INTERACTIVE ARITHMETIC

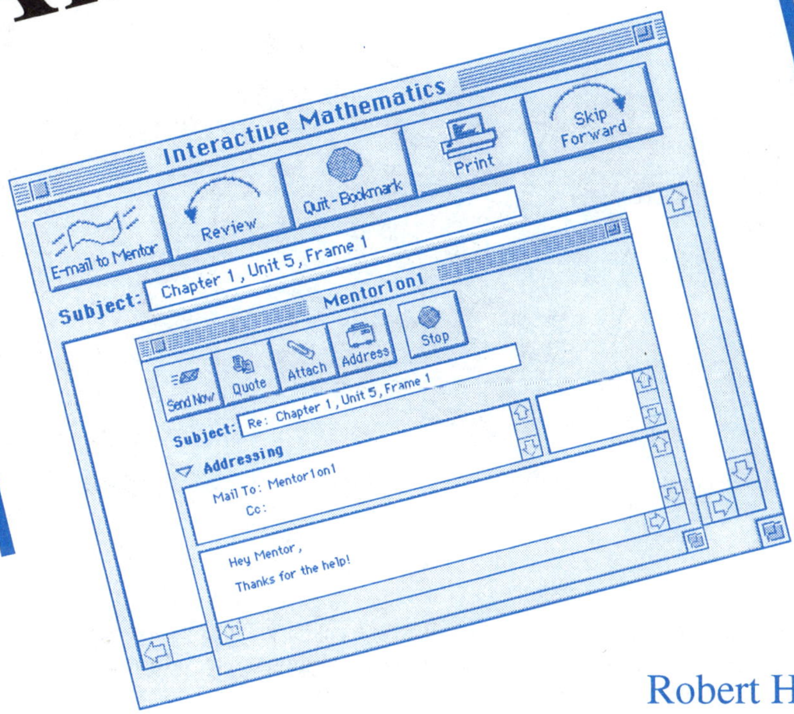

Robert H. Alwin
Robert D. Hackworth
Joseph Howland

H&H INTERACTIVE MATHEMATICS SERIES

H&H Publishing Company, Inc.
Clearwater, Florida

H&H Publishing Company, Inc.

1231 Kapp Drive
Clearwater, FL 33765
(813) 442-7760
(800) 366-4079
FAX (813) 442-2195
email: info@HHPublishing.com
web site: www.HHPublishing.com

INTERACTIVE ARITHMETIC

Robert H. Alwin
Robert D. Hackworth
Joseph Howland

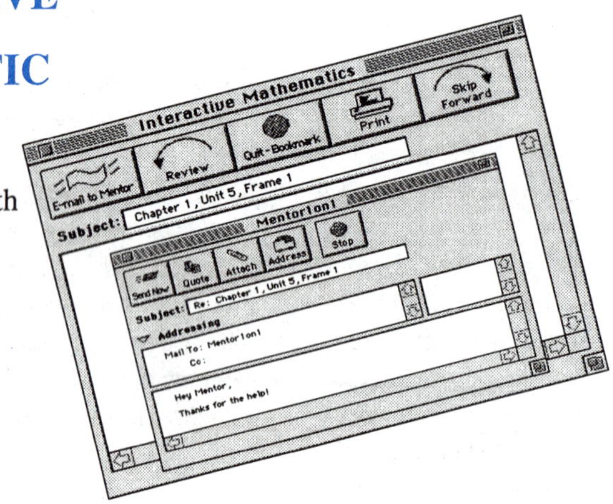

ISBN 0-943202-68-X

This text in previous versions was originally titled **Programmed Arithmetic**.

Printing is the lowest number: 10 9 8 7 6 5 4 3 2 1

PREFACE

Interactive Arithmetic is designed to aid the student in learning its material.

The text possesses strong teaching qualities that virtually assure the conscientious student of steady, confident progress. At the same time, the text's interactive processes assure the student of continuous opportunities to get help where and when it is needed. Solid mathematics, quality instruction, and personal assistance are the hallmark of this series of texts.

The text is designed in a format that constantly creates an interaction between the student, the mathematics to be learned, and an experienced teacher either in a classroom or in cyberspace. The majority of the book is a series of questions/answers that achieves a true dialogue with its student reader. Most of the time, the dialogue teaches the mathematics effectively, and students learn easily without benefit of outside assistance. Sometimes, however, the dialogue may be insufficient; it is in such cases where the student can find extra, personal assistance through conversation with a friend or teacher. These helpful conversations can occur in person, but also are available with a cyberspace instructor who is always as close as the nearest computer.

The quality of the mathematical content and instructional use of this text have been thoroughly documented in a wide range of situations and with great numbers of students. Specifically, there are no units or chapters where common difficulties can be expected. When a student has a problem understanding the material, the difficulties will be unique to that student. Something of importance was overlooked, misunderstood, etc. That is the reason for providing personal assistance in those instances and getting the student back into the flow of the book's dialogue. The personal assistance will both answer the questions of individual students and give guidance in the best ways to study mathematics.

Because of the design of this book, the student can expect other dramatic improvements in the study and learning of mathematics. One of those advantages is the student's ability to vary the pace of instruction to match their ability to learn. At times, a student needs to work slowly through the instruction and carefully ponder each new concept. At other times, the student grasps the material quickly and can accelerate their learning rapidly through the material. As a result, most students will move through this material more quickly than in a regular class situation.

Another great advantage of this book is the fact that students have an opportunity to learn to ask intelligent, specific questions about the mathematics in a context that encourages and does not threaten the student. Study skill deficiencies can be corrected within the context of these materials while maintaining traditional levels of achievement and rigor.

The content of this book is comparable to an accumulation of the arithmetic topics normally taught prior to an algebra course. Chapters 1-4 deal only with the arithmetic of the counting numbers, and those who are already good at adding, subtracting, multiplying, and dividing whole numbers will finish these chapters quickly with a little review. Chapters 5-7 are devoted to fractions, and most students will find that these chapters need to be carefully studied and learned. Chapters 8-10 deal with decimal fractions, and some students will be adept at most of the skills and need a limited amount of review. Chapter 11 presents ratios and proportions, and this material will usually require concentrated study. Chapter 12 on Percent, like Chapter 11, is likely to require careful study. The final two chapters, Chapters 13-14, pave the way for entering a Beginning Algebra course. Throughout the book, a spiraling effect on learning is used. That is, the skills gained in early chapters are again and again revisited as they are extended to other numbers and real life situations. Each chapter ends with an Application Unit teaching word problem-solving processes. Many of the applications involve geometry so that the student completing the text will have developed the necessary knowledge and skills for dealing well with most two-dimensional figures.

The book contains many opportunities for the student to make wise decisions about the level of understanding achieved. Each chapter begins with an Objectives Test that illustrates, by example, all of the concepts and skills which are to be presented. Each problem is accompanied by a designation showing where it will be taught in the chapter. Each chapter is broken into units in a way that divides the content into topics of such length that it gives the student a way of organizing their own studying to make it most effective. Most units are designed to be completed in a single study session and, for best results, should be handled in that way. The unit ends with a Feedback which is an opportunity for the student to test the level of understanding achieved. It is strongly suggested that the Feedback be taken at the beginning of the next study session so that some time intervenes between the learning and assessment. The chapter ends with a Mastery Test that is similar to the Objectives Test. When a student can do all the problems correctly on a Mastery Test, it indicates a complete understanding of all the material in the chapter. Again, each problem is accompanied by a designation showing where it was taught in the chapter. Answers for all problems in the Tests and Feedbacks are included at the back of the book. Good learning can occur only when the student is aware of what is known and what still needs to be learned.

We are greatly indebted to thousands of teachers and hundreds of thousands of students who have successfully used our previous materials and given us the benefit of their own experiences and ideas. The strength of this new book is a result of that special kind of study of the work and challenges we face in teaching mathematics to students — some in desperate need for overcoming past experiences that were markedly unsuccessful and have interfered with motivations to try again.

Robert H. Alwin
Robert D. Hackworth

To The Student

You are about to begin a new, successful experience in learning mathematics because this textbook can be of tremendous assistance to you. It was written to provide true instruction and it will teach you all you need to know if you use it as described.

First, however, you must start a process of making sense out of mathematics. Students who ask "WHY" questions in mathematics become good at the subject because it really isn't just a "bunch of meaningless rules." Students who skip the "WHY" questions and try to learn "HOW" to do each type of problem find there are more types of problems than they can memorize successfully. The more you understand, the better you will remember and the less you will need to memorize. Every arithmetic process has an answer to any "WHY" question you want to ask and this book tries to provide those answers, but if you don't find them here then ask anyone who will listen. Just the process of asking "WHY" questions will, by itself, improve your ability to learn mathematics. Make a habit of asking "WHY" questions.

Now to the specific directions for this book. The most important part of this book is its frames. If you look at page 5 of the text, you will see an entire page of frames. Each frame is a numbered item. The left column of a frame gives you some information and a question (problem) to answer. Do the question and then look at the book's answer which is printed in color in the right column of the frame. If your answer is the same as the book's, then you have learned the lesson of the frame and you should continue to the next frame. If your answer is different from the book's, then you have missed the lesson of the frame and need to get help before proceeding further. That help may be simply re-reading the frame, but if that is not sufficient then seek out a teacher or friend to get you back on the right track.

The frames are the most important part of the book because they are the only part of the text where you **learn** the mathematics. All other parts of the book are intended to give you information about how well you have learned the mathematics, but only in the frames do you learn mathematics. If you do the frames well, everything else will follow accordingly. If you skip the frames because you "kind of" understand them, then you will make learning unnecessarily difficult.

Some frames are of special importance and are indicated by a small hand (✍). Pay special attention to the frames that are marked in this manner. When you are reviewing, you can skip from one hand-marked frame to another.

Each chapter of the text begins with an Objectives Test like that on page 2 that previews the chapter for you. You are not expected to do the problems before you study the chapter. Those problems simply show you the types of problems you will study in the chapter. Each problem is marked so you can tell in which unit it will be studied. Answers for all of these problems are given at the back of the book.

Each chapter ends with a Mastery Test like the one on page 24 and when you finish the chapter, most, hopefully all, of the problems can be correctly completed. Each problem of the Mastery Test has been completely explained in the chapter. If any problem seems difficult you will find that its unit has been indicated and you will need to restudy that section. Answers for all of these problems are given at the back of the book.

Feedback Exercises, like the one on page 23, are at the end of each unit of a chapter. The problems of a Feedback Exercise give you a method of evaluating your work each day. Answers for all of these problems are given at the back of the book. If you have any trouble with a Feedback Exercise you will find help in the unit just completed.

Every problem in this book has an answer given for it. Your learning successfully depends upon knowing, as you go along, whether you are right or wrong. You must accept the responsibility for determining the accuracy of your work. Looking at the answer is one way for you to meet that responsibility. Learning how to check an answer is another way to meet the same responsibility. Regardless of your method, you are expected to always know whether your work has been done correctly.

TABLE OF CONTENTS

Focus On Study Skills For Mathematics

Starting college and taking an Arithmetic Course is a downer. The student is studying, one more time, material that was not learned before. That is discouraging, and the negative feelings of some students interfere greatly with opportunities for success.

Many students in these situations think something is inherently wrong with them. They tend to give up before they start and make statements like:

"I've never been good at math."
"I don't have a math mind."
"I don't like math."
"My mother can't do math either."

All of these statements, and others like them, really mean that the student doesn't believe he/she can be successful in a math course. Not surprisingly, once a student accepts that belief, the likelihood of doing well in mathematics is slight indeed.

Attitudes Of Students Who Do Well In Mathematics

Students that do well in mathematics believe they will succeed and then do the common sense things needed to support that belief.

1. They attend class without fail.
2. They listen carefully and take good notes.
3. They always do all of their assignments.
4. They ask questions — lots of questions.
5. They show all their work and they arrange it very neatly so they can refer to it later and read it easily.
6. They work hard because nothing worth while is ever gained without a struggle.

Attitudes Of Students Who Don't Do Well In Mathematics

Students who expect to fail in mathematics don't believe they can succeed and frequently do such silly things that their belief comes true.

1. They frequently miss class and make little effort to find out what was missed.
2. They don't listen carefully and rarely take notes that would help them remember.
3. They give up on any assignment that is difficult and skip some problems because they seem easy.
4. They write far less than their teacher would and they write so small that hardly anyone could read it anyway.
5. They hardly ever ask questions and are usually afraid that they will ask a "stupid" question. There really is no such thing as a "stupid" question.
6. They believe they are destined to do poorly anyway so they don't work hard. What's the use?

Today Is A New Day For You And Mathematics

The first objective you need to accomplish in this course is to establish the belief that you can, and will, succeed. This is not an easy task. You need to carefully consider your past experiences in mathematics and make plans to change those behaviors which did not work for you in other courses. Today is a new day in your mathematics life, a chance to start over and do it right, and an opportunity to find the easy ways to study and learn mathematics.

The authors of this book do not believe you are destined to fail. In fact, our many years experience with students like you have convinced us that there is nothing biologically or intellectually deficient about you. More commonly, our students have come to us with beliefs about mathematics and study habits in the subject which have made the subject very hard to learn. The aim of this section of the book is to present efficient and effective ways of dealing with your learning of mathematics.

Many study skills have been preached constantly at you over the years. You have probably been told:

> "Read mathematics carefully."
>
> "Keep paper and pencil handy so you can constantly try the problems."
>
> "Arrange your work neatly so you or someone else can find any errors."
>
> "Do all your assignments and never get behind."

All of these statements are excellent advice. If you have ignored them in the past, make certain that you follow them in this course.

In this section, however, we want to go beyond some of the past advice you have been given and expand on four major themes for organizing your study of mathematics. Our aim here is to encourage you to change the ways you study mathematics. It is the way you have studied in the past that is the problem. The mathematics will be far easier if you find new ways to study it.

The four themes for organizing your study of mathematics are from the work of Claire E. Weinstein, Professor of Educational Psychology, University of Texas at Austin. These four themes are:

1. Work and study in quality environments.
2. Learn what is to be memorized and what must not be memorized.
3. Keep your mind active.
4. Learn to evaluate your own work and study habits.

1. Always Work And Study In A Quality Environment

A student does better work, stays at it longer, and feels better about the learning when that work is done in a setting in which the student is comfortable and physically relaxed. Your body and your mind should not be the source of any distractions. You want to be able to devote full energies, physical and mental, to your study.

A good restaurant earns its reputation on much more than the quality of its food. A restaurant is judged as much on its atmosphere (ambiance) as on its menu. Customers expect fine food to be served in a comfortable environment.

Make a list of the physical attributes of a fine restaurant. Some of these attributes may be different than you want in a good study. The important new idea to incorporate in your behavior is the selection, by you, of an environment that will allow you to study more effectively. Don't settle for less in your study environment than you demand in your other activities. Otherwise, you will not be your best when studying.

Physical Attributes Of Quality Environments

Physical attributes such as lighting, temperature, and noise level are obvious factors to be carefully regulated by the student. Yet these obvious factors are sometimes ignored by students who are having trouble in a subject. Serious study cannot occur while working the drive-in window at MacDonald's or watching the Super Bowl on television. Although these examples may seem extreme, you need to look at your own study sessions and make certain they eliminate all possible physical distractions.

The sounds in your study environment would be a good place for you to start altering the way you work. Complete silence is not necessarily the best condition, but loud rock and roll is definitely not helpful. You should try to find sound that is soft enough not to be intrusive and with a slow beat. Try playing classical musical at such a low volume that you cannot pay attention to it. That music will over-ride other sounds in your environment that might be distracting.

Psychological Attributes Of A Quality Environment

Besides the obvious physical attributes of your learning environment, but equally important, are qualities which are psychological in nature. Fear of mathematics or anxiety over mathematics interferes greatly with the study habits of some students. Those fears or anxieties must be dealt with prior to a study session if they are to be overcome. Your mind, as well as your body, must be free from distractions. A mind that is full of fear or worry has no capacity to also learn mathematics. If your mind is distracted by other thoughts find ways to clear it before studying. Otherwise, it will make your study time inefficient and distasteful. The result will be a lessening of your efforts and eventual failure.

Relaxation techniques can be learned and they will lower the level of fear or anxiety. You can learn methods like Benson's Relaxation Technique from books in the school library. You can also ask for some personal assistance from a school counselor.

All students should learn, as a minimum, to pay attention to their breathing. Do not attempt to alter your breathing. Just pay attention to it and its rhythm. That simple focusing will, by itself, have beneficial results in any anxiety producing situation. Try it. The more you try it, the better will be your results.

Play The Role Of A Scholar

A final aspect of your environment needs close attention. Education is more than simply learning a list of facts. It is a different lifestyle with a value system that respects books, music, art, and beauty from an altered perspective. Put good literature, classical music, fine art, and flowers into your environment. Live with these elements in your study environment.

Imagine the private study of a learned professor at Yale and try to create the same study conditions for yourself. Playing the role of a scholar is helpful in becoming a good student.

2. Learn What Part Of Mathematics Needs To Be Memorized And Those Portions Of The Subject That Must Be Figured Out.

A major problem for many students is the belief that all mathematics should be memorized. This is false and trying to memorize everything always leads to failure. If you have been memorizing much in mathematics, that is not only inefficient – it is destructive. Read this section carefully and develop new ways of dealing with mathematics information.

Memorize Definitions And Symbols

In general, the **only** material that should be memorized in a mathematics course is the meaning of words and symbols. A good mathematics student must know the meaning of a word like "percent" and its symbol "%." **Memorize words and symbols** because their meanings are crucial to working out problems that contain them.

Words and symbols play a crucial role in mathematics which is often overlooked by students. At the same time, words and symbols do not have meanings that can be figured out or discovered. Someone, your teacher or your textbook, needs to give you clear definitions for all the words that are relevant and clear descriptions of any new symbols that are introduced.

Words like "percent" and symbols like exponents, 10^7, must be clearly understood before problems that contain them can be learned. Such words and symbols must be memorized. Don't skip past them and attempt to do problems without first understanding them.

Do Not Memorize Rules, Laws, Or Processes

In general, do not memorize the steps involved in working a problem. **Do not memorize** rules or laws. Such memorization clutters your mind unnecessarily and tends to misdirect all your study. Blindly memorizing a "bunch of rules" is not the same as learning mathematics. Find reasons why rules, laws, processes, or steps work. If you understand why, there will be no need to memorize more than words or symbols.

Treat rules or laws in mathematics just as you treat the legal laws in your community. Nobody memorizes all the state laws. Instead, the good citizen tries to make sense of them and learns how they apply to his/her needs.

Rules and laws of mathematics also need to make sense. Your study of them needs to uncover the logic that supports them. If you learn the reasons for different steps in a problem then you will find it is both unnecessary and unwise to learn the steps themselves!

Focus

Finally, and very importantly, you need to direct your attention to the particular idea or skill that is to be learned. **FOCUS** on each new idea or skill.

You need to **FOCUS** specifically on the purpose of each new idea or skill as it is presented. Do not be distracted by other parts of a problem or its accompanying material.

Each new idea or skill in mathematics will be presented to you in a context which contains many other ideas with other skills. It is essential that you recognize both the new ideas and the older ones. You must learn to concentrate on the new ideas. This is what is meant by **focusing**.

Poor mathematics students generally try to learn with only a vague idea of the particular new skill being taught at any moment. Good mathematics students practice the habit of giving their undivided attention to the one new aspect of a problem that needs to be learned while treating the other parts of the problem as background information.

3. Keep Your Mind Active

Learning is an activity of the mind and no one is learning when they aren't thinking.

There is no way to divorce learning from thinking.

You can improve your ability to learn by constantly making certain that you are thinking. Good learning of mathematics occurs when the student is thinking about the mathematics. Poor learning of mathematics occurs when the student's mind is elsewhere.

It may come as quite a surprise, but even the best learners encounter difficulties at trying to maintain their thinking. All students, good and bad, have their minds wander. Good students have learned methods for constantly bringing their minds back to the activity. Poor learners have frequently seemed pleased when they could do a lot of problems without thinking.

If you find yourself doing mathematics without thinking, stop and bring yourself back to learning.

Inner Dialogues Keep Your Mind Active

An excellent way to keep your mind active is to carry on a conversation with yourself as you study. This "inner dialogue" might be structured by always asking the following two questions:

"What is the meaning of this problem?"
"Why is the problem worked this way?"

Neither of these questions is simple. Both of them force the learner to think and that is their purpose. The first question will keep the student on track in constantly knowing the meaning of words and symbols. The second question will emphasize the need to understand the reasoning behind a problem rather than memorizing the steps used to solve it.

An "*inner dialogue*" is a form of talking to oneself and such conversations are valuable. It is also valuable to talk with someone else when studying. Find another student who will participate in a common learning experience. Again, the two major questions you can build into your study sessions are:

> "What is the meaning of this problem?"
> "Why is the problem worked this way?"

You want to constantly look at both meaning and reasoning. The questions, whether in an "inner dialogue" or with a learning "buddy," will structure your thinking and keep you on track for learning mathematics effectively.

4. Learn To Evaluate Your Own Learning

Many students believe that it is the task of the teacher to give them a grade and, consequently, accept little responsibility for making accurate judgements about their own progress in a mathematics course. When such students are asked, "How did you do on the test?" they are likely to give very vague or inaccurate responses. This is a characteristic that is drastically different from good students. When good students are asked, "How did you do on the test?" they are likely to be very specific in their comments and will accurately predict the grade they will receive.

Most math programs and texts provide students with opportunities for evaluating the extent to which learning has occurred. For example, most texts contain answers for many of the problem sets. Students who check the answers to problems carefully and restudy those problems that are missed gained valuable insight into both the mathematics and their learning of the mathematics.

To make any evaluation of your work pay its best dividends, be certain that enough is written to provide a clear trail of your thinking on each problem. Finding wrong answers is not nearly as helpful as finding the point in a problem where you went wrong.

Two Kinds Of Knowledge

Knowing how well you are learning in a mathematics course is very different from knowing the mathematics. In other words, there are always two types of knowledge for the mathematics student:

1. Knowledge of the mathematics, and
2. Knowledge about how well the mathematics has been learned.

Knowledge about the mathematics is not enough. The most effective student needs to have a knowledge about his/her learning of the mathematics. In studying mathematics, always pay attention to both the subject and the learning of that subject.

A simple example involves the checking of a problem. The student who does a problem but skips the check of the problem is accepting responsibility for the mathematics, but ignoring the equally important responsibility for evaluating their learning.

The best of students make mistakes, misunderstand directions, and encounter real difficulties with some topics. But these students have developed mechanisms for telling them when those errors in learning occur.

Knowing when the required material has not been learned gives direction and meaning to further study. Excellence is achieved for the mathematics student when he/she knows the mathematics and also knows that he/she knows the mathematics.

Failures In Learning Are Correctable

A major difference between good and poor mathematics students is the fact that poor students do not know whether they know the mathematics. After studying for a mathematics test, poor students are unable to determine correctly whether they know or don't know the material. This is truly tragic because it is almost always correctable. The tragedy is that poor students generally do not take advantage of the opportunities they have for accurately assessing how much they know.

Use your textbook, your instructor, and other sources for help on and off campus. If you look for them there are ample opportunities for you to determine how much and how well you know the mathematics. Problem sets, review exercises, practice tests, supplementary teaching aids (television, computers, etc.) and learning centers are usually available. Good students take advantage of them. Poor students may not realize their value in the learning process and either ignore them or give them only slight attention.

If you want to improve your knowledge of mathematics, begin by raising your awareness of the ways you use to learn mathematics.

All students have troubles, at times, learning the mathematics. Good students realize that and take the necessary steps to overcome the difficulties.

Learn the mathematics, but use every tool available to accurately tell you how well you have learned the mathematics.

A good relationship with yourself, your textbook and your instructor will make learning mathematics a very important part of your total educational and real life experiences.

CHAPTER 1 OBJECTIVES

The following problems describe, by example, the skills and knowledge studied in this chapter. Each problem is accompanied by a symbol, (U), which indicates the Unit where the problem is studied. Answers are at the back of the book.

1. 7 + 2 = _____ (1)

2. 2 + 7 + 3 = _____ (1)

3. What is the ten's digit of 4,639? (2)

4. What is the thousand's digit of 836,417? (2)

5. 347 + 296 = _____ (2)

6. 583 + 489 = _____ (2)

7. 276 + 148 = _____ (2)

8. 319 + 693 = _____ (2)

9. A druggist filled 432 prescriptions in July and 856 in August. What was the sum for the two months? (3)

10. Hospitals in Town A had 913 beds and in Town B 476 beds. What was the total number of beds in the two towns? (3)

11. 27,946 + 3,049 + 46,905 = _____ (3)

12. 21,419 + 5,004 + 8,510 = _____ (3)

CHAPTER 1
ADDITION OF WHOLE NUMBERS

UNIT 1: ADDING BY COUNTING

1
The box ☆☆☆☆☆ contains 5 stars.
How many stars are in ☆☆☆? ______ 3

2
How many hearts are in ♥♥♥♥♥♥♥♥? ______ 8

3 ✍
Addition is a form of counting. The addition of 5 and 3 is the same as counting 5 and then ________ 3 more. counting

4
To add 5 + 3 think of 5 stars ☆☆☆☆☆ and 3 stars ☆☆☆. Count all the stars. How many stars are there? ______ 8

5
Add 9 + 4 by counting 4 more than 9. ______ 9 + 4 = 13

6
6 + 3 = ______ 9

7
8 + 7 = ______ 15

8
5 + 9 = ______ 14

9
The answer to an addition problem is its **sum**. Find the sum for 4 + 7 = ______. 11

10
Find the sum for 9 + 9 = ______. 18

11
The answer to an addition problem is its sum or **total**. Find the total for 8 + 4 = ______. 12

12
Find the total for 8 + 9 = ______. 17

ADDITION TERMS

The numbers in the addition problem at the right are **addends**. 5,469, 395, and 4,067 are **addends**.

The answer of an addition problem is its **sum**. 9,931 is the **sum** of the addition.

```
  5469
   395
+ 4067
  9931
```

13
The problem 6 + 9 can be written in vertical form as shown at the right. Write 8 + 3 in vertical form. _____

$$\begin{array}{r} 6 \\ +9 \\ \hline \end{array}$$

$$\begin{array}{r} 8 \\ +3 \\ \hline \end{array}$$

14
Add.

$$\begin{array}{r} 5 \\ +3 \\ \hline \end{array}$$

8

15
Add.

$$\begin{array}{r} 2 \\ +9 \\ \hline \end{array}$$

11

16
Find the sum.

$$\begin{array}{r} 3 \\ +6 \\ \hline \end{array}$$

9

17
Find the total.

$$\begin{array}{r} 4 \\ +5 \\ \hline \end{array}$$

9

18
Add.

$$\begin{array}{r} 1 \\ +8 \\ \hline \end{array}$$

9

19
Find the sum.

$$\begin{array}{r} 3 \\ +7 \\ \hline \end{array}$$

10

20
Find the sum of 15 and 8 by counting 8 more than 15. _____

15 + 8 = 23

21
Find the total of 16 and 5 by counting 5 more than 16. ______ 21

22
12 + 3 = _______ 15

23
13 + 9 = _______ 22

24
17 + 4 = _______ 21

25 ✍
The numbers in an addition problem are **addends**. What are the addends in the problem 16 + 7? ______ 16 and 7

26
In the problem 14 + 2 the addends are ______ and ______. 14, 2

27
In the problem 11 + 7 the addends are ______ and ______ 11, 7
which have a sum of ______. 18

28 ✍
To add three numbers (addends) add the first two, get an answer (sum), and then add the third addend to the sum. 6 + 3 + 4 = 9 + 4 = _______ 13

29
Add 7 + 4 + 6 by first adding 7 and 4. _____ 17

30
6 + 5 + 8 = _____ 19

31

4 + 9 + 6 = _____

19

32

2 + 8 + 7 = _____

17

33

6 + 1 + 8 = _____

15

34

The problem 4 + 5 + 9 written in vertical form is shown at the right.

$$\begin{array}{r} 4 \\ 5 \\ +\ 9 \\ \hline \end{array}$$

Write 2 + 7 + 3 in vertical form.

$$\begin{array}{r} 2 \\ 7 \\ +3 \\ \hline \end{array}$$

35

Find the sum of the addends 4, 3, and 8 using the vertical form.

$$\begin{array}{r} 4 \\ 3 \\ +8 \\ \hline 15 \end{array}$$

36

Find the total of the addends 2, 6, and 9 using the vertical form.

$$\begin{array}{r} 2 \\ 6 \\ +9 \\ \hline 17 \end{array}$$

37

Find the sum of 6, 3, and 5 using the vertical form.

$$\begin{array}{r} 6 \\ 3 \\ +5 \\ \hline 14 \end{array}$$

38

Find the total of 7, 4, and 9 using the vertical form.

$$\begin{array}{r} 7 \\ 4 \\ +9 \\ \hline 20 \end{array}$$

39
To add 56 and 23 the addends are written in vertical form.

$$\begin{array}{r} 56 \\ +\underline{23} \end{array}$$

Write 63 + 25 in vertical form.

$$\begin{array}{r} 63 \\ +\underline{25} \end{array}$$

40
Write 725 + 42 in vertical form.

$$\begin{array}{r} 725 \\ +\underline{42} \end{array}$$

41
To add 56 + 23 the columns are added from right to left. Add only the right hand column of the problem.

$$\begin{array}{r} 56 \\ +\underline{23} \end{array}$$

$$\begin{array}{r} 56 \\ +\underline{23} \\ 9 \end{array}$$

42
Complete the addition shown at the right by adding the left hand column.

$$\begin{array}{r} 56 \\ +\underline{23} \\ 9 \end{array}$$

$$\begin{array}{r} 56 \\ +\underline{23} \\ 79 \end{array}$$

43
Complete the problem at the right by adding the columns from right to left. _____

$$\begin{array}{r} 73 \\ +\underline{25} \end{array}$$

$$\begin{array}{r} 73 \\ +\underline{25} \\ 98 \end{array}$$

44
Add 437 + 551 in vertical form. _____

$$\begin{array}{r} 437 \\ +\underline{551} \\ 988 \end{array}$$

45
Add 618 + 270 in vertical form. ______

$$\begin{array}{r} 618 \\ +\underline{270} \\ 888 \end{array}$$

FEEDBACK UNIT 1

This quiz reviews the preceding unit. Answers are at the back of the book. Begin the next unit only after all problems of this unit are understood.

1. 6 + 8 = 14
2. 56 + 12 = 68
3. 13 + 8 = 21
4. 84 + 14 = 98
5. 7 + 5 + 9 = 21
6. 15 + 43 = 58
7. 6 + 8 + 4 = 18
8. 814 + 152 = 966
9. The answer to an addition problem is its sum or total.
10. The numbers to be added in an addition problem are its addends.

UNIT 2: BASE 10 NUMERATION SYSTEM

1
In the numeral 562, the **digits** are 5, 6, and 2.
In the numeral 946, the digits are ___, ___, and ___. 9, 4, 6

2
What are the digits of 54? _____ 5 and 4

3
Each digit in 562 has a **place value**. The digit 5 is in the hundred's place and means 5 hundreds. The digit 6 is in the ten's place and means 6 ______ and the digit 2 is in the ______ place and means 2 ______. tens one's ones

4
Each digit in 8,102 has a **place value**. The digit 8 is in the thousand's place and means 8 thousands. The digit 1 is in the hundred's place and means 1 ______. hundred

PLACE VALUES

Every **digit** in a whole number has a **place value**. The **place value** of a **digit** is determined by its column.

The figure at the right shows the **place values** of the **digits** in 8,943,625. 6 is in the **hundreds** place and is called the **hundreds digit**. 4 is the **ten thousands digit**.

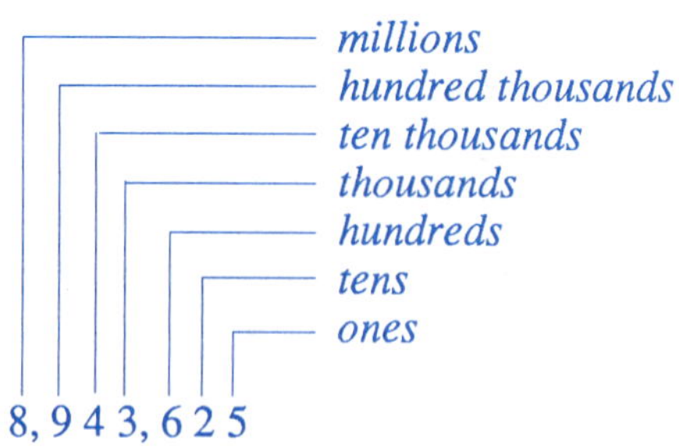

5
In the numeral 5,632 the digit "2" means 2 ones, the digit "3" means 3 tens, and the digit "6" means 6 ___________. hundreds

6
In the numeral 4,157,986 the digit "5" means 5 ten thousands and the digit "1" means 1 ________ ___________. hundred thousands

7
What is the millions digit in 6,498,075? ___________ 6

8
What is the hundreds digit in 5,938,407? ___________ 4

9
What is the ones digit in 8,697,753? ___________ 3

10
What is the tens digit in 45,687? ___________ 8

11
What is the hundreds digit in 8,697,453? ___________ 4

12
Write a three-digit numeral with ones digit 5, tens digit 9, and hundreds digit 7. ___________ 795

13
Write a three-digit numeral with hundreds digit 8, tens digit 6, and ones digit 0. ___________ 860

14
Write a three-digit numeral with tens digit 7, ones digit 8, and hundreds digit 9. ___________ 978

A BASE 10 NUMERATION SYSTEM

Each column of a base 10 numeration system is a power of 10. Adjacent columns are related by multiples of ten.

Each ten is equal to ten ones. Each hundred is equal to ten tens. Each thousand is equal to ten hundreds, etc.

ones
10 ones = ten
10 tens = hundred
10 hundreds = thousand
10 thousands = ten thousand
10 ten thousands = hundred thousand
10 hundred thousands = million
10 millions = ten million

15

10 is 10 ones.
100 is 10 tens.
1,000 is 10 _________.

hundreds

16

10,000 is 10 thousands.
100,000 is 10 ten thousands.
1,000,000 is 10 __________________.

hundred thousands

17 ✍

The numeral 47 means: 4 tens, and 7 ones.
Is 47 also equal to 3 tens, and 17 ones? _____

Yes

18

If the digit in the tens column is reduced by one, the number of ones is increased by ten.

47 = 4 tens, 7 ones = 3 tens, 17 ones
79 = 7 tens, 9 ones = 6 tens, ___ ones

19

19

96 means 9 tens, 6 ones or 8 tens, ___ ones.

16

20

65 means 6 tens, 5 ones or 5 tens, ___ ones.

15

21
The numeral 360 means: 3 hundreds, and 6 tens.
Is 360 also equal to 2 hundreds, and 16 tens? _____ Yes

22
If the digit in the hundreds column is reduced by one, the number of tens is increased by ten.
917 = 9 hundreds, 1 ten, 7 ones
= 8 hundreds, 11 tens, 7 ones
760 = 7 hundreds, 6 tens
= 6 hundreds, ___ tens 16

23
936 means: 9 hundreds, 3 tens, 6 ones
or 8 hundreds, ___ tens, 6 ones 13

24
807 means: 8 hundreds, 0 tens, 7 ones
or 7 hundreds, ___ tens, 7 ones 10

25
The numeral 5,800 means: 5 thousands and 8 hundreds.
Is 5,800 also equal to 4 thousands and 18 hundreds? _____ Yes

26
If the digit in the thousand's column is reduced by one, the number of hundreds is increased by ten.
5,800 = 5 thousands, 8 hundreds
= 4 thousands, 18 hundreds
2,600 = 2 thousands, 6 hundreds
= 1 thousand, ___ hundreds 16

27
6,400 means: 6 thousands, 4 hundreds
or 5 thousands, ___ hundreds 14

28 ✍

800 means:	8 hundreds, 0 tens, 0 ones	
or	7 hundreds, 10 tens, 0 ones	
or	7 hundreds, 9 tens, ___ ones	10

29

6,000 means:	6 thousands, 0 hundreds	
or	5 thousands, 10 hundreds	
or	5 thousands, 9 hundreds, ___ tens	10

30

900 means:	9 hundreds, 0 tens, 0 ones	
or	8 hundreds, ___ tens, 0 ones	10

31

700 means:	7 hundreds, 0 tens, 0 ones	
or	6 hundreds, ___ tens, 10 ones	9

FEEDBACK UNIT 2

This quiz reviews the preceding unit. Answers are at the back of the book. Begin the next unit only after all problems of this unit are understood.

1. In the numeral 6,589,143 the digit "9" means 9 ____________.
2. The ten thousands digit in 4,367,598 is _____.
3. 67 means 6 tens, 7 ones or 5 tens, _____ ones.
4. 43 means 4 tens, 3 ones or 3 tens, _____ ones.
5. 917 means 9 hundreds, 1 ten, 7 ones or 8 hundreds, _____tens, 7 ones.
6. 7,600 has 7 thousands, 6 hundreds or 6 thousands, _____ hundreds.
7. 200 has 2 hundreds, 0 tens, 0 ones or 1 hundred, _____ tens, 10 ones.
8. 500 has 5 hundreds, 0 tens, 0 ones or 4 hundreds, 9 tens, _____ ones.

UNIT 3: THE "CARRYING" PROCESS IN ADDITION

1
Write 4,063 + 2,914 for vertical addition.
____

Answer:

```
  4063
+ 2914
```

2
Add 4,063 + 2,914.
____

Answer:

```
  4063
+ 2914
  6977
```

3
Vertical addition always begins with the digits in the ones column. In adding 2,427 and 8,396 the vertical addition begins with the digits ____ and ____.

Answer: 7 and 6

4
Adding 2,427 and 8,396 begins with the ones digits. The sum of 7 and 6 is ____.

```
  2427
+ 8396
```

Answer: 13

5 ✍
Whenever the sum of any column exceeds 9, the process of "carrying" is used. Notice, that 3 is placed in the ones column and ___ is carried to the _____ column.

```
     1
  2427
+ 8396
     3
```

Answer: 1; tens

6
After adding and carrying from the ones column, the sum of the digits in the tens column exceeds 9. Is carrying necessary? _____

```
     1
  2427
+ 8396
     3
```

Answer: Yes

7
The carrying process for addition of the tens column is shown at the right. Because the sum of the tens column is 12, a ___ is placed in the tens column and 1 is carried to the __________ column.

$$\begin{array}{r} {\scriptstyle 1\,1} \\ 2427 \\ +\underline{8396} \\ 23 \end{array}$$

2

hundreds

VERTICAL ADDITION REQUIRING "CARRYING"

To add 4,093 and 2,398 the numerals are arranged vertically so that **digits with the same place value are in the same column.**

$$\begin{array}{r} 4093 \\ +\underline{2398} \\ 6491 \end{array}$$

Each column from right to left is separately added. Whenever the answer to any column exceeds nine, the process of carrying is used.

The "carrying" marks from the additions of the ones and the tens columns are shown in the figure at the right. Use these "carrying" marks if they improve your accuracy.

$$\begin{array}{r} {\scriptstyle 1\,1} \\ 4093 \\ +\underline{2398} \\ 6491 \end{array}$$

8
To add 68 and 24, carrying is necessary because the sum of the digits in the ones column exceeds 9. Add 47 and 35 and show any carrying mark. _____

$$\begin{array}{r} {\scriptstyle 1} \\ 68 \\ +\underline{24} \\ 92 \end{array}$$

$$\begin{array}{r} {\scriptstyle 1} \\ 47 \\ +\underline{35} \\ 82 \end{array}$$

9
Add 438 and 927 using vertical addition. Show clearly any carrying that is necessary.

____

$$\begin{array}{r} {\scriptstyle 1\;\;1} \\ 438 \\ +\underline{927} \\ 1365 \end{array}$$

10
Add 9,036 and 5,281 using vertical addition. Show clearly any carrying that is necessary.

____

```
1  1
  9036
 +5281
 14317
```

11
Add 543 and 628 using vertical addition. Show clearly any carrying that is necessary.

____

```
1  1
  543
 +628
 1171
```

12
487 + 252 = ________

```
  1
  487
 +252
  739
```

13
654 + 359 = ________

```
 111
  654
 +359
 1013
```

14
495 + 358 = ________

```
 11
 495
+358
 853
```

15
294 + 6,308 = _____

```
   11
   294
 +6308
  6602
```

16
8,638 + 958 = _____

```
 1  1
 8638
 +958
 9596
```

17
Arrange the three addends 635, 43, and 8,136 in vertical columns and find the sum of only the one's column. ______

$$\begin{array}{r} {\scriptstyle 1} \\ 635 \\ 43 \\ +8136 \\ \hline 4 \end{array}$$

18
Arrange the three addends 4,315, 216, and 483 in vertical columns and find the sum of only the one's column. ______

$$\begin{array}{r} {\scriptstyle 1} \\ 4315 \\ 216 \\ +483 \\ \hline 4 \end{array}$$

19
Add 905 + 1,483 + 384 using column addition. Show all carrying.

$$\begin{array}{r} {\scriptstyle 1\ 1\ 1\ \ } \\ 905 \\ 1483 \\ +384 \\ \hline 2772 \end{array}$$

20
Add 674 + 8,905 + 655 using column addition. Show all carrying.

$$\begin{array}{r} {\scriptstyle 1\ 2\ 1\ 1\ \ } \\ 674 \\ 8905 \\ +655 \\ \hline 10234 \end{array}$$

21
426 + 5,270 + 8,654 = __________ 14,350

22
7,179 + 869 + 7,658 = __________ 15,706

23
6,845 + 97,403 + 6,510 = __________ 110,758

24
44,509 + 81,654 + 3,124 = __________ 129,287

25
4,326,913 + 58,496 = __________ 4,385,409

26
67,315 + 412,684 + 5,094 + 678 = __________ 485,771

27
575,903 + 96,318 + 5,585 + 987,865 = ______ 1,665,671

FEEDBACK UNIT 3

This quiz reviews the preceding unit. Answers are at the back of the book. Begin the next unit only after all problems of this unit are understood.

1. 374 + 2,331 + 47 = ______
2. 10,003 + 412 + 78 + 39 + 112 = ______
3. 2,398 + 97 + 783 + 79 = ______
4. 18 + 4,374 + 399 + 415 + 10,473 = ______
5. 6,279 + 848 + 3,958 = ______
6. 28,145 + 46,703 + 2,110 = ______

UNIT 4: WORD PROBLEMS REQUIRING ADDITION

1
The numbers in an addition problem are **addends**. In the problem 8 + 3 = 11 the numbers 8 and 3 are ________.

addends

2
The answer to an addition problem is its **sum** or **total**. In the problem 8 + 3 = 11 the number 11 is the _______.

sum or total

3
The answer to an addition problem is its **sum** or **total**. Find the sum of 87,306 and 422,312.

509,618

4
A rancher has three pastures measuring 4,352 acres, 518 acres, and 47 acres. To find her total pasture land the numbers should be ________.

added

5
Find the total number of acres for the problem in frame 4. _____

4,917 acres

6
Sam Jones owes two bills of $512 and $87. To find the sum of money Sam owes the numbers should be _________.

added

7
Find the sum Sam owes for the problem in frame 6. _____

$599

8
A druggist prepared 45 prescriptions on Monday, 32 on Tuesday, and 54 Wednesday. What was the total number of prescriptions those three days? _____

131 prescriptions

9 ✍
It makes no sense to add 3 cats and 4 dogs because the numbers have different labels - cats and dogs. Is it sensible to add 57 apples and 12 peaches? ______

No

10
Is it sensible to find the sum of 47 Fords and 53 Dodges? ___

No

11
To add two numbers they must have the same label. Is it possible to find the total of 19 cows and 483 cows? _____

Yes

12
To add 15 inches to 1 foot, both measurements need to be expressed as inches. 1 foot equals 12 inches. What is the sum of 15 inches and 1 foot? ______

27 inches

13 ✍
The **perimeter** of a geometric figure is the sum of the lengths of its sides. To find the perimeter of a triangle, _____ the lengths of its three sides.

add

14
The **perimeter** of the quadrilateral shown at the right is the sum of the lengths of the four sides. Find the perimeter. ______

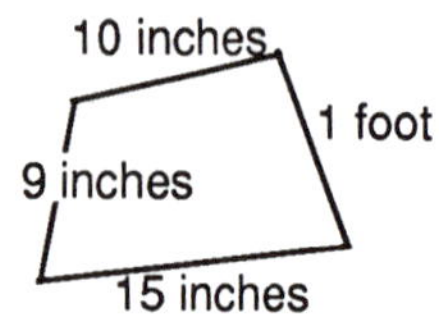

[Note: A **quadrilateral** is a four-sided figure.]

1 foot = 12 inches

12 + 15 + 9 + 10
46 inches is the perimeter

15
The **perimeter** of the **trapezoid** shown at the right is the sum of the lengths of the four sides. Find the perimeter. ______

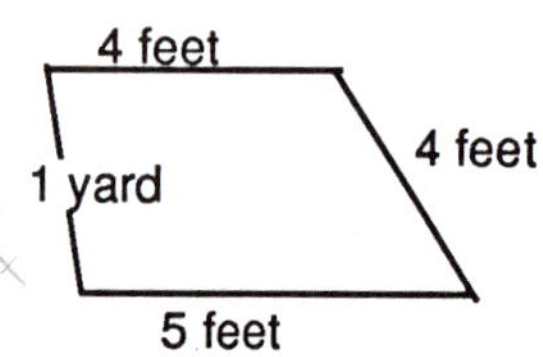

[Note: A **trapezoid** is a quadrilateral with one pair of parallel sides.]

1 yard = 3 feet

3 + 4 + 4 + 5
16 feet is the perimeter

16
The perimeter of the **parallelogram** shown at the right is the sum of the lengths of the four sides. Find the perimeter. ______

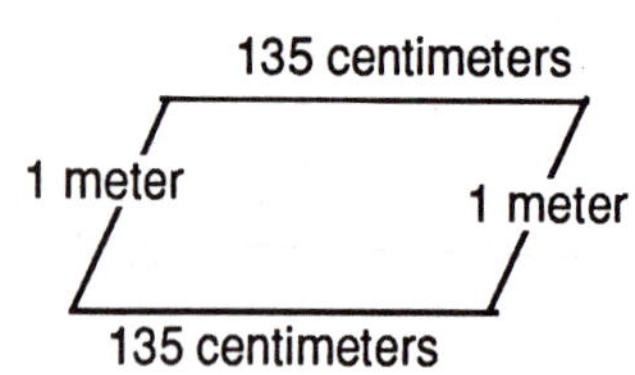

[Note: A **parallelogram** is a quadrilateral with two pairs of parallel sides.]

100 cm = 1 meter

100 + 135 + 100 + 135
470 cm is the perimeter

17
Find the perimeter of the **rectangle** shown at the right.

______

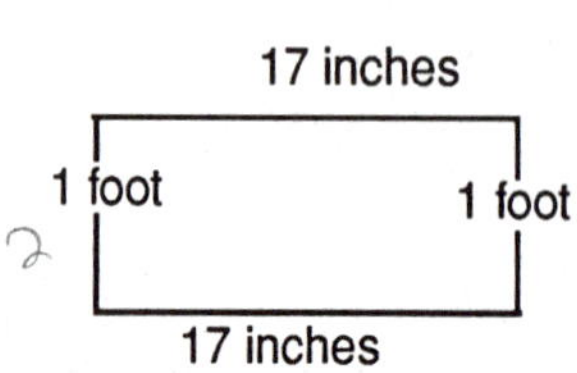

[Note: A **rectangle** is a parallelogram with four right angles.]

12 + 17 + 12 + 17
58 inches is the perimeter

FEEDBACK UNIT 4

This quiz reviews the preceding unit. Answers are at the back of the book.
Take the Chapter Mastery Test only after all problems of this unit are understood.

1. Numbers must not be added if they have ____________(the same, different) labels.

2. A nurse worked 20 days in June, 24 days in July, 19 days in August, and 12 days in September. What was the sum of days worked those four months?

3. Bill loaded the following amounts of concrete onto a truck: 100 lbs, 50 lbs, 25 lbs, 93 lbs. Find the total.

4. Find the sum of Melanie's investments if she has separately invested $46,724 and $9,764.

5. Find the perimeter of the quadrilateral shown at the right.

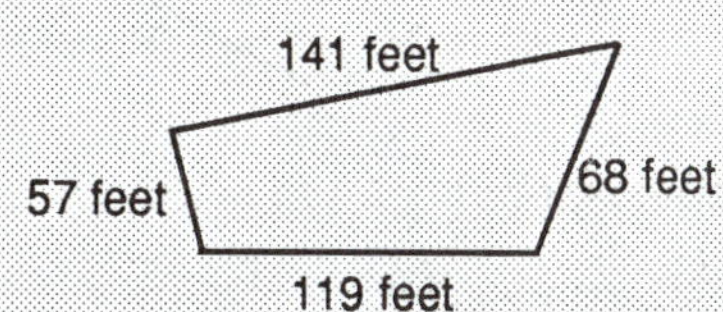

6. Find the perimeter of the trapezoid shown at the right.

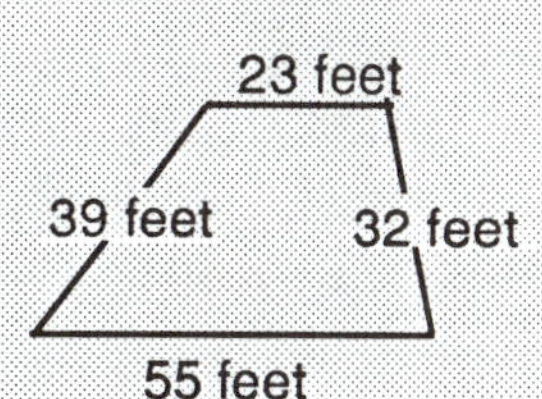

7. Find the perimeter of the parallelogram shown at the right.

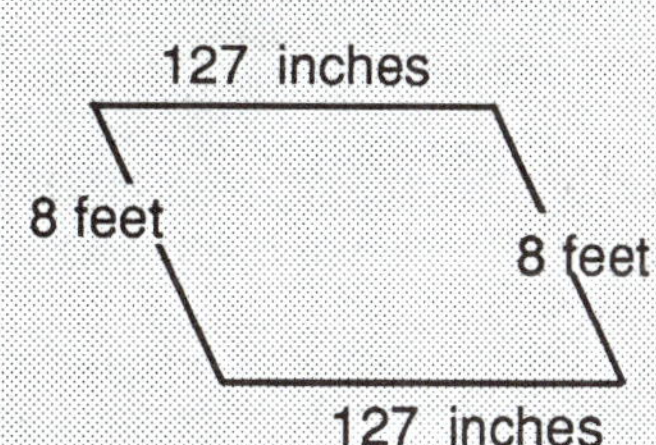

8. Find the perimeter of the rectangle shown at the right.

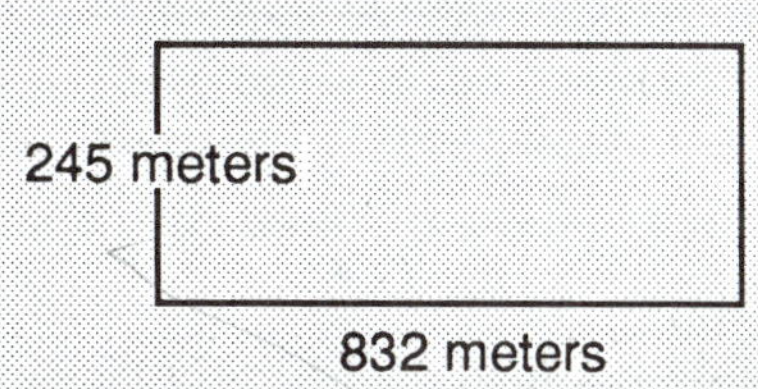

CHAPTER 1 MASTERY TEST

This test provides an opportunity for assuring that **all** of the objectives of Chapter 1 have been learned. Each problem is accompanied by the symbol (U) which indicates the Unit in which that type of problem was studied. If any problems are missed, they need to be corrected. Answers are at the back of the book.

1. 4 + 5 + 8 = ______ (1)

2. 9 + 3 + 7 = ______ (1)

3. What is the tens digit of 4,369? (2)

4. What is the thousands digit of 741,638? (2)

5. The words ________ and ________ mean the answer to an addition problem. (1)

6. The word ________ means the numbers to be added in an addition problem. (1)

7. 743 + 269 = ______ (2)

8. 538 + 894 = ______ (2)

9. 12,458 + 1,893 + 56,732 = __________ (3)

10. 9,245 + 48,407 + 32,612 = __________ (3)

11. The distance from Town A to Town B is 435 miles and from Town B to Town C is 658 miles. Find the sum of the distances. (4)

12. A carpenter earned $319 one week and $67 the next week. What were the total earnings for the two week period? (4)

13. Find the perimeter of the trapezoid shown at the right. (4)

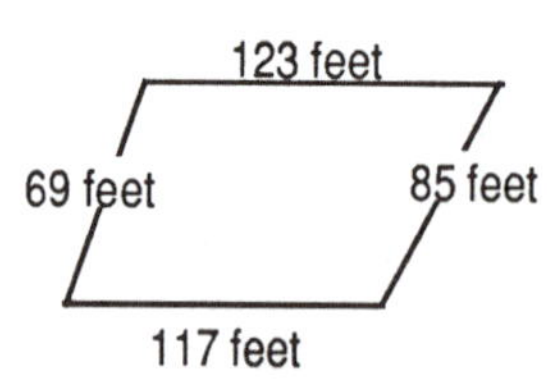

14. Find the perimeter of the rectangle shown at the right. (4)

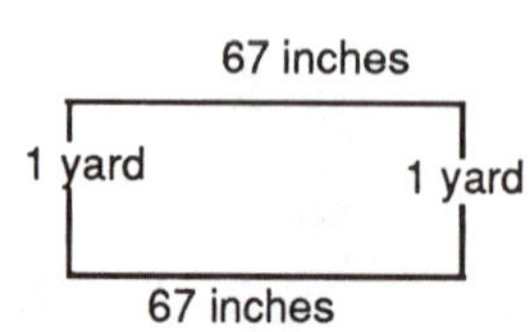

Chapter 2 Objectives

The following problems describe, by example, the skills and knowledge studied in this chapter. Each problem is accompanied by a symbol, (U), which indicates the Unit where the problem is studied. Answers are at the back of the book.

1. Arrange the following numbers in size from smallest to greatest. (1)
   87, 512, 3062, 913, 439, 67

2. 8 − 5 = _____ (2)

3. 7 − 2 = _____ (2)

4. 6 − 8 = _____ (2)

5. 13 − 7 = _____ (2)

6. 4,179 − 253 = ________ (3)

7. 6,367 − 5,143 = ________ (3)

8. 7,006 − 3,142 = ________ (3)

9. 5,000 − 2,963 = ________ (3)

10. 63,408 − 21,937 = ________ (3)

11. School A had 825 sophomores this year and School B had 573. How many more sophomores did School A have than School B? (4)

12. A manufacturer delivered 1358 orders in April and 902 in May. What was the difference in orders delivered for the two months? (4)

Chapter 2
Subtraction Of Whole Numbers

Unit 1: The Distance Between Whole Numbers

The Number Ray

The figure below is a number ray. The dot at the left end of the number ray shows the position of zero (0). The positions of the numbers 0, 1, 2, 3, 4, 5, 6, 7, 8, 9, 10, 11, and 12 are indicated by marks on the number ray.

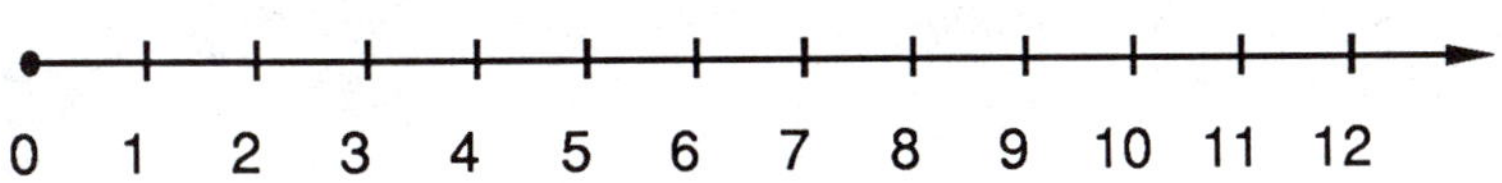

The arrow at the right end of the number ray indicates that the ray continues to the right and has positions for the other whole numbers, such as 13, 14, 15, etc.

1
The figure below shows a number ray. Which capital letter (A) or (B) indicates the position of 3? ______ B

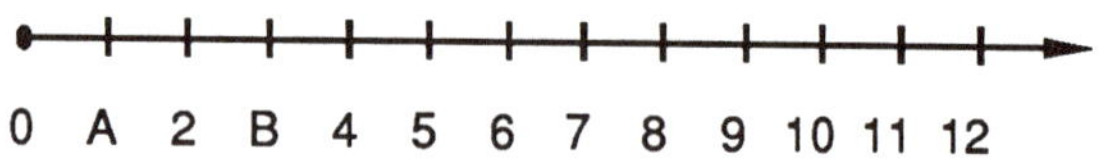

2
On the number ray below, which capital letter shows the position of 4? _____ C

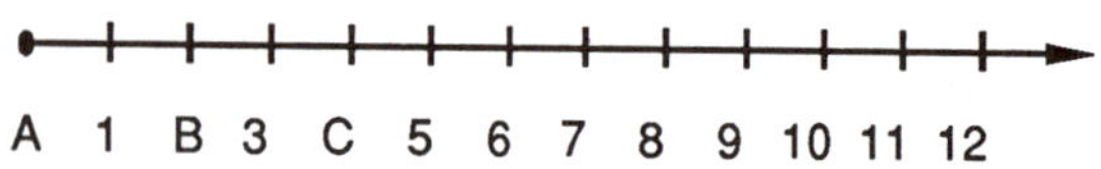

3
On the number ray below, which capital letter shows the position of 7? _____ C

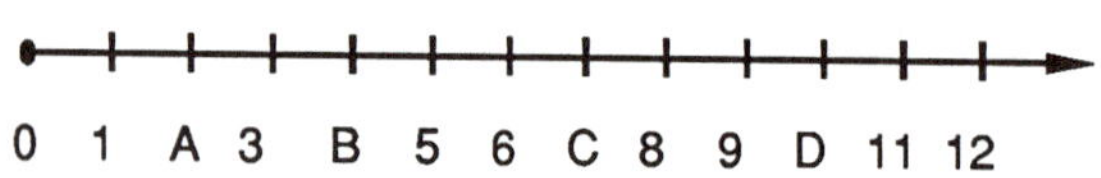

4
On the number ray below, which capital letter shows the position of 5? _____ F

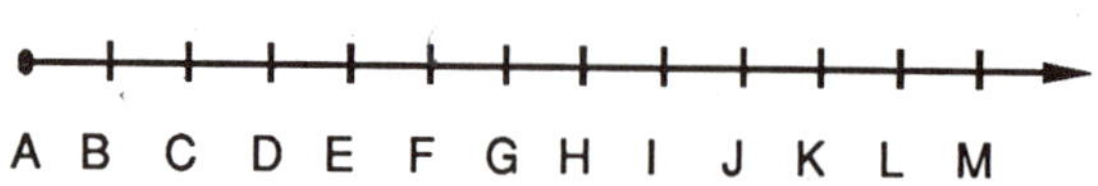

5
The arrow at the right end of the number ray shows that the whole numbers continue. 17 is on the continuation of the number ray. Is 18 on the number ray? _____ Yes

6
47 and 912 are on the number ray. Is 4,677 on the number ray? _____ Yes

7

67 and 849 are on the number ray. Is 93,609 on the number ray? _____ Yes

8 ✍

5 is to the right of 2 on the number ray. This means that 5 is **greater** than 2. Is 6 to the right of 4 on the number ray? _____ (Is 6 greater than 4?) Yes

9

15 is to the right of 8 on the number ray. 15 is greater than 8. Is 17 both to the right of 12 and greater than 12? _____ Yes

10

The size of a number is determined by its position on the number ray. The further to the right a number is, the greater its size. Which number is greater, 19 or 23? _____ 23

11

Which number, 59 or 108, is further to the right on the number ray and therefore greater? _____ 108

12

Which number, 813 or 459, is further to the right on the number ray and therefore greater? _____ 813

13

Which number, 693 or 7,015, is further to the right on the number ray and therefore greater? _____ 7,015

14

25 is greater than 19 because 25 is further to the _____ on the number ray. right

15

Is 28 greater than 12? _____ Yes

16
Is 45 greater than 63? _____ No

17
Is 102,456 greater than 97,412? _____ Yes

18
Is 57,645 greater than 9,999? _____ Yes

19
The figure below shows that 7 is 4 units from 3.
How far is 8 from 5? _____ 3

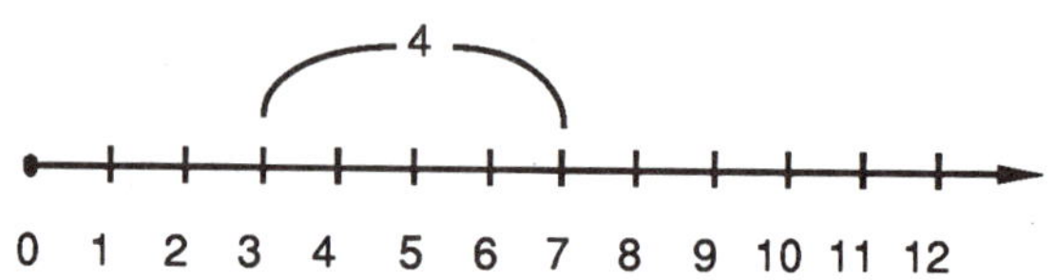

20
6 is 5 units from 1. How far is 9 from 5? _____ 4

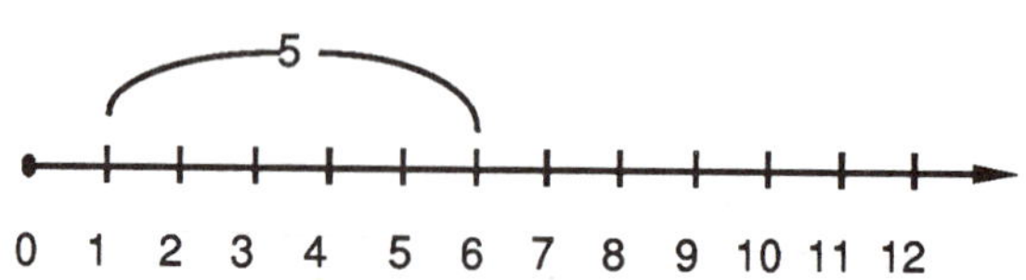

21
How far is 7 from 1? _____ 6

22
How far is 8 from 3? _____ 5

23
How far is 8 from 8? _____ 0

24
How far is 9 from 3? _____ 6

FEEDBACK UNIT 1

This quiz reviews the preceding unit. Answers are at the back of the book. Begin the next unit only after all problems of this unit are understood.

1. On the number ray below what number is indicated by:
   a) C b) J
   c) F d) A

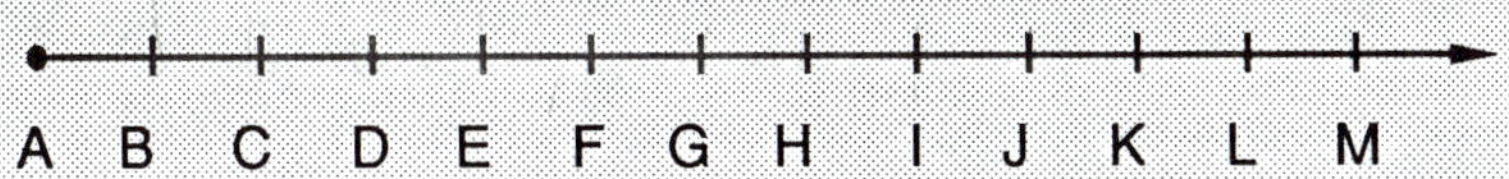

2. Arrange the following numbers in order of size listing the smallest first and the greatest last. 17, 49, 763 39, 8012, 5, 94, 212, 934, 8

3. How far is 8 from 6?

4. How far is 7 from 1?

5. How far is 13 from 9?

6. How far is 15 from 7?

7. How far is 9 from 5?

8. How far is 11 from 3?

UNIT 2: SUBTRACTING WHOLE NUMBERS

SUBTRACTION AND DISTANCE

The question, "How far is it from 2 to 7?" is equivalent to asking, "What is the distance from 2 to 7?"

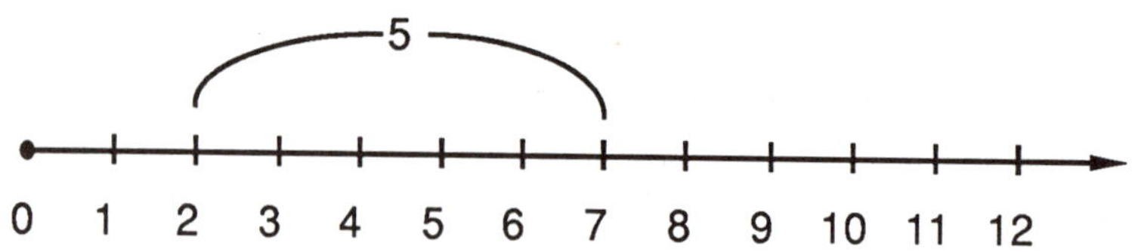

The subtraction problem $7 - 2 =$ _____ has the answer 5. The distance from 2 to 7 is 5.

1
Find the distance from 5 to 8 and answer the subtraction problem $8 - 5 =$ _____. 3

2
Find the answer to $4 - 1 =$ _____ by finding the distance from 1 to 4. 3

3
Find the answer to $12 - 5 =$ _____ by finding the distance from 5 to 12. 7

4
Find the distance from 4 to 6 and find the answer to $6 - 4 =$ _____. 2

5
Find the distance from 5 to 14 and find the answer to $14 - 5 =$ _____. 9

6

7 − 4 = _____

3

7

3 − 1 = _____

2

8

8 − 0 = _____

8

9

6 − 4 = _____

2

10

13 − 8 = _____

5

11

14 − 4 = _____

10

12

16 − 9 = _____

7

13

9 − 7 = _____

2

14

18 − 9 = _____

9

15

12 − 8 = _____

4

16

The answer to a subtraction problem is its **difference**. Find the difference between 7 and 3. _____

4; because 7 − 3 = 4

17
In the subtraction problem $8 - 5 = 3$, the difference is ______. 3

18 ✍
In the subtraction problem $12 - 7 = 5$, the number before the minus sign is the **minuend**, 12. What is the minuend for $14 - 8 = 6$? ______ 14

19
What is the minuend for $7 - 3 = 4$? ______ 7

20 ✍
In the subtraction problem $12 - 7 = 5$, the number after the minus sign is the **subtrahend**, 7. What is the subtrahend for $14 - 8 = 6$? ______ 8

21
What is the minuend for $7 - 3 = 4$? ______ 7

22 ✍
The subtraction problem $4 - 7 =$ ___ has no whole number answer because the subtrahend, 7, is greater than the minuend, 4. Does the problem $5 - 9 =$ ___ have a whole number answer? ______ No

23
When a subtraction problem has no answer because its subtrahend is greater than its minuend use a capital letter X. $4 - 5 = X$ because 5 is greater than ______. 4

24
Which problem below has no whole number answer?

$4 - 1 =$ ___ $1 - 4 =$ ___ $1 - 4 = X$

25

$6 - 10 =$ ______ X

26
$9 - 4 =$ ______ 5

27
$4 - 11 =$ ______ X

28
$8 - 5 =$ _____ 3

29
$4 - 2 =$ ______ 2

30
$0 - 8 =$ ______ X

31
$7 - 5 =$ ______ 2

32
$14 - 5 =$ ______ 9

33
The subtraction problem $6 - 3$ has been written vertically as shown at the right.

$$\begin{array}{r} 6 \\ \underline{-3} \end{array}$$

Write $16 - 8$ vertically.

$$\begin{array}{r} 16 \\ \underline{-8} \end{array}$$

34
Write $7 - 2$ vertically and find the difference.

$$\begin{array}{r} 7 \\ \underline{-2} \\ 5 \end{array}$$

35
Write $12 - 6$ vertically and find the difference.

$$\begin{array}{r} 12 \\ \underline{-6} \\ 6 \end{array}$$

36
Write 14 – 8 vertically and find the difference.

$$\begin{array}{r} 14 \\ \underline{-8} \\ 6 \end{array}$$

37
Write 8 – 15 vertically and find the difference..

$$\begin{array}{r} 8 \\ \underline{-15} \\ \text{X} \end{array}$$

FEEDBACK UNIT 2

This quiz reviews the preceding unit. Answers are at the back of the book. Begin the next unit only after all problems of this unit are understood.

1. 11 – 7 = _____
2. 15 – 7 = _____
3. 5 – 9 = _____
4. 7 – 14 = _____
5. 13 – 6 = _____
6. 12 – 7 = _____
7. 5 – 3 = _____
8. 0 – 6 = _____
9. 2 – 0 = _____
10. 6 – 3 = _____

UNIT 3: SUBTRACTING REQUIRING "BORROWING"

COLUMN SUBTRACTION

The subtraction problem 8,437 − 205 = _____ is written vertically at the right.

The subtraction is completed by separately subtracting each column from right to left. Always begin any vertically arranged subtraction problem by subtracting the column at the far right.

$$\begin{array}{r} 8437 \\ -205 \\ \hline 8232 \end{array}$$

1
8,437 − 205 = _____ is written vertically at the right.

$$\begin{array}{r} 8437 \\ -205 \\ \hline \end{array}$$

Subtract only in the one's column. _____

$$\begin{array}{r} 8437 \\ -205 \\ \hline 2 \end{array}$$

2
Subtraction begins in the one's column. Complete the subtraction problem shown at the right by subtracting the columns right to left. _____

$$\begin{array}{r} 8437 \\ -205 \\ \hline 2 \end{array}$$

$$\begin{array}{r} 8437 \\ -205 \\ \hline 8232 \end{array}$$

3
Write the subtraction problem 265 − 42 = ____ vertically. _____

$$\begin{array}{r} 265 \\ -42 \\ \hline \end{array}$$

4
Subtract only in the one's column in the problem shown at the right. _____

$$\begin{array}{r} 265 \\ -42 \\ \hline \end{array}$$

$$\begin{array}{r} 265 \\ -42 \\ \hline 3 \end{array}$$

5

Complete the subtraction shown at the right by subtracting from right to left.

$$\begin{array}{r} 265 \\ -\underline{42} \\ 3 \end{array}$$

$$\begin{array}{r} 265 \\ -\underline{42} \\ 223 \end{array}$$

6

Write 437 − 125 = _____ vertically and find the answer. _____

$$\begin{array}{r} 437 \\ -\underline{125} \\ 312 \end{array}$$

7

Write 286 − 143 = _____ vertically and find the answer. _____

$$\begin{array}{r} 286 \\ -\underline{143} \\ 143 \end{array}$$

8

Write 637 − 407 = _____ vertically and find the answer. _____

$$\begin{array}{r} 637 \\ -\underline{407} \\ 230 \end{array}$$

9

Write 155 − 31 = _____ vertically and find the answer. _____

$$\begin{array}{r} 155 \\ -\underline{31} \\ 124 \end{array}$$

10 ✍

The problem 584 − 846 = _____ has no whole number answer because 846 is greater than 584. Does the problem 647 − 34 = _____ have a whole number answer? ___

Yes, 34 is less than 647

11

Subtraction with a whole number answer can only occur when the minuend is greater than the subtrahend. Does 497 − 812 = ____ have a whole number answer? ___

No

12
If a subtraction problem has no whole number answer (the subtrahend is greater than the minuend) it should be indicated by an X. 558 – 913 = X because 913, the subtrahend, is ________ than 558, the minuend.

greater

13
989 – 101 = _____

$$\begin{array}{r} 989 \\ -101 \\ \hline 888 \end{array}$$

14
183 – 291 = _____

X

15
397 – 104 = _____

$$\begin{array}{r} 397 \\ -104 \\ \hline 293 \end{array}$$

16 ✍
The problem 483 – 219 = ____ requires "borrowing." 9 cannot be subtracted from 3 in the one's column because 9 is ________ than 3.

greater

17
483 is 4 hundreds, 8 tens, 3 ones but it can also be written as: 4 hundreds, 7 tens, ___ ones.

13

18
The borrowing process is shown at the right.

$$\begin{array}{r} 483 \\ -219 \\ \hline \end{array} \qquad \begin{array}{r} 4\,\overset{7}{\not{8}}\,{}^{1}3 \\ -\,2\;1\;9 \\ \hline \end{array}$$

Can each column now be subtracted? ______

Yes

19
In subtracting the problem shown at the right, which column requires borrowing? ______

$$\begin{array}{r} 546 \\ -\underline{127} \end{array}$$

Ones column
7 is greater than 6

20
In subtracting 546 − 127, borrowing requires that 546 be written as: 5 hundreds, ___ tens, ___ ones.

3, 16

21
The borrowing process is shown at the right.

$$\begin{array}{r} 546 \\ -\underline{127} \end{array} \qquad \begin{array}{r} \scriptstyle 3 \\ 5\,\not{4}\,{}^{1}6 \\ -\underline{1\;2\;7} \end{array}$$

Find the answer (difference). ______

419

22
Will the problem shown at the right require borrowing?

$$\begin{array}{r} 6157 \\ -\underline{4032} \end{array}$$

______

No

23
Will the problem shown at the right require borrowing?

$$\begin{array}{r} 4512 \\ -\underline{2042} \end{array}$$

______

Yes

24
Which column requires borrowing in the problem shown at the right?

$$\begin{array}{r} 4512 \\ -\underline{2042} \end{array}$$

______

Tens, 4 is greater than 1

25
Show the borrowing necessary to complete the subtraction shown at the right. ______

$$\begin{array}{r} 4512 \\ -\underline{2042} \end{array}$$

$$\begin{array}{r} \scriptstyle 4 \\ 4\,\not{5}\,{}^{1}1\,2 \\ -\underline{2\;0\;4\;2} \end{array}$$

26
Subtract 4,512 − 2,042. ______

$$\begin{array}{r} 4\ \overset{4}{\not 5}\ {}^{1}1\ 2 \\ -\ 2\ 0\ 4\ 2 \\ \hline 2\ 4\ 7\ 0 \end{array}$$

27
Will the problem shown at the right require borrowing? ______

$$\begin{array}{r} 4663 \\ -4381 \\ \hline \end{array}$$

Yes

28
Which column requires borrowing in the problem shown at the right? ______

$$\begin{array}{r} 4663 \\ -4381 \\ \hline \end{array}$$

Tens, 8 is greater than 6

29
Show the borrowing necessary to complete the subtraction shown at the right. ______

$$\begin{array}{r} 4663 \\ -4381 \\ \hline \end{array}$$

$$\begin{array}{r} 4\ \overset{5}{\not 6}\ {}^{1}6\ 3 \\ -\ 4\ 3\ 8\ 1 \\ \hline \end{array}$$

30
Subtract 4,663 − 4,381. ______

$$\begin{array}{r} 4\ \overset{5}{\not 6}\ {}^{1}6\ 3 \\ -\ 4\ 3\ 8\ 1 \\ \hline 2\ 8\ 2 \end{array}$$

31
Subtract 4,765 − 1,538 and clearly show any borrowing. ______

$$\begin{array}{r} 4\ 7\ \overset{5}{\not 6}\ {}^{1}5 \\ -\ 1\ 5\ 3\ 8 \\ \hline 3\ 2\ 2\ 7 \end{array}$$

32
Subtract 6,184 − 2,048 and clearly show any borrowing. ______

$$\begin{array}{r} 6\ 1\ \overset{7}{\not 8}\ {}^{1}4 \\ -\ 2\ 0\ 4\ 8 \\ \hline 4\ 1\ 3\ 6 \end{array}$$

33
Subtract 5,436 − 4,254 and clearly show any borrowing. ______

$$\begin{array}{r} 3\phantom{\;{}^{1}3\;6} \\ 5\;\not{4}\;{}^{1}3\;6 \\ -\;4\;2\;5\;4 \\ \hline 1\;1\;8\;2 \end{array}$$

34
Subtract 4,738 − 1,835 and clearly show any borrowing. ______

$$\begin{array}{r} 3\phantom{\;{}^{1}7\;3\;8} \\ \not{4}\;{}^{1}7\;3\;8 \\ -\;1\;8\;3\;5 \\ \hline 2\;9\;0\;3 \end{array}$$

35
Subtract 1,747 − 1,651 and clearly show any borrowing. ______

$$\begin{array}{r} 6\phantom{\;{}^{1}4\;7} \\ 1\;\not{7}\;{}^{1}4\;7 \\ -\;1\;6\;5\;1 \\ \hline 9\;6 \end{array}$$

36
Subtract 5,618 − 4,906 and clearly show any borrowing. ______

$$\begin{array}{r} 4\phantom{\;{}^{1}6\;1\;8} \\ \not{5}\;{}^{1}6\;1\;8 \\ -\;4\;9\;0\;6 \\ \hline 7\;1\;2 \end{array}$$

BORROWING IN TWO COLUMNS

The problem 6,518 − 2,544 requires borrowing in two columns. 6,518 is rewritten for the borrowing in the ten's column. Then another borrowing situation occurs in the hundred's column.

$$\begin{array}{r} 6518 \\ -\;2544 \\ \hline \end{array} \quad \text{becomes} \quad \begin{array}{r} 4\phantom{\;{}^{1}1\;8} \\ 6\;\not{5}\;{}^{1}1\;8 \\ -\;2\;5\;4\;4 \\ \hline \end{array}$$

$$\text{and then it becomes} \quad \begin{array}{r} 5\;14\phantom{\;{}^{1}1\;8} \\ \not{6}\;\not{5}\;{}^{1}1\;8 \\ -\;2\;5\;4\;4 \\ \hline \end{array}$$

37
The problem at the right requires borrowing in ____ (how many) columns?

$$\begin{array}{r} 4756 \\ -\;1829 \\ \hline \end{array}$$

two

38

Write 4,756 for the first borrowing situation that occurs in the subtraction shown at the right.

$$\begin{array}{r} 4756 \\ -\,1829 \\ \hline \end{array}$$

$$\begin{array}{cccc} & & 4 & \\ 4 & 7 & \not{5} & {}^{1}6 \\ -1 & 8 & 2 & 9 \\ \hline \end{array}$$

39

Complete the subtraction shown at the right by clearly showing the second borrowing situation.

$$\begin{array}{cccc} & & 4 & \\ 4 & 7 & \not{5} & {}^{1}6 \\ -1 & 8 & 2 & 9 \\ \hline & & 2 & 7 \end{array}$$

$$\begin{array}{cccc} 3 & & 4 & \\ \not{4} & {}^{1}7 & \not{5} & {}^{1}6 \\ -1 & 8 & 2 & 9 \\ \hline 2 & 9 & 2 & 7 \end{array}$$

40

Complete the subtraction shown at the right by clearly showing both of the necessary borrowings.

$$\begin{array}{cccc} 9 & 5 & 1 & 6 \\ -4 & 2 & 6 & 7 \\ \hline \end{array}$$

$$\begin{array}{cccc} & 4 & 10 & \\ 9 & \not{5} & \not{1} & {}^{1}6 \\ -4 & 2 & 6 & 7 \\ \hline 5 & 2 & 4 & 9 \end{array}$$

41

Complete the subtraction shown at the right by clearly showing both of the necessary borrowings.

$$\begin{array}{cccc} 9 & 3 & 1 & 6 \\ -5 & 6 & 0 & 9 \\ \hline \end{array}$$

$$\begin{array}{cccc} 8 & & 0 & \\ \not{9} & {}^{1}3 & \not{1} & {}^{1}6 \\ -5 & 6 & 0 & 9 \\ \hline 3 & 7 & 0 & 7 \end{array}$$

42

Complete the subtraction shown at the right by clearly showing both of the necessary borrowings.

$$\begin{array}{cccc} 4 & 6 & 8 & 2 \\ -3 & 9 & 4 & 7 \\ \hline \end{array}$$

$$\begin{array}{cccc} 3 & & 7 & \\ \not{4} & {}^{1}6 & \not{8} & {}^{1}2 \\ -3 & 9 & 4 & 7 \\ \hline & 7 & 3 & 5 \end{array}$$

43

Complete the subtraction shown at the right. Clearly show all necessary borrowings.

$$\begin{array}{cccc} 9 & 4 & 5 & 3 \\ -2 & 2 & 8 & 1 \\ \hline \end{array}$$

$$\begin{array}{cccc} & 3 & & \\ 9 & \not{4} & {}^{1}5 & 3 \\ -2 & 2 & 8 & 1 \\ \hline 7 & 1 & 7 & 2 \end{array}$$

44

Complete the subtraction shown at the right. Clearly show all necessary borrowings.

$$\begin{array}{r} 4\ 9\ 2\ 7 \\ \underline{-3\ 1\ 8\ 2} \end{array}$$

$$\begin{array}{rrrr} & 8 & & \\ 4 & \not{9} & {}^{1}2 & 7 \\ \hline -3 & 1 & 8 & 2 \\ \hline 1 & 7 & 4 & 5 \end{array}$$

45

Complete the subtraction shown at the right. Clearly show all necessary borrowings.

$$\begin{array}{r} 6\ 7\ 3\ 8 \\ \underline{-4\ 9\ 7\ 3} \end{array}$$

$$\begin{array}{rrrr} 5 & 16 & & \\ \not{6} & \not{7} & {}^{1}3 & 8 \\ \hline -4 & 9 & 7 & 3 \\ \hline 1 & 7 & 6 & 5 \end{array}$$

Borrowing In Several Columns

To subtract 8,000 – 6,257 borrowing is done in three stages because borrowing is necessary in the ones, tens, and hundreds columns.

These three successive borrowings are shown below.

$$\begin{array}{r} 8000 \\ \underline{-6257} \end{array} \qquad \begin{array}{rrrr} 7 & & & \\ \not{8} & {}^{1}0 & 0 & 0 \\ \hline -6 & 2 & 5 & 7 \end{array} \qquad \begin{array}{rrrr} 7 & 9 & & \\ \not{8} & {}^{1}\not{0} & {}^{1}0 & 0 \\ \hline -6 & 2 & 5 & 7 \end{array} \qquad \begin{array}{rrrr} 7 & 9 & 9 & \\ \not{8} & {}^{1}\not{0} & {}^{1}\not{0} & {}^{1}0 \\ \hline -6 & 2 & 5 & 7 \\ \hline 1 & 7 & 4 & 3 \end{array}$$

46

6,000 is 6 thousands, 0 hundreds, 0 tens, and 0 ones, but is also 5 thousands, 10 hundreds, 0 tens, 0 ones. Is it also 5 thousands, 9 hundreds, 9 tens, and 10 ones?

Yes

47

The four problems below show the borrowing that is needed to subtract 6,000 – 3,179. Complete the subtraction.

$$\begin{array}{r} 6000 \\ \underline{-3179} \end{array} \qquad \begin{array}{rrrr} 5 & & & \\ \not{6} & {}^{1}0 & 0 & 0 \\ \hline -3 & 1 & 7 & 9 \end{array} \qquad \begin{array}{rrrr} 5 & 9 & & \\ \not{6} & {}^{1}\not{0} & {}^{1}0 & 0 \\ \hline -3 & 1 & 7 & 9 \end{array} \qquad \begin{array}{rrrr} 5 & 9 & 9 & \\ \not{6} & {}^{1}\not{0} & {}^{1}\not{0} & {}^{1}0 \\ \hline -3 & 1 & 7 & 9 \end{array}$$

2,821

48
Complete the subtraction shown at the right. Clearly show all necessary borrowings.

$$\begin{array}{r} 5000 \\ -2043 \\ \hline \end{array}$$

```
  4  9  9
  5 ¹0 ¹0 ¹0
- 2  0  4  3
  2  9  5  7
```

49
Complete the subtraction shown at the right. Clearly show all necessary borrowings.

$$\begin{array}{r} 4000 \\ -1564 \\ \hline \end{array}$$

```
  3  9  9
  4 ¹0 ¹0 ¹0
- 1  5  6  4
  2  4  3  6
```

50
Subtract.

$$\begin{array}{r} 2000 \\ -1547 \\ \hline \end{array}$$

```
  1  9  9
  2 ¹0 ¹0 ¹0
- 1  5  4  7
     4  5  3
```

51
Subtract.

$$\begin{array}{r} 50,406 \\ -15,698 \\ \hline \end{array}$$

```
  4  9 13  9
  5 ¹0 ¹4 ¹0 ¹6
- 1  5  6  9  8
  3  4  7  0  8
```

52
Subtract.

$$\begin{array}{r} 63,819 \\ -25,947 \\ \hline \end{array}$$

```
  5 12 17
  6 ¹3 ¹8 ¹1  9
- 2  5  9  4  7
  3  7  8  7  2
```

53
Subtract.

$$\begin{array}{r} 47,832 \\ -26,278 \\ \hline \end{array}$$

```
        7 12
  4  7  8 ¹3 ¹2
- 2  6  2  7  8
  2  1  5  5  4
```

54
Subtract.

$$\begin{array}{r} 40,340 \\ -\ \ 6,186 \\ \hline \end{array}$$

```
  3     2 13
  4 ¹0  3 ¹4 ¹0
-    6  1  8  6
  3  4  1  5  4
```

55
Subtract.

$$\begin{array}{r} 50{,}035 \\ -\ 19{,}478 \\ \hline \end{array}$$

$$\begin{array}{r} \overset{4}{\cancel{5}}\ \overset{9}{{}^{1}\cancel{0}}\ \overset{9}{{}^{1}\cancel{0}}\ \overset{12}{{}^{1}\cancel{3}}\ {}^{1}5 \\ -\ 1\ \ 9\ \ 4\ \ 7\ \ 8 \\ \hline 3\ \ 0\ \ 5\ \ 5\ \ 7 \end{array}$$

FEEDBACK UNIT 3

This quiz reviews the preceding unit. Answers are at the back of the book. Begin the next unit only after all problems of this unit are understood.

Subtract. Use an X to answer any problem that has no whole number answer.

1. 405 − 128 = ______
2. 852 − 347 = ______
3. 560 − 259 = ______
4. 801 − 583 = ______
5. 6,415 − 7,318 = ______
6. 4,819 − 2,742 = ______
7. 7,000 − 1,693 = ______
8. 10,806 − 8,597 = ______
9. 40,037 − 6,859 = ______
10. 62,507 − 15,894 = ______

Unit 4: Word Problems Requiring Subtraction

1
The numbers in a subtraction problem are the **minuend** and **subtrahend**.

4653	**minuend**
$\underline{-3197}$	**subtrahend**

In the problem above, the **minuend** is 4,653 and the **subtrahend** is ______.

3,197

2
The **subtrahend** of the problem at the right is 3,197 and the **minuend** is ______.

$$\begin{array}{r} 4653 \\ \underline{-3197} \end{array}$$

4,653

3
The **minuend** of 5 – 3 is 5 and the **subtrahend** is 3. What is the **minuend** of 56 – 47?______

56

4
What is the **subtrahend** of 963 – 673? ______

673

5
The **minuend** of 6,715 – 2,438 is ______.

6,715

6
The **subtrahend** of 4,693 – 1,946 is ______.

1,946

7
In an addition problem the numbers being added are called **addends**. In a subtraction problem the numbers are the ______ and ______.

minuend, subtrahend

8
An addition problem has two or more addends. Does a subtraction problem have two or more minuends?______

No, only one

9
The answer to an addition problem is called its sum. What is the sum of 476 + 51? ______

527

10 ✍

The answer to a subtraction problem is called its **difference**. What is the **difference** of $57 - 41 = 16$? ______ 16

11

In the subtraction problem $12 - 5 = 7$ the 7 is called the ______. difference

12

Use the numbers from the problem $83 - 32 = 51$. Find the sum of the subtrahend and the difference. ______ $32 + 51 = 83$

13

At the right are shown a subtraction and a related addition problem.

$459 - 127 = 332$

$127 + 332 = 459$

The sum of the addition problem is the ______ of the subtraction problem. minuend

14 ✍

To check a subtraction problem, find the sum of the subtrahend and the difference.
To check $9 - 2 = 7$, add ______ and ______. 2 and 7

15

To check a subtraction, add the subtrahend and the difference to see if it equals the ______. minuend

16

Does the subtraction below check? ______
$16 - 9 = 5$ and $9 + 5 = 14$. No, 14 is not the minuend

17

Subtract and check. $573 - 268 =$ ______

$573 - 268 = 305$
$268 + 305 = 573$

18
Subtract and check. 917 − 432 = ______

917 − 432 = 485
432 + 485 = 917

19
Subtract and check. 465 − 199 = ______

465 − 199 = 266
199 + 266 = 465

20
Subtract and check. 600 − 362 = ______

600 − 362 = 238
362 + 238 = 600

21 ✍
Subtraction is used to solve problems which ask "how much more" or "how many more." Would subtraction be used to solve: If Sara has $15 and Mike has $38, how many more dollars does Mike have? ______

Yes, 38 − 15 = 23

22
If Dan weighs 158 pounds and Joe weighs 141 pounds, how much more does Dan weigh?
Would subtraction be used to solve? ______

Yes, 158 − 141 = 17

23
Subtraction is used to solve problems that ask "how much was left." Penny took 78 cents to the store and bought a 43 cent pen. How much money was left?
Would subtraction be used to solve? ______

Yes, 78 − 43 = 35

24
Would subtraction be used to solve? ______
Farmer Smith sold 930 acres from her 2,536 acre ranch. How many acres were left?

Yes,
2,536 − 930 = 1,606

25
In an addition problem should the numbers have the same label? ______

Yes

26 ✍
In both addition and subtraction problems the numbers must have the same labels. Does it make sense to add 5 dogs and 3 cats? ______

No, cats and dogs are different labels.

27
Does it make sense to subtract 47 cows from 58 pigs?
______

No, cows and pigs are different labels.

28
Does it make sense to subtract 97 boxes from 482 boxes? ______

Yes, both labels are the same.

29
If Sam worked 25 days in January and 18 days in February, how many more days did he work in January?
______

25 − 18
7 days

30
A sculptor started a statue from a block of granite weighing 942 pounds. She estimated chipping off 150 pounds. How much granite was left?

942 − 150
792 pounds

FEEDBACK UNIT 4

This quiz reviews the preceding unit. Answers are at the back of the book. Take the Chapter Mastery Test only after all problems of this unit are understood.

1. In the problem 17 − 9 = 8 the minuend is ______.
2. In the problem 47 − 35 = 12 the difference is ______.
3. In the problem 56 − 11 = 45 the subtrahend is ______.
4. Solve and check.
   a. 5,614 − 2,163
   b. 40,936 − 15,487
   c. 6,149 − 2,847
   d. 50,000 − 19,306
5. If a nurse earns $147 a week and a cab driver earns $205 a week, how much more does a cab driver earn each week?
6. If a woman has read 387 pages of a 520 page book, how many pages are left?

CHAPTER 2 MASTERY TEST

This test provides an opportunity for assuring that **all** of the objectives of Chapter 2 have been learned. Each problem is accompanied by the symbol (U) which indicates the Unit in which that type of problem was studied. If any problems are missed, they need to be corrected. Answers are at the back of the book.

1. Arrange the following numbers in size from smallest to largest. (1)
   78, 215, 6032, 139, 394, 76

Subtract.

2. 9 − 5 = _____ (2)

3. 8 − 3 = _____ (2)

4. 5 − 7 = _____ (2)

5. 14 − 9 = _____ (2)

6. 1,749 − 532 = _______ (2)

7. 3,667 − 3,415 = _______ (2)

8. 8,009 − 413 = _______ (2)

9. 7,000 − 3,692 = _______ (3)

10. 57,603 − 28,512 = ________ (3)

11. 9,392 − 478 = ________ (3)

12. 43,681 − 57,314 = _________ (3)

13. It is 486 miles from Town A to Town B and only 195 miles from Town A to Town C. How much further is it from Town A to Town B than from A to C? (4)

14. Farmer A has 853 acres and Farmer B has 290 acres. What is the difference in the size of their farms? (4)

CHAPTER 3 OBJECTIVES

The following problems describe, by example, the skills and knowledge studied in this chapter. Each problem is accompanied by a symbol, (U), which indicates the Unit where the problem is studied. Answers are at the back of the book.

1. The numbers of a multiplication problem are the multiplicand and the multiplier or both may be called ______. (5)

2. The answer to a multiplication problem is called its ______. (5)

3. 47 x 8 (3)

4. 53 x 6 (3)

5. 5,173 x 7 (3)

6. 6,047 x 9 (3)

7. $\begin{array}{r} 342 \\ \underline{\times\ 18} \end{array}$ (4)

8. $\begin{array}{r} 502 \\ \underline{\times\ 37} \end{array}$ (4)

9. $\begin{array}{r} 2835 \\ \underline{\times\ 602} \end{array}$ (4)

10. $\begin{array}{r} 4137 \\ \underline{\times\ 540} \end{array}$ (4)

11. A grocer received 34 cartons of peas with 18 cans in each carton. How many cans of peas were there? (5)

12. Jane Smith pays $185 each month for rent. How much rent will she pay in 12 months? (5)

Chapter 3
Multiplication Of Whole Numbers

Unit 1: Multiples Of Whole Numbers

Multiples Of 6

The list of numbers 6, 12, 18, 24, 30 shows five **multiples** of 6. The first number in the list is 6 and each other number is 6 more than the preceding one.

$$1 \times 6 = 6$$

Addition	Multiplication
6 + 6 = 12	2 x 6 = 12
12 + 6 = 18	3 x 6 = 18
18 + 6 = 24	4 x 6 = 24
24 + 6 = 30	5 x 6 = 30

After 30 the next **multiple** of 6 is 36 because 30 + 6 = 36.

1 ✍
The list 5, 10, 15, 20, 25, 30 shows **multiples** of 5.
The first number in the list is 5 and each other number is
______ more than the preceding one. — 5

2
The list 5, 10, 15, 20, 25, 30 shows **multiples** of 5.
The next number, after 30, in the list is ______. — 35 (30 + 5)

3
The list 5, 10, 15, 20, 25, 30, 35 shows **multiples** of 5.
The next number, after 35, in the list is ______. — 40

4
The list 5, 10, 15, 20, 25, 30, 35, 40 shows eight
multiples of ______. — 5

5
The list 5, 10, 15, 20, 25, 30, 35, 40 shows eight
multiples of 5. After 40, what would be the next
two multiples of 5?______ — 45, 50

6
The list 2, 4, 6, 8, 10 shows five multiples of 2.
After 10, what would be the next multiple of 2? ____ — 12 (10 + 2)

7
The list 2, 4, 6, 8, 10, 12 shows six multiples of ______. — 2

8
Make a list, starting with 2, of eight multiples of 2. ______ — 2,4,6,8,10,12,14,16

9
The list 3, 6, 9, 12, 15 shows five multiples of 3.
After 15, what would the next multiple of 3? ______ — 18 (15 + 3)

10
3, 6, 9, 12, 15 is a list of multiples of 3. What are the next five multiples of 3? ______ — 18,21,24,27,30

11
The list 4, 8, 12, 16, 20 shows five multiples of ______. — 4

12
4, 8, 12, 16, 20. After 20 what is the next multiple of 4? ______ — 24 (20 + 4)

13
4, 8, 12, 16, 20 is a list of five multiples of 4. After 20, what are the next five multiples of 4? ______ — 24,28,32,36,40

14
Start with 2 and make a list of ten multiples of 2. ____________ — 2,4,6,8,10,12, 14,16,18,20

15
Start with 3 and make a list of ten multiples of 3. ____________ — 3,6,9,12,15,18, 21,24,27,30

16
Start with 4 and make a list of ten multiples of 4. ____________ — 4,8,12,16,20,24, 28,32,36,40

17
Start with 5 and make a list of ten multiples of 5. ____________ — 5,10,15,20,25,30, 35,40,45,50

18
Multiples of 6 are 6, 12, 18, 24, 30. After 30, what are the next five multiples of 6? ____________ — 36,42,48,54,60

19
The numbers in the list 7, 14, 21, 28, 35 are multiples of ______. — 7

20
What multiple of 7 comes after 35? ______ 42

21
7, 14, 21, 28, 35 are multiples of 7. After 35, what are the next five multiples of 7? ______ 42,49,56,63,70

22
The numbers 8, 16, 24, 32, 40 are multiples of ______. 8

23
What multiple of 8 comes after 40? ______ 48

24
8, 16, 24, 32, 40 are multiples of 8. After 40, what are the next five multiples of 8? ______ 48,56,64,72,80

25
The numbers in the list 9, 18, 27, 36, 45 are multiples of ______. 9

26
What multiple of 9 comes after 45? ______ 54

27
9, 18, 27, 36, 45 are multiples of 9. After 45, what are the next five multiples of 9? ______ 54,63,72,81,90

28
Start with 6 and make a list of ten multiples of 6. 6,12,18,24,30,36, 42,48,54,60
____

29
Start with 7 and make a list of ten multiples of 7. 7,14,21,28,35,42, 49,56,63,70
____

30
Start with 8 and make a list of ten multiples of 8.
____

8,16,24,32,40,48,
56,64,72,80

31
Start with 9 and make a list of ten multiples of 9.
____

9,18,27,36,45,54,
63,72,81,90

32
4, 8, 12, 16, etc. is a list of multiples of ______.

4

33
7, 14, 21, 28 etc. is a list of multiples of ______.

7

34 ✍
The numbers 30, 35, 40 are three **consecutive** multiples of 5. The numbers 16, 20, 24 are three consecutive multiples of ______.

4

35
32, 40, 48 are three **consecutive** multiples of 8.
24, 30, 36 are three consecutive multiples of ______.

6

36
21, 28, 35 are three consecutive multiples of ______.

7

37
27, 36, 45, 54 are four consecutive multiples of ______.

9

38
18, 21, 24, 27 are four consecutive multiples of ______.

3

FEEDBACK UNIT 1

This quiz reviews the preceding unit. Answers are at the back of the book. Begin the next unit only after all problems of this unit are understood.

1. Make a list of ten multiples of:
   a. 2 c. 7
   b. 5 d. 8

2. 12, 18, 24, 30, 36 are consecutive multiples of ______.

3. 36, 45, 54, 63 are consecutive multiples of ______.

4. 12, 15, 18, 21 are consecutive multiples of ______.

5. After 28, what is the next multiple of 7?

6. After 32, what is the next multiple of 4?

UNIT 2: BASIC MULTIPLICATION FACTS

THE MULTIPLICATION FACTS (ZERO THROUGH NINE)

In this section the basic multiplication facts for the whole numbers 0 through 9 are reviewed.

The table of "facts" is shown at the right. Use the table when necessary, but try to remember its entries.

X	0	1	2	3	4	5	6	7	8	9
0	0	0	0	0	0	0	0	0	0	0
1	0	1	2	3	4	5	6	7	8	9
2	0	2	4	6	8	10	12	14	16	18
3	0	3	6	9	12	15	18	21	24	27
4	0	4	8	12	16	20	24	28	32	36
5	0	5	10	15	20	25	30	35	40	45
6	0	6	12	18	24	30	36	42	48	54
7	0	7	14	21	28	35	42	49	56	63
8	0	8	16	24	32	40	48	56	64	72
9	0	9	18	27	36	45	54	63	72	81

1

0 x 7 = 0
0 x 4 = 0
0 x 0 = 0

Whenever 0 is multiplied by a whole number the answer is ______. 0

2

0 x 8 = ______. 0

3

0 x 3 = ______. 0

4

5 x 0 = 0
2 x 0 = 0

Any whole number multiplied by zero gives ______. 0

5

6 x 0 = ______. 0

6

Whenever zero is a factor in a multiplication problem the answer is ______. 0

7

1 x 8 = 8
1 x 6 = 6
1 x 3 = ______. 3

8

When one (1) is multiplied with any whole number the answer will be that same whole number.

1 x 4 = ______. 4

9

1 x 9 = ______. 9

10

$5 \times 1 = 5$
$2 \times 1 = 2$
$7 \times 1 =$ ______. 7

11

$0 \times 5 =$ ______. 0

12

$1 \times 9 =$ ______. 9

13

$6 \times 0 =$ ______. 0

14

$7 \times 1 =$ ______. 7

15

Five multiples of 2 are: 2, 4, 6, 8, 10. These multiples show that:

$1 \times 2 = 2$
$2 \times 2 = 4$
$3 \times 2 = 6$
$4 \times 2 = 8$
$5 \times 2 =$ ______. 10

16

List nine multiples of 2 starting with 2. ____________ 2,4,6,8,10,12, 14,16,18

17

2, 4, 6, 8, 10, 12, 14, 16, 18

The list of multiples of 2 can be used to answer

$2 \times 6 =$ ______.

Find the sixth number in the list of multiples.______ 12

18

$2 \times 6 = 12$ and 12 is the sixth number in the list of multiples: 2, 4, 6, 8, 10, 12, 14, 16, 18.

Find $2 \times 8 =$ ______ using the list of multiples. 16

19

2 x 7 = ______ and it is the ______ number in the list 2, 4, 6, 8, 10, 12, 14, 16, 18.

14, seventh

20

2 x 3 = ______ and it is the ______ number in the list 2, 4, 6, 8, 10, 12, 14, 16, 18.

6, third

21

2 x 9 = ______ and it is the ______ number in the list 2, 4, 6, 8, 10, 12, 14, 16, 18.

18, ninth

22

3, 6, 9, 12, 15, 18, 21, 24, 27 is a list of the multiples of ______.

3

23

To multiply by 3 think of the multiples of 3.

3, 6, 9, 12, 15, 18, etc.

3 x 4 = ______ and it is the ______ number in the list of multiples.

12, fourth

24

3, 6, 9, 12, 15, 18, 21, 24, 27

3 x 7 = ______ and it is the ______ number in the list of multiples.

21, seventh

25

If necessary, use the multiples of 3.

Find 3 x 8 = ____.

24

26

If necessary, use the multiples of 3.

Find 3 x 5 = _____.

15

27

0 x 6 = ______.

0

28
1 x 7 = ______. 7

29
2 x 4 = ______. 8

30
3 x 9 = ______. 27

31
Write a list of nine multiples of 4 beginning with 4. ____________ 4,8,12,16,20,24,28,32,36

32
Use a list of multiples of 4 to find 4 x 3 = ______. 12

33
Use the multiples of 4 to find 4 x 7 = ______. 28

34
Use the multiples of 4 to find 4 x 5 = ______. 20

35
Use the multiples of 4 to find 4 x 9 = ______. 36

36
0 x 9 = ______ 0

37
2 x 6 = ______ 12

38
3 x 6 = ______ 18

39
1 x 8 = ______ 8

40
4 x 8 = ______ 32

41
Write a list of nine multiples of 5 beginning with 5. ____________ 5,10,15,20,25,30, 35,40,45

42
Use the multiples of 5 to find 5 x 4 = ______. 20

43
Use the multiples of 5 to find 5 x 7 = ______. 35

44
Use the multiples of 5 to find 5 x 9 = ______. 45

45
5 x 6 = ______ 30

46
3 x 7 = ______ 21

47
4 x 8 = ______ 32

48
0 x 7 = ______ 0

49
2 x 9 = ______ 18

50
1 x 1 = ______ 1

51
5 x 5 = ______ 25

52
Make a list of nine multiples of 6 beginning with 6. ____________ 6,12,18,24,30,36, 42,48,54

53
Use the multiples of 6 to find 3 x 6 = ______. 18

54
Use the multiples of 6 to find 6 x 8 = ______. 48

55
Use the multiples of 6 to find 6 x 7 = ______. 42

56
Use the multiples of 6 to find 6 x 6 = ______. 36

57
Use the multiples of 6 to find 6 x 9 = ______. 54

58
3 x 8 = _____ 24

59
4 x 7 = ______ 28

60
2 x 2 = ______ 4

61
6 x 7 = ______ 42

62
Make a list of nine multiples of 7 beginning with 7. ____________ 7,14,21,28,35,42, 49,56,63

63
Use the multiples of 7 to find 7 x 2 = ______.

14

64
Use the multiples of 7 to find 7 x 7 = ______.

49

65
Use the multiples of 7 to find 7 x 6 = ______.

42

66
Use the multiples of 7 to find 7 x 9 = ______.

63

67
0 x 8 = ______

0

68
4 x 4 = ______

16

69
5 x 8 = ______

40

70
6 x 5 = ______

30

71
7 x 8 = ______

56

72
Make a list of nine multiples of 8 beginning with 8. ____________

8,16,24,32,40,48,
56,64,72

73
Use the multiples of 8 to find 8 x 3 = ______.

24

74
Use the multiples of 8 to find 8 x 8 = ______.

64

75
Use the multiples of 8 to find $8 \times 6 =$ ______. 48

76
Use the multiples of 8 to find $8 \times 9 =$ ______. 72

77
$5 \times 9 =$ ______ 45

78
$6 \times 8 =$ ______ 48

79
$3 \times 7 =$ ______ 21

80
$7 \times 6 =$ ______ 42

81
$8 \times 7 =$ ______ 56

82
Make a list of nine multiples of 9 beginning with 9. ____________ 9,18,27,36,45,54,63,72,81

83
Use the multiples of 9 to find $9 \times 4 =$ ______. 36

84
Use the multiples of 9 to find $9 \times 8 =$ ______. 72

85
Use the multiples of 9 to find $9 \times 9 =$ ______. 81

86
$4 \times 8 =$ ______ 32

87
7 x 8 = ______ 56

88
6 x 9 = ______ 54

89
8 x 8 = ______ 64

90
9 x 7 = ______ 63

91 ✍
In multiplying **do not guess** the answer. Make a list of multiples when you are not certain. If you don't know the answer to a multiplication problem you ______ (should, shouldn't) guess. shouldn't

92
If you haven't memorized the multiplication facts, ______ (use, don't use) a list of multiples. use

FEEDBACK UNIT 2

This quiz reviews the preceding unit. Answers are at the back of the book. Begin the next unit only after all problems of this unit are understood.

Multiply.

1. 4 x 7	2. 6 x 9	3. 9 x 7
4. 8 x 6	5. 3 x 8	6. 5 x 9
7. 9 x 4	8. 6 x 6	

UNIT 3: MULTIPLICATION REQUIRING "CARRYING"

MULTIPLYING FACTORS WITH TWO OR MORE DIGITS

The material of this unit requires the use of basic multiplication facts with speed and accuracy. If you are unsure of any multiplication facts review the lists of multiples until they are mastered.

```
  6 3 0 2 1
x         2
1 2 6 0 4 2  = 126,042
```

To multiply 63,021 by 2, each digit of 63,021 is multiplied by 2. This process is shown at the left.

- 2 x 1 one = 2 ones
- 2 x 2 tens = 4 tens
- 2 x 0 hundreds = 0 hundreds
- 2 x 3 thousands = 6 thousands
- 2 x 6 ten thousands = 12 ten thousands or 1 hundred thousand and 2 ten thousands

---

1
Multiply 1,302 by 3 by multiplying each digit of 1,302 by 3.

```
1302
x  3
```

```
1302
x  3
3906
```

2
Multiply 1,304 by 2 by multiplying each digit of 1,304 by 2.

```
1304
x  2
```

```
1304
x  2
2608
```

3
Multiply.

```
4310
x  2
```

```
4310
x  2
8620
```

4
Multiply.

$$\begin{array}{r} 2012 \\ \underline{\times \quad 4} \end{array} \qquad \begin{array}{r} 2012 \\ \underline{\times \quad 4} \\ 8048 \end{array}$$

5
Multiply.

$$\begin{array}{r} 3104 \\ \underline{\times \quad 2} \end{array} \qquad \begin{array}{r} 3104 \\ \underline{\times \quad 2} \\ 6208 \end{array}$$

6
Multiply.

$$\begin{array}{r} 1021 \\ \underline{\times \quad 3} \end{array} \qquad \begin{array}{r} 1021 \\ \underline{\times \quad 3} \\ 3063 \end{array}$$

7
Multiply.

$$\begin{array}{r} 5022 \\ \underline{\times \quad 4} \end{array} \qquad \begin{array}{r} 5022 \\ \underline{\times \quad 4} \\ 20088 \end{array}$$

8
Multiply.

$$\begin{array}{r} 9013 \\ \underline{\times \quad 2} \end{array} \qquad \begin{array}{r} 9013 \\ \underline{\times \quad 2} \\ 18026 \end{array}$$

9
Multiply.

$$\begin{array}{r} 7123 \\ \underline{\times \quad 3} \end{array} \qquad \begin{array}{r} 7123 \\ \underline{\times \quad 3} \\ 21369 \end{array}$$

10
Multiply.

$$\begin{array}{r} 8221 \\ \underline{\times \quad 4} \end{array} \qquad \begin{array}{r} 8221 \\ \underline{\times \quad 4} \\ 32884 \end{array}$$

"CARRYING" IN MULTIPLICATION

$$\begin{array}{r} {\scriptstyle 1\,2} \\ 37 \\ \underline{\times \quad 4} \\ 148 \end{array} = 148$$

4 x 7 ones = 28 ones or 2 tens, 8 ones

4 x 3 tens = 12 tens plus 2 (carried) tens
= 14 tens or 1 hundred, 4 tens

The multiplication example shown at the left requires "carrying." The "carrying" in multiplication is shown. Show your "carrying" in the problems that follow.

11
At the right is shown the "carrying" in the multiplication of 19 by 6.

```
 1 5
   1 9
 x   6
 1 1 4
```

Multiply 43 by 6 and show the "carrying." ______

```
 2 1
   4 3
 x   6
 2 5 8
```

12
Multiply.

```
  85
 x 9
```

```
 7 4
   8 5
 x   9
 7 6 5
```

13
Multiply.

```
  76
 x 7
```

```
 5 4
   7 6
 x   7
 5 3 2
```

```
 4 3
   5 4 1
 x     8
 4 3 2 8  = 4,328
```

- 8 x 1 one
- 8 x 4 tens = 3 hundreds (carried) plus 2 tens
- 8 x 5 hundreds plus 3 (carried) hundreds = 43 hundreds or = 4 thousands plus 3 hundreds

In multiplying 541 by 8, each digit of 541 is multiplied by 8. Because 8 x 4 tens = 32 tens, carrying is necessary in the hundred's column. The multiplication is shown at the left.

14
The problem at the right requires "carrying" in the hundred's column. Do the problem and show your "carrying." ______

```
   2 4 1
 x     6
```

```
 1 2
   2 4 1
 x     6
 1 4 4 6
```

15
Multiply and clearly show your "carrying."

$$\begin{array}{r} 572 \\ \times\ \ 4 \\ \hline \end{array}$$

$$\begin{array}{l} \mathbf{2\ 2} \\ \quad 572 \\ \underline{\times \qquad 4} \\ 2288 \text{ or } 2{,}288 \end{array}$$

16
Multiply.

$$\begin{array}{r} 351 \\ \times\ \ 7 \\ \hline \end{array}$$

$$\begin{array}{l} \mathbf{2\ 3} \\ \quad 351 \\ \underline{\times \qquad 7} \\ 2457 \text{ or } 2{,}457 \end{array}$$

17
Multiply.

$$\begin{array}{r} 480 \\ \times\ \ 9 \\ \hline \end{array}$$

$$\begin{array}{l} \mathbf{4\ 7} \\ \quad 480 \\ \underline{\times \qquad 9} \\ 4320 \text{ or } 4{,}320 \end{array}$$

$$\begin{array}{l} \mathbf{1\ 1} \\ \quad 6532 \\ \underline{\times \qquad\ \ 3} \\ 19596 = 19{,}596 \end{array}$$

In the problem at the left, "carrying" is necessary in the thousand's column.

- 3 x 2 ones
- 3 x 3 tens
- 3 x 5 hundreds = 15 hundreds or 5 hundreds and 1 thousand
- 3 x 6 thousands plus 1 (carried) thousand = 19 thousands or = 1 ten thousand and 9 thousands

---

18
Multiply.

$$\begin{array}{r} 5802 \\ \times\ \ \ 4 \\ \hline \end{array}$$

$$\begin{array}{l} \mathbf{2\ 3} \\ \quad 5802 \\ \underline{\times \qquad\ \ 4} \\ 23208 \text{ or } 23{,}208 \end{array}$$

19
Multiply.

```
  6812
x    3
```

```
 2 2
  6812
x    3
 20436 or 20,436
```

20
Multiply.

```
  9801
x    7
```

```
 6 5
  9801
x    7
 68607 or 68,607
```

```
2 3 1
  463
x   5
 2315 = 2,315
```

The multiplication shown at the left requires "carrying" in three columns.

- 5 x 3 ones
- 5 x 6 tens plus one (carried) ten
- 5 x 4 hundreds plus 3 (carried) hundreds

21
Multiply.

```
  362
x   7
```

```
2 4 1
  362
x   7
 2534 or 2,534
```

22
Multiply.

```
  6284
x    3
```

```
1  2 1
  6284
x    3
 18852 or 18,852
```

23
Multiply.

```
 5 7 3 1
x      6
```

```
3 4 1
  5 7 3 1
x       6
3 4 3 8 6 or 34,386
```

24
The "carrying" in multiplication can be done in your head. Try to do the problem at the right without writing any of the carrying symbols.

```
 5783
x   6
```

34,698

25
Multiply without writing any carrying symbols.

```
 4156
x   4
```

16,624

26
Multiply without writing any carrying symbols.

```
 2023
x   6
```

12,138

27
Multiply.

```
 9619
x   4
```

38,476

28
Multiply.

```
 7549
x   6
```

45,294

29
Multiply.

```
 6287
x   9
```

56,583

FEEDBACK UNIT 3

This quiz reviews the preceding unit. Answers are at the back of the book. Begin the next unit only after all problems of this unit are understood.

Multiply.

1. $\begin{array}{r} 5173 \\ \underline{\times\ 2} \end{array}$

2. $\begin{array}{r} 4536 \\ \underline{\times\ 2} \end{array}$

3. $\begin{array}{r} 6154 \\ \underline{\times\ 7} \end{array}$

4. $\begin{array}{r} 3157 \\ \underline{\times\ 6} \end{array}$

5. $\begin{array}{r} 4936 \\ \underline{\times\ 5} \end{array}$

6. $\begin{array}{r} 8159 \\ \underline{\times\ 9} \end{array}$

7. $\begin{array}{r} 9317 \\ \underline{\times\ 3} \end{array}$

8. $\begin{array}{r} 4939 \\ \underline{\times\ 4} \end{array}$

9. $\begin{array}{r} 2607 \\ \underline{\times\ 7} \end{array}$

10. $\begin{array}{r} 4883 \\ \underline{\times\ 9} \end{array}$

UNIT 4: TWO-DIGIT MULTIPLIERS

To multiply 67 x 485, each digit of 67 is separately multiplied by 485. This multiplication is shown at the right.

$$\begin{array}{r l} 485 & \\ \underline{\times\ 67} & \\ 3395 & = 7 \text{ ones} \times 485 \\ \underline{2910} & = 6 \text{ tens} \times 485 \\ 32495 & \end{array}$$

The numbers 3395 and 2910 are **partial products**. Because the 6 in 67 is the tens digit, the partial product 2910 is put directly under the 6.

The partial products are added in columns to give the final answer, 32,495. This final answer is the **product** of 67 x 485.

1

To multiply 54 x 914 each of the digits of 54 (5 and 4) is multiplied by 914. The example at the right shows part of the multiplication.

```
 914
 x 54
 3656
```

Complete the multiplication.______

```
  914
  x 54
  3656
 4570
 49356
```

2

To multiply 753 by 64, each digit of 64 is separately multiplied by ______.

753

3

Find the first partial product for:

```
 753
 x 64
```

______

```
  753
  x 64
 3012
```

4

Find the second partial product for:

```
 753
 x 64
 3012
```

______

```
  753
  x 64
  3012
 4518
```

5

Find the final answer for the multiplication problem at the right by adding the partial products in columns. ______

```
  753
  x 64
  3012
 4518
```

```
  753
  x 64
  3012
 4518
 48192
```

6

Multiply 28 x 943 by separately multiplying the digits of 28. ______

```
 943
 x 28
```

```
  943
  x 28
  7544
 1886
 26404
```

7

Multiply 57 x 818 by finding the two partial products and adding them in columns. ______

```
 818
 x 57
```

```
  818
  x 57
 5726
4090
46626
```

8

Multiply.

```
 493
 x 46
```

```
  493
  x 46
 2958
1972
22678
```

9

Multiply.

```
 758
 x 38
```

```
  758
  x 38
 6064
2274
28804
```

```
   619
 x 408
  4952
  000
2476
252552
```

In the example at the left, 619 is multiplied by 408. Notice that three partial products are required; one for each digit of 408. Notice that the partial product when multiplying by zero is zero.

4952 = 8 ones x 619
000 = 0 tens x 619
2476 = 4 hundreds x 619

10

Multiply.

```
 571
 x 305
```

```
   571
 x 305
  2855
  000
1713
174155
```

11
Multiply.

```
  368
x 480
```

```
    368
  x 480
    000
  2944
 1472
 176640
```

12
Multiply.

```
  695
x 908
```

```
    695
  x 908
   5560
   000
 6255
 631060
```

FEEDBACK UNIT 4

This quiz reviews the preceding unit. Answers are at the back of the book. Begin the next unit only after all problems of this unit are understood.

Multiply.

1. 5072 x 48
2. 312 x 96
3. 496 x 58
4. 294 x 87
5. 5890 x 40
6. 2693 x 68
7. 543 x 67
8. 687 x 48
9. 269 x 508
10. 487 x 438

UNIT 5: WORD PROBLEMS REQUIRING MULTIPLICATION

MULTIPLICATION TERMINOLOGY AND APPLICATIONS

In a multiplication problem the numbers being multiplied are **factors**. For example, in the problem 6 x 9 = 54, the numbers 6 and 9 are **factors**. 54 is the product.

6	**factor**
x 9	**factor**
54	**product**

When two numbers are multiplied using a vertical arrangement of the factors, the factor on top is the **multiplicand** and the factor on the bottom is the **multiplier**.

438	**multiplicand**
x 69	**multiplier**
3942	partial product
2628	partial product
30222	**product**

1 ✍

The numbers in a multiplication problem are **factors**. In the problem 6 x 7 = 42, the factors are ___ and ___.

6, 7

✍

2

In the problem 8 x 7 = 56, the **product** is 56 and the factors are ______ and ______.

7, 8

3

What are the factors in 3 x 6 = 18?

______

3, 6

4

In the problem 10 x 73 = 730, the product is 730 and the factors are ______ and ______.

10, 73

5 ✍
When one factor of a multiplication problem is 10, the product will always be the other factor with a zero affixed.

10 x 98 = 980
10 x 45 = ______ — 450

6

10 x 543 = 5,430
10 x 764 = ______ — 7,640

7 ✍
There is a shortcut for multiplying by 100. Affix two zeros to the other factor.

100 x 47 = 4,700
100 x 83 = ______ — 8,300

8 ✍
There is a short cut for multiplying by 1000. Affix three zeros to the other factor.

1000 x 675 = 675,000
1000 x 432 = ______ — 432,000

9 ✍
Word problems requiring multiplication frequently contain the word "each." If the word "each" appears in a word problem then it probably indicates that the numbers should be ______. — multiplied

10
What word often indicates multiplication in a word problem? ______ — each

11
If candy bars cost 15 cents each, then the price of 7 bars is found by __________ 15 and 7. — multiplying

12
What word in the phrase, "Six boys, each weighing 74 pounds" indicates multiplication may be needed? ______

each

13
Six boys, each weighing 74 pounds, weigh a total of ______ pounds.

6 x 74
444 pounds

14
If an orderly works 8 days and each day works 7 hours, how many hours were worked? ______

8 x 7
56 hours

15 ✍
Numbers in a multiplication problem do not have to have the same label. Do numbers in an addition problem have to have the same label? ______

Yes

16
Numbers in a multiplication problem may have different labels. Can numbers in a subtraction problem have different labels? ______

No

17
A therapist needed to spend 14 hours with each of 13 patients. Find the total number of hours needed. _____

182 hours

18
A shipment of sports cars consists of 27 units priced at $9,700 each. What is the value of the shipment? ______

$261,900

19
A geological survey team brought back 548 specimens that weighed an average of 29 kilograms each. Find the total weight of the specimens. ______

15,892 kilograms

20
A bottle of medicine contains 83 pills. Each pill weighs 27 milligrams. What is the weight of the pills? ______

2,241 milligrams

21
If each ticket in a bundle of 2,000 tickets costs 76 cents, what is the cost of the bundle? ______

152,000 cents
or $1,520

22
A case holds 64 bottles. If each bottle contains 55 grams of mercury, how much mercury is in the case? ______

3,520 gm

23
The **area** of the **rectangle** shown at the right is the product of its length and width. Find the area. ______

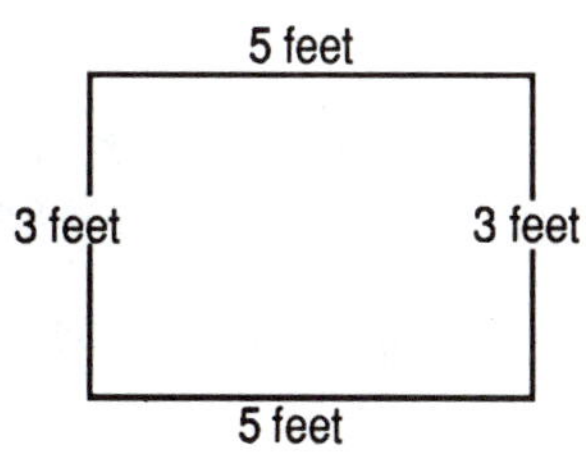

5 x 3 = 15
15 square feet
is the area

24
The **area** of a rectangle is measured in terms of the number of small squares that it can contain. How many 1-inch squares are contained in the rectangle shown at the right? _____

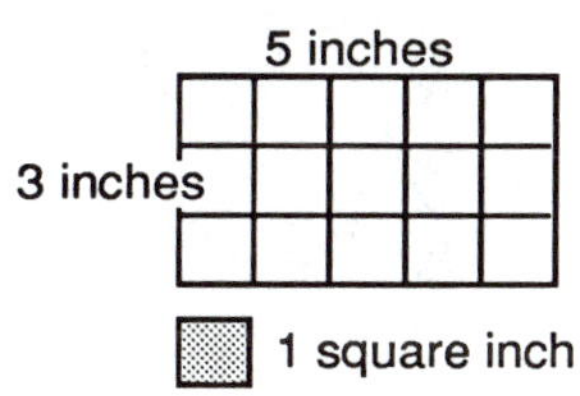

15

25
The formula for finding the area of a **rectangle** is A = LW where L is the length of the rectangle and W is the width. Use the formula to find the area of the rectangle shown at the right. ______

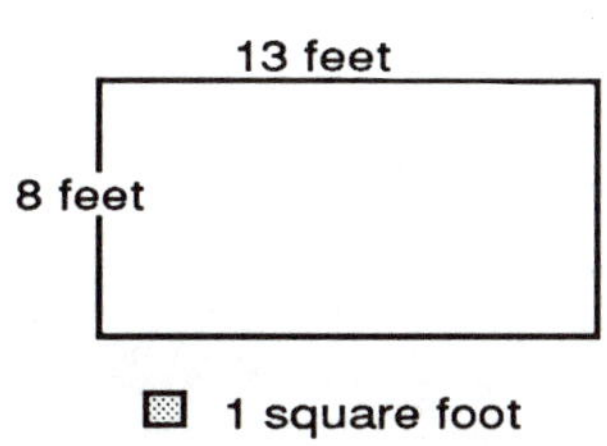

A = L x W
= 13 x 8
104 square feet is
the area

26
Area is always measured in square units such as square inches, square feet, or square centimeters. Find the area of a rectangle with length 20 yards and width 9 yards.

______

A = L x W
= 20 x 9
180 square yards is the area

27
The formula for finding the area of a **parallelogram** is A = BH where B is the base of the parallelogram and H is the height. For the parallelogram shown at the right, the base is 13 meters and the height is _______.
(Note: Height is perpendicular to the base.)

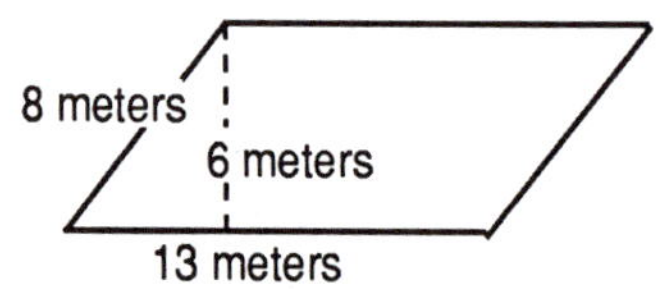

6 meters
8 meters is not the height; it is the length of a side

28
Use the formula A = BH to find the area of the parallelogram shown at the right. ______

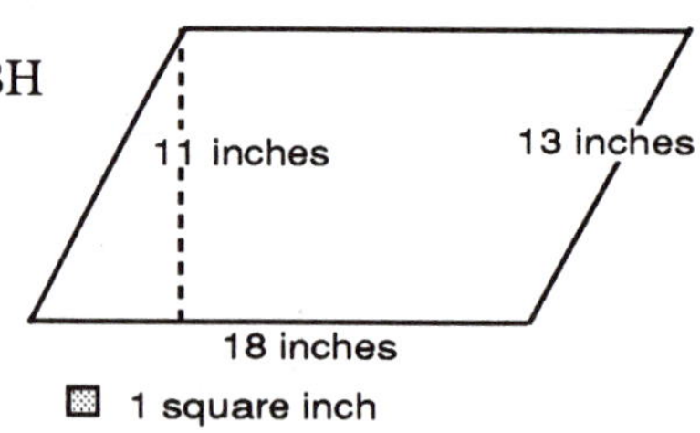

A = B x H
= 18 x 11
198 square inches is the area

29
The distance around a circle can be approximated by multiplying 3 times the length of its diameter. For the circle shown at the right, the diameter is _______.

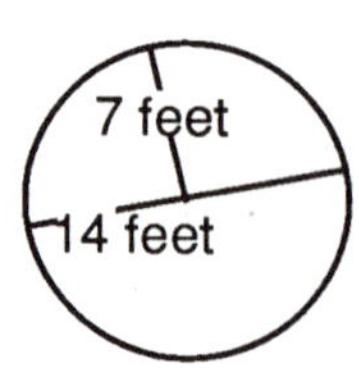

14 feet
7 feet is the radius

30
The distance around a circle is its **circumference**. The circumference can be approximated by multiplying 3 times its diameter. Approximate the circumference of the circle shown at the right.

____

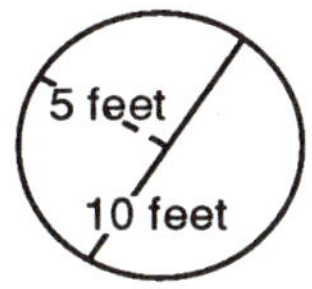

C is approximately
3 x 10 or 30 feet

31
Approximate the circumference of a circle by multiplying its diameter by 3. What is the approximate **circumference** of the circle shown at the right?

____

(d = 2 X 24)
C is approximately
3 x 48 or 144 inches

FEEDBACK UNIT 5

This quiz reviews the preceding unit. Answers are at the back of the book.
Take the Chapter Mastery Test only after all problems of this unit are understood.

1. The numbers in a multiplication problem are called the multiplier and the multiplicand. Another word for either number in a multiplication problem is ______.

2 Use a multiplication shortcut to find the answers for:
   a. 10 x 56 = ______
   b. 10 x 5,678 = ______
   c. 100 x 61 = ______
   d. 100 x 48 = ______

3. Each test tube in a box weighs 57 grams. Find the total weight of the test tubes if there are 144 tubes in the box.

4. Seventeen girl scouts went on a camping trip. Each scout carried a 26 pound backpack. What was the total weight in the backpacks?

5. Find the area of the rectangle shown below.

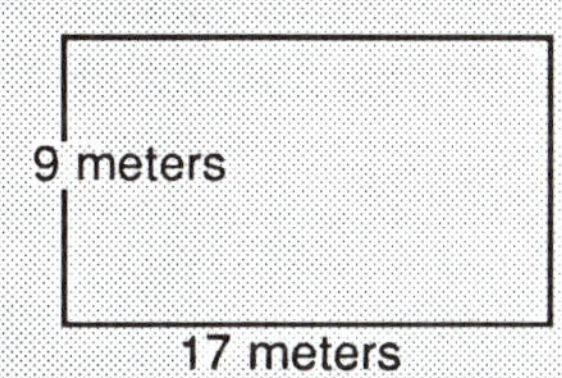

6. Find the area of the parallelogram below.

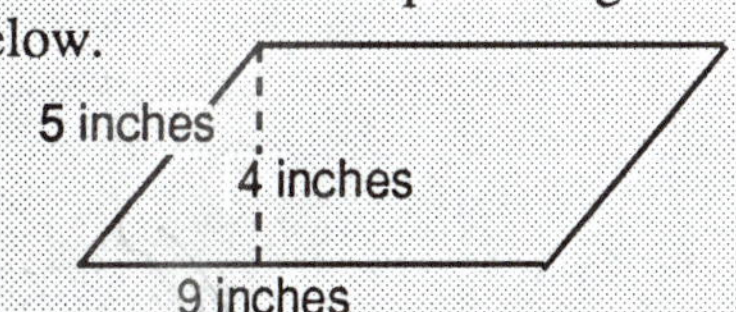

CHAPTER 3 MASTERY TEST

This test provides an opportunity for assuring that **all** of the objectives of Chapter 3 have been learned. Each problem is accompanied by the symbol (U) which indicates the Unit in which that type of problem was studied. If any problems are missed, they need to be corrected. Answers are at the back of the book.

1. The numbers of a multiplication problem are the multiplicand and the multiplier or both may be called ______. (5)

2. The answer to a multiplication problem is its ______. (5)

3.	65 x 8	(3)	4.	35 x 6	(3)	5.	3175 x 7	(3)	6.	4076 x 9	(3)
7.	234 x 18	(4)	8.	205 x 73	(4)	9.	3852 x 602	(4)	10.	3417 x 450	(4)
11.	458 x 10	(5)	12.	508 x 100	(5)	13.	6139 x 100	(5)	14.	432 x 1000	(5)

15. A druggist received 28 cartons with 24 jars of vitamins in each carton. How many vitamin jars were there? (5)

16. Rosa James is paid $204 each week. How much does she earn in 15 weeks? (5)

17. Find the area of the rectangle shown below. (5)

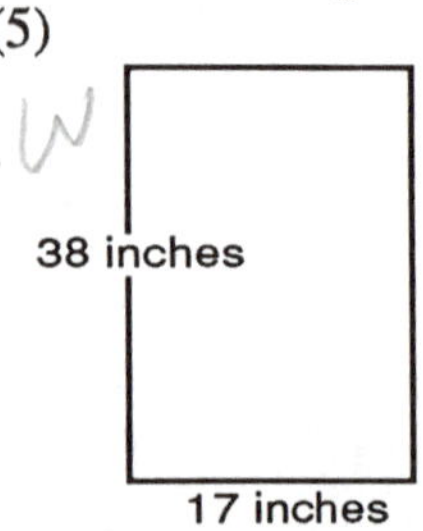

18. Find the area of the parallelogram shown below. (5)

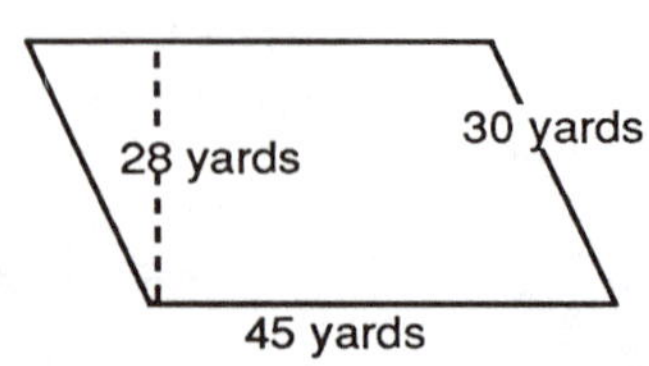

Chapter 4 Objectives

The following problems describe, by example, the skills and knowledge studied in this chapter. Each problem is accompanied by a symbol, (U), which indicates the Unit where the problem is studied. Answers are at the back of the book.

1. In the problem at the right (5)
   the divisor is ______,
   the dividend is ______,
   and the quotient is ______.

$$47\overline{)2726}\quad\text{quotient } 58$$

Divide. (3)

2. $4\overline{)3,126}$
3. $7\overline{)5,263}$
4. $8\overline{)5,063}$
5. $5\overline{)9,146}$

Divide. (4)

6. $6\overline{)1,235}$
7. $23\overline{)5,128}$
8. $37\overline{)36,541}$
9. $57\overline{)41,365}$
10. $93\overline{)72,810}$

11. Find the average of 57, 65, 72, and 66. (5)

12. 480 fluid ounces of saline solution is to be distributed equally into 15 containers. How many fluid ounces should be poured into each container? (5)

CHAPTER 4
DIVISION OF WHOLE NUMBERS

UNIT 1: MULTIPLES OF WHOLE NUMBERS

1

2,4,6,8,10, etc. are multiples of 2.
Is 12 a multiple of 2? ______ Yes

2

2,4,6,8,10,12, etc. are counting number multiples of 2.

$2 \times 0 = 0.$

Is zero (0) also a multiple of 2? ______ Yes

3

0,2,4,6,8,10,12, etc. are whole number multiples of 2. Is 18 a whole number multiple of 2? ______ Yes, $2 \times 9 = 18$

4

0,2,4,6,8, etc. are multiples of 2.
Is 13 a multiple of 2? ______ No

5
3,6,9,12,15, etc. are multiples of ______. 3

6
3,6,9,12,15, etc. are multiples of 3.
Is zero (0) a multiple of 3? ______ Yes, 3 x 0 = 0

7
0,3,6,9,12, etc. are multiples of 3. Yes, 3 x 6 = 18
Is 18 a multiple of 3? ______

8
0,3,6,9,12,15, etc.are multiples of 3. No
Is 26 a multiple of 3? ______

9
0,4,8,12,16, etc. are multiples of ______. 4

10
20 is a multiple of 4 because 4 x 5 = 20.
Is zero (0) a multiple of 4? ______ Yes, 4 x 0 = 0

11
Is 28 a multiple of 4? ______ Yes, 4 x 7 = 28

12
Is 18 a multiple of 4? ______ No

13
0,5,10,15,20, etc. are multiples of ______. 5

14
Is 34 a multiple of 5? ______ No

15
Is zero a multiple of 5? ______ Yes, 5 x 0 = 0

16
0,6,12,18,24, etc. are multiples of ______. | 6

17
Is 42 a multiple of 6? ______ | Yes, 6 x 7 = 42

18
0,7,14,21,28, etc. are multiples of ______. | 7

19
Is 56 a multiple of 7? ______ | Yes, 7 x 8 = 56

20
Is zero a multiple of 7? ______ | Yes 7 x 0 = 0

21
0,8,16,24,32, etc. are multiples of ______. | 8

22
Is 60 a multiple of 8? ______ | No

23
Is zero a multiple of 8? ______ | Yes, 8 x 0 = 0

24
0,9,18,27,36, etc. are multiples of ______. | 9

25
Is 56 a multiple of 9? ______ | No

26
Is zero a multiple of 9? ______ | Yes, 9 x 0 = 0

27
What number is a multiple of every
whole number? ______ | Zero

28
What number can be multiplied by 4 to give 20? ______ 5; $4 \times 5 = 20$

29
What number can be multiplied by 9 to give 36? ______ 4; $9 \times 4 = 36$

30
24 divided by 6 is a division problem. The answer is 4 because $24 = 6 \times$ ______. 4

31
32 divided by 4 is a division problem. The answer is 8 because $32 = 4 \times$ ______. 8

32
$40 \div 8$ is a division problem. The answer is 5 because $40 = 8 \times$ ______. 5

33
$0 \div 7 = 0$ because $0 = 7 \times$ ______. 0

34
$12 \div 4 = 3$ because $12 = 4 \times$ ______. 3

35
$64 \div 8 = 8$ because $64 = 8 \times$ ______. 8

36
$18 \div 9 =$ ______ 2; $9 \times 2 = 18$

37
$12 \div 6 =$ ______ 2; $6 \times 2 = 12$

38
$35 \div 7 =$ ______ 5

39

$54 \div 9 =$ ______

6

40

$6 \div 6 =$ ______

1

41

$0 \div 8 =$ ______

0; $8 \times 0 = 0$

42

$9 \div 3 =$ ______

3

43

$12 \div 4 =$ ______

3

FEEDBACK UNIT 1

This quiz reviews the preceding unit. Answers are at the back of the book. Begin the next unit only after all problems of this unit are understood.

1. Which of the following numbers are multiples of 6? 12, 28, 32, 42, 48
2. Which of the following numbers are multiples of 4? 12, 18, 27, 28, 32
3. Which of the following numbers are multiples of 9? 0, 27, 32, 45, 56
4. Which of the following numbers are multiples of 7? 7, 17, 21, 42, 56
5. $63 \div 7$
6. $0 \div 6$
7. $18 \div 3$
8. $8 \div 8$
9. $36 \div 9$
10. $24 \div 4$

UNIT 2: INTRODUCTION TO DIVISION

1
Is 2 a multiple of 3? ______

No

2
Is 27 a multiple of 4? ______

No

3
Is 33 a multiple of 5? ______

No

ASSOCIATING MULTIPLIES WITH DIVISION

Multiples of 6 are: 0, 6, 12, 18, 24, 30, 36, 42, 48, 54, etc.

The greatest multiple of 6 that can be subtracted from 34 is 30.

34 − 30 = 4 and 30 = 6 x 5

30 is less than 34 and it is the greatest multiple of 6 that is less than 34.

4
What is the greatest multiple of 3 that can be subtracted from 23? ______

21 = 3 x 7,
23 − 21 = 2

5
What is the greatest multiple of 3 that can be subtracted from 2? ______

0 = 3 x 0,
2 − 0 = 2

6
What is the greatest multiple of 4 that is less than 27? ______

24 = 4 x 6

7
What is the greatest multiple of 5 that is less than 32? ______

$30 = 5 \times 6$

DIVISION SYMBOLS

The division of 8 by 3 is indicated by

$$8 \div 3 \quad \text{or} \quad 3\overline{)8}$$

To divide 8 by 3 it is necessary to ask:

What is the greatest multiple of 3 that can be subtracted from 8?

The answer, 6 ($6 = 3 \times 2$), means that 3 goes into 6 two times. When 6 is subtracted from 8 it gives a remainder (R) of 2.

$$\begin{array}{r} 2 \\ 3\overline{)8} \\ \underline{-6} \\ 2 \end{array} \qquad 8 \div 3 = 2\text{ R}2$$

8 ✍
The division problem $12 \div 5$ is written at the right using the long division symbol. Write $6 \div 7$ using the long division symbol. ______

$5\overline{)12}$

$7\overline{)6}$

9
Write $13 \div 4$ using the long division symbol. ______

$4\overline{)13}$

10
Write $9 \div 6$ using the long division symbol. ______

$6\overline{)9}$

STEPS IN DIVISION

To do the division problem $7 \div 3$ $\quad 3\overline{)7}$

Ask: What is the greatest multiple of 3 that can be subtracted from 7?

Answer: 6 because 3×2 is less than 7 and 3×3 is greater than 7.

Place: 2 is placed directly above the 7 and 3×2 or 6 is placed directly below 7.

$$\begin{array}{r} 2 \\ 3\overline{)7} \\ 6 \end{array}$$

Subtract: Subtract the multiple of 3 ($3 \times 2 = 6$) from 7 to find the remainder.

$$\begin{array}{rl} 2 & \text{R1} \\ 3\overline{)7} & \\ \underline{-6} & \\ 1 & \end{array}$$

11
To complete the division problem $9 \div 4$, the first step is to find the greatest multiple of 4 that can be subtracted from 9. What is it? ______

$8 = 4 \times 2$

12
For the problem at the right we know that $4 \times 2 = 8$ and 8 is the greatest multiple of 4 that is less than 9. Place the numbers 2 and 8 in their correct positions in the division problem. _____ $\quad 4\overline{)9}$

$$\begin{array}{r} 2 \\ 4\overline{)9} \\ 8 \end{array}$$

13
The answer to $9 \div 4$ is 2 R1 as shown by the problem shown at the right.

$$\begin{array}{rl} 2 & \text{R1} \\ 4\overline{)9} & \\ \underline{-8} & \\ 1 & \end{array}$$

Show all the work in completing the problem $7 \div 2$. ______

$$\begin{array}{rl} 3 & \text{R1} \\ 2\overline{)7} & \\ \underline{-6} & \\ 1 & \end{array}$$

14
For the problem $6 \div 7$, find the greatest multiple of 7 that is less than 6. What is it? ______

$0 = 7 \times 0$

15
The problem at the right shows the work of $6 \div 7$. What is the remainder?

$$\begin{array}{r} 0 \\ 7\overline{)6} \\ \underline{-0} \\ 6 \end{array}$$

______

6

16
Show all your work for $9 \div 5$.

______

$$\begin{array}{r} 1 \text{ R4} \\ 5\overline{)9} \\ \underline{-5} \\ 4 \end{array}$$

17
Show all your work for $4 \div 6$.

______

$$\begin{array}{r} 0 \text{ R4} \\ 6\overline{)4} \\ \underline{-0} \\ 4 \end{array}$$

18
Show all your work for $6 \div 3$.

______

$$\begin{array}{r} 2 \text{ R0} \\ 3\overline{)6} \\ \underline{-6} \\ 0 \end{array}$$

19

Show all your work for $6 \div 9$.

______

$$\begin{array}{r} 0 \text{ R6} \\ 9\overline{)6} \\ \underline{-0} \\ 6 \end{array}$$

FEEDBACK UNIT 2

This quiz reviews the preceding unit. Answers are at the back of the book. Begin the next unit only after all problems of this unit are understood.

1. What is the greatest multiple of 6 that can be subtracted from 51?
2. What is the greatest multiple of 5 that can be subtracted from 42?
3. What is the greatest multiple of 9 that can be subtracted from 38?
4. What is the greatest multiple of 3 that can be subtracted from 8?
5. What is the greatest multiple of 7 that can be subtracted from 4?

6. $5 \div 4$
7. $4 \div 8$
8. $1 \div 2$
9. $8 \div 4$
10. $9 \div 8$

UNIT 3: SINGLE DIGIT DIVISORS

To divide $37 \div 4$ the following steps are used:

1. Begin as if the problem were: $3 \div 4$ or $4\overline{)3}$
2. Divide $3 \div 4$.

$$\begin{array}{r} 0 \\ 4\overline{)37} \\ \underline{-0} \\ 3 \end{array}$$

3. Bring down the next digit of the dividend, 37.

$$\begin{array}{r} 0 \\ 4\overline{)37} \\ \underline{-0} \\ 37 \end{array}$$

4. Since 36 is the greatest multiple of 4 that can be subtracted from 37, the problem is completed as shown at the right.

$$\begin{array}{r} 09 \\ 4\overline{)37} \\ \underline{-0} \\ 37 \\ \underline{-36} \\ 1 \end{array}$$

$37 \div 4 = 9\text{ R}1$

1
To divide $37 \div 5$, first divide as if the problem were $3 \div 5$. What is the result? ______

$$\begin{array}{r} 0 \\ 5\overline{)3} \\ \underline{-0} \\ 3 \end{array}$$

2
For the problem at the right, bring down the 7 and complete the division by finding the greatest multiple of 5 that is less than 37. ______

$$\begin{array}{r} 0 \\ 5\overline{)37} \\ \underline{-0} \\ 3 \end{array}$$

$$\begin{array}{r} 07 \text{ R}2 \\ 5\overline{)37}\phantom{\text{ R}2} \\ \underline{-0}\phantom{\text{ R}2} \\ 37\phantom{\text{ R}2} \\ \underline{-35}\phantom{\text{ R}2} \\ 2\phantom{\text{ R}2} \end{array}$$

3
To divide $85 \div 6$, first divide as if the problem were $8 \div 6$. What is the result of this first step? ______

$$6\overline{)85}$$

$$\begin{array}{r} 1 \\ 6\overline{)85} \\ \underline{-6} \\ 2 \end{array}$$

4
Complete the division of $85 \div 6$ by bringing down the 5 and finding the greatest multiple of 6 that is less than 25.

$$\begin{array}{r} 1 \\ 6\overline{)85} \\ \underline{-6} \\ 2 \end{array}$$

$$\begin{array}{r} 14 \text{ R}1 \\ 6\overline{)85}\phantom{\text{ R}1} \\ \underline{-6}\phantom{\text{ R}1} \\ 25\phantom{\text{ R}1} \\ \underline{-24}\phantom{\text{ R}1} \\ 1\phantom{\text{ R}1} \end{array}$$

5
Divide 83 ÷ 5 and show all your work.

$5\overline{)83}$

```
   1 6   R3
5 ) 8 3
   -5
    3 3
   -3 0
      3
```

6
Divide 59 ÷ 7 and show all your work.

$7\overline{)59}$

```
   0 8   R3
7 ) 5 9
   -0
    5 9
   -5 6
      3
```

7
Divide 43 ÷ 8 and show all your work.

$8\overline{)43}$

```
   0 5   R3
8 ) 4 3
   -0
    4 3
   -4 0
      3
```

8
Divide 75 ÷ 4 and show all your work.

$4\overline{)75}$

```
   1 8   R3
4 ) 7 5
   -4
    3 5
   -3 2
      3
```

9
The division of 4,395 by 6 requires that each digit of 4,395 be divided by 6. This process begins at the left of 4,395 and moves to the right.

The division is shown at the right. What is the answer when 4,395 is divided by 6? _________

```
   0 7 3 2   R3
6 ) 4 3 9 5
   -0
    4 3
   -4 2
      1 9
     -1 8
        1 5
       -1 2
          3
```

732 R3

10
Complete the long division of 4,738 by 8 that has been started at the right.

```
   05
8)4738
 -0
  47
 -40
   73
```

```
   0592   R2
8)4738
 -0
  47
 -40
   73
  -72
    18
   -16
     2
```

11
To divide 7,269 by 5, divide 5 into each digit of 7,269 from left to right. Show all your work.

```
5)7269
```

```
   1453   R4
5)7269
 -5
  22
 -20
   26
  -25
    19
   -15
     4
```

12
Divide.

```
6)387
```

```
   064   R3
6)387
 -0
  38
 -36
   27
  -24
    3
```

13
Divide.

```
4)597
```

```
   149   R1
4)597
 -4
  19
 -16
   37
  -36
    1
```

THE NECESSITY FOR NEATNESS IN LONG DIVISION

In dividing 2,514 by 7, each digit of 2,514 has an "answer" digit directly above it.

Writing the problem neatly and maintaining the columns will, by itself, greatly improve your long division accuracy.

```
    0359  R1
 7 )2514
   -0
    25
   -21
     41
    -35
      64
     -63
        1
```

14
Divide.

```
 8 )4937
```

```
    0617  R1
 8 )4937
   -0
    49
   -48
     13
    -08
      57
     -56
        1
```

15
Divide.

```
 6 )7934
```

```
    1322  R2
 6 )7934
   -6
    19
   -18
     13
    -12
      14
     -12
        2
```

16
Divide.

$8\overline{)8143}$

$$\begin{array}{r} 1017 \quad R7 \\ 8\overline{)8143} \\ -8 \\ \hline 01 \\ -0 \\ \hline 14 \\ -08 \\ \hline 63 \\ -56 \\ \hline 7 \end{array}$$

MAINTAIN VERTICAL COLUMNS CAREFULLY

In dividing 2,741 by 6, it is important to remember that one digit of the answer goes directly above each digit of 2,741.

In practice, the first digit of the answer that is written is not zero. This common practice is shown in the example at the right.

$$\begin{array}{r} 456 \quad R5 \\ 6\overline{)2741} \\ -24 \\ \hline 34 \\ -30 \\ \hline 41 \\ -36 \\ \hline 5 \end{array}$$

17
Divide.

$9\overline{)3845}$

$$\begin{array}{r} 427 \quad R2 \\ 9\overline{)3845} \\ -36 \\ \hline 24 \\ -18 \\ \hline 65 \\ -63 \\ \hline 2 \end{array}$$

18
Divide.

$6\overline{)1973}$

$$\begin{array}{r} 328 \quad R5 \\ 6\overline{)1973} \\ -18 \\ \hline 17 \\ -12 \\ \hline 53 \\ -48 \\ \hline 5 \end{array}$$

19
Divide.

$3\overline{)1826}$

$$\begin{array}{r} 608 \\ 3\overline{)1826} \\ -18 \\ 02 \\ -0 \\ 26 \\ -24 \\ 2 \end{array} \quad R2$$

20
Divide.

$5\overline{)6537}$

$$\begin{array}{r} 1307 \\ 5\overline{)6537} \\ -5 \\ 15 \\ -15 \\ 03 \\ -0 \\ 37 \\ -35 \\ 2 \end{array} \quad R2$$

FEEDBACK UNIT 3

This quiz reviews the preceding unit. Answers are at the back of the book. Begin the next unit only after all problems of this unit are understood.

Divide and show all the work.

1. $4\overline{)5173}$
2. $6\overline{)3715}$
3. $2\overline{)8146}$
4. $3\overline{)8831}$
5. $5\overline{)1536}$
6. $7\overline{)2523}$
7. $8\overline{)5176}$
8. $9\overline{)1836}$
9. $6\overline{)4038}$
10. $8\overline{)5002}$

UNIT 4: TWO AND THREE DIGIT DIVISORS

REVIEWING THE LONG DIVISION PROCESS

The long division shown at the right follows exactly the same process as the previous units.

To maintain the process, remember:

1. A digit of the answer will appear directly above each digit of 437,628 with the understanding that the zeros over 4 and 3 will not be written.

2. Maintain the columns carefully.

```
          1261  R61
347 ) 437628
     -347
       906
      -694
       2122
      -2082
         408
        -347
          61
```

1
Which of the problems at the right has a quotient of zero? ______

$8\overline{)6}$ $8\overline{)9}$

$8\overline{)6}$

2
The problem $6 \div 8$ has zero as its quotient because 6 is less than 8. Which problem below has zero as its quotient? _____

$11\overline{)13}$ $13\overline{)11}$

$13\overline{)11}$

3
Which of the problems below has zero as the quotient? ______

$24\overline{)7}$ $24\overline{)72}$ $24\overline{)725}$

$24\overline{)7}$

4
Which of the problems below have zero as the quotient? ______

$34\overline{)2}$ $34\overline{)27}$ $34\overline{)275}$

$34\overline{)2}$, $34\overline{)27}$

ESTIMATE THE SIZE OF A DIVISION ANSWER

In dividing whole numbers two important facts need to be noted before any computation is attempted.

1. Determine the position of the first non-zero digit in the answer.
2. Determine the number of digits in the completed answer.

In the problem at the right, the position of the first non-zero digit in the quotient is indicated by the capital letter X. The answer to the division will have four digits.

```
        0 0 X
463 ) 8 1 4 3 7 6
     -0
      8 1
     - 0
      8 1 4
```

5
For the problem at the right, place a capital letter X in the correct position for the first non-zero digit of the quotient. ______

$57\overline{)4835}$

```
      0 0 X
57 ) 4 8 3 5
    - 0
     4 8 3
```

6
In dividing the problem shown at the right, each capital letter X indicates a position for one digit in the quotient. How many digits will be in the quotient? ______

```
      0 0 X X
57 ) 4 8 3 5
    - 0
```

2

7
Use an X to mark the position of the first non-zero digit in the quotient of the problem shown at the right. ______

92) 43871

X
92) 4 3 8 7 1

8
How many digits will be in the division quotient for the problem at the right? _____

92) 43871

3 digits
XXX
92) 4 3 8 7 1

9
Use an X to mark the position of the first non-zero digit in the quotient of the problem shown at the right. ______

52) 98372

X
52) 9 8 3 7 2

10
How many digits will be in the division quotient for the problem at the right? _____

52) 98372

4 digits
XXXX
52) 9 8 3 7 2

11
Use an X to mark the position of the first non-zero digit in the quotient of the problem shown at the right. ______

83) 72516

X
83) 7 2 5 1 6

12
How many digits will be in the division quotient for the problem at the right? _____

83) 72516

3 digits
XXX
83) 7 2 5 1 6

13
The example at the right shows 3 as the first non-zero digit of the answer. Multiply 3 x 576 and write the answer in its correct position.

3
5 7 6)1 6 7 3 9

3
5 7 6)1 6 7 3 9
−1 7 2 8

14 ✍

The work shown at the right clearly shows that 3 is not the first non-zero digit of the answer. Why? ____

```
         3
576)16739
   -1728
```

```
 1673
-1728
     X (the subtrahend
        is larger than
        the minuend)
```

15 ✍

The first non-zero digit of the answer is not 3. The correct first digit will be ______ (less, greater) than 3.

```
         3
576)16739
   -1728
```

less

16

Multiply 2 x 576 and place the product in its correct position in the problem.

```
         2
576)16739
```

```
         2
576)16739
   -1152
```

17

Is 2 the correct first digit of the answer for the problem shown at the right? Why? ___________

```
         2
576)16739
   -1152
     521
```

Yes, 521 is less than 576

18

What is the correct next step in completing the division shown at the right? ______

```
         2
576)16739
   -1152
     521
```

"Bring Down" the 9 of 16739

19

Multiply 8 x 576 and place the product in its correct position in the problem at the right. ______

```
        28
576)16739
   -1152
     5219
```

```
        28
576)16739
   -1152
     5219
    -4608
```

20 ✍

For the partially completed problem at the right, subtract and determine whether 8 is the correct second digit of the answer. ______

```
           2 8
576 )1 6 7 3 9
    -1 1 5 2
       5 2 1 9
      -4 6 0 8
```

No, 611 is greater than 576

21 ✍

The correct second digit of the answer, for the problem at the right, is not 8. The correct second digit is ______ (less, greater) than 8.

```
           2 8
576 )1 6 7 3 9
    -1 1 5 2
       5 2 1 9
      -4 6 0 8
         6 1 1
```

greater

22

Finish the division problem shown at the right.

```
           2
576 )1 6 7 3 9
    -1 1 5 2
       5 2 1 9
```

```
           2 9  R35
576 )1 6 7 3 9
    -1 1 5 2
       5 2 1 9
      -5 1 8 4
           3 5
```

23

Find and place the first non-zero digit of the division at the right.

```
35 ) 9 3 6 2
```

```
       2
35 ) 9 3 6 2
     7 0
     2 3
```

24

Find and place the first non-zero digit of the division at the right.

```
78 ) 5 7 3 6 2
```

```
         7
78 ) 5 7 3 6 2
     5 4 6
       2 7
```

25

Find and place the first non-zero digit of the division at the right.

```
57 ) 2 5 7 3 8
```

```
         4
57 ) 2 5 7 3 8
     2 2 8
       2 9
```

26
Find and place the first non-zero digit of the division at the right.

```
83 ) 47362
```

```
        5
83 ) 47362
     415
      58
```

27
Find and place the first non-zero digit of the division at the right.

```
65 ) 19836
```

```
        3
65 ) 19836
     195
       3
```

28
Complete the division shown at the right.

```
       26
35 ) 9362
    -70
     236
    -210
      262
```

```
       267  R17
35 ) 9362
    -70
     236
    -210
      262
      245
       17
```

29
Complete the division shown at the right. Caution: There is a zero in the answer. Be certain to place it correctly.

```
58 ) 17635
```

```
        304   R3
58 ) 17635
    -174
       23
       -0
       235
      -232
         3
```

30
Complete the division shown at the right.

```
92 ) 40635
```

```
        441  R63
92 ) 40635
    -368
      383
     -368
       155
       -92
        63
```

31
Complete the division shown at the right.

87)44134

```
      507  R25
 87)44134
   -435
      63
     -0
     634
    -609
      25
```

32
Complete the division shown at the right.

84)37605

```
      447  R57
 84)37605
   -336
     400
    -336
     645
    -588
      57
```

FEEDBACK UNIT 4

This quiz reviews the preceding unit. Answers are at the back of the book. Begin the next unit only after all problems of this unit are understood.

Divide and show all the work.

1. 65)8143
2. 59)29367
3. 43)9214
4. 28)14632
5. 17)5736
6. 85)9437
7. 72)14314
8. 96)25438
9. 15)9145
10. 71)6308
11. 473)81479
12. 276)31257

UNIT 5: WORD PROBLEMS REQUIRING DIVISION

DIVISION NOMENCLATURE AND APPLICATIONS

In a division problem each of the numbers has a special name.

In the problem at the right,
628 is the **divisor,**
85 is the **quotient,**
53,672 is the **dividend,**
292 is the **remainder.**

```
         8 5  R292
628 ) 5 3,6 7 2
     -5 0 2 4
        3 4 3 2
       -3 1 4 0
          2 9 2
```

1
In the problem at the right, 55 is the divisor and 948 is the dividend.

55) 948

In the problem at the right, the divisor is _____.

78) 9136

78

2
In the problem at the right, the dividend is _____.

724) 41632

41,632

3
In the problem at the right,
the **divisor** is 35,
the **dividend** is 1936,
the **quotient** is 55,
and the **remainder** is _____.

```
       5 5   R11
35 ) 1 9 3 6
    -1 7 5
       1 8 6
      -1 7 5
         1 1
```

11

4
In the problem at the right,
the **quotient** is _______.

```
      5 4
14 ) 7 5 6
    -7 0
      5 6
     -5 6
        0
```

54

5
The divisor of the problem at the right is _____.

$$6\overline{)93}\quad 15 \text{ R3}$$

6

6
The dividend of the problem at the right is _____.

$$8\overline{)436}\quad 54 \text{ R4}$$

436

7
The quotient of the problem at the right is _____.

$$7\overline{)434}\quad 62$$

62

8 ✍
To check the accuracy of a division answer, multiply the quotient and the divisor. This product is added to the remainder. If the sum equals the dividend the division is correct.

Check. $5\overline{)73}\quad 14 \text{ R3}$

$(14 \times 5) + 3 = 73$

9
Show how to check the division problem at the right. __________ $45\overline{)195}\quad 4 \text{ R15}$

$(4 \times 45) + 15 = 195$

10
In a division problem with remainder zero, the product of the ______ and ______ must equal the ___________.

quotient, divisor
dividend

11 ✍
To find the **average** of the numbers 59, 68, and 62 add the numbers and divide the sum by 3.

$59 + 68 + 62 = 189$

Find the average of: 59, 68, and 62._____

$3\overline{)189}\quad 63$ is the average

12
Find the average of 257, 364, 295, and 308 by adding the numbers and dividing the sum by 4. ______

$$4\overline{)1224}\ \ 306$$

13
Find the average of 19, 25, 30, 36, 25, and 27 by counting the numbers and dividing their sum by that count. ______

$$6\overline{)162}\ \ 27$$

14
Find the average of 261 and 313. ______

$$2\overline{)574}\ \ 287$$

15
Find the average of 679, 513, and 521. ______

$$3\overline{)1713}\ \ 571$$

16
When a word problem uses the word "**each**" it usually indicates a multiplication or a division. Multiplication or division is often indicated by the word "______."

each

17
A word problem with the word "each" usually indicates multiplication or ______.

division

18
Solve the problem: If 6 people earn a total of $864, what is the average earned by each person? ______
Hint: Divide 864 into 6 equal parts.

$144

19
Solve the problem: If a chemist has 845 milligrams of medicine in 5 test tubes, what is the average in each tube? ______
Hint: Divide 845 into 5 equal parts.

169 mg

20
If 5 nurses have a total of 65 vacation days, what is the average vacation due each nurse? ______
Hint: Divide 65 into 5 equal parts.

13 days

21
If 8,624 fluid ounces is to be equally distributed into 7 containers, what amount should be put in each container?
______

1232 fl oz

22
Each tire in a sale is priced at $37. The total value of the tires is $2,775. Find the number of tires. ______
Hint: The answer will be the number of 37's there are in 2,775.

75 tires

23
A total of 7,101 grams of gold is to be made into ornaments that weigh 9 grams each. How many ornaments can be made?
______

789 ornaments

24
If 893 pearls are to be made into necklaces containing 19 pearls each, how many necklaces can be obtained? _____.

47 necklaces

25
How many statues weighing 23 tons each can be poured with 2,116 tons of concrete?
______

92 statues

FEEDBACK UNIT 5

This quiz reviews the preceding unit. Answers are at the back of the book. Take the Chapter Mastery Test only after all problems of this unit are understood.

1. In the problem at the right, the divisor is ______, the dividend is ______, the quotient is ______, and the remainder is ______.

```
      1 9  R38
46 ) 9 1 2
   - 4 6
     4 5 2
   - 4 1 4
       3 8
```

2. Show how to check the division problem at the right.

```
      507  R17
18 ) 9143
```

3. Find the average of 57, 46, 52, and 49.

4. If 16 employees have total wages of $5712, find the average pay of each employee.

5. 9,378 cards are to be separated into 6 equal piles. How many cards will be in each pile?

CHAPTER 4 MASTERY TEST

This test provides an opportunity for assuring that **all** of the objectives of Chapter 4 have been learned. Each problem is accompanied by the symbol (U) which indicates the Unit in which that type of problem was studied. If any problems are missed, they need to be corrected. Answers are at the back of the book.

1. In the problem at the right, (5)
   the divisor is ______,
   the dividend is ______,
   and the quotient is ______.

$$46\overline{)2990}^{\;65}$$

Divide (3)

2. $4\overline{)3615}$ 3. $7\overline{)6213}$ 4. $8\overline{)1946}$ 5. $5\overline{)6053}$

Divide (4)

6. $28\overline{)5213}$ 7. $37\overline{)36541}$ 8. $63\overline{)65314}$ 9. $58\overline{)91326}$

10. Find the average of 52, 60, 67, and 61. (5)

11. 640 customers are to be equally distributed to 16 salespersons. How many customers will each salesperson receive? (5)

Chapter 5 Objectives

The following problems describe, by example, the skills and knowledge studied in this chapter. Each problem is accompanied by a symbol, (U), which indicates the Unit where the problem is studied. Answers are at the back of the book.

1. Which of the following fractions is the smallest? $\frac{2}{3}$ or $\frac{3}{4}$ (4)

2. In the fraction $\frac{4}{3}$, the numerator is ______. (1)

3. In $\frac{7}{11}$, the denominator is ______. (1)

4. Which fraction in the following list is equal to $\frac{3}{4}$? $\frac{1}{2}, \frac{6}{10}, \frac{9}{12}, \frac{3}{5}, \frac{9}{16}, \frac{8}{16}, \frac{2}{3}, \frac{8}{12}$ (3)

5. Is 9 a divisor of 375? (2)

6. Is 4 a divisor of 1128? (2)

7. Is 5 a divisor of 551? (2)

8. Is 10 a divisor of 720? (2)

9. Is 3 a divisor of 81? (2)

Simplify the fractions in problems 10-14. (3)

10. $\frac{35}{45}$

11. $\frac{24}{45}$

12. $\frac{93}{102}$

13. $\frac{28}{42}$

14. $\frac{12}{12}$

In problems 15-19 write each fraction as a mixed number. (5)

15. $\frac{10}{9}$

16. $\frac{44}{5}$

17. $\frac{56}{14}$

18. $\frac{44}{13}$

19. $\frac{62}{7}$

In problems 20-24 write each mixed number as an improper fraction. (6)

20. $2\frac{1}{2}$

21. $5\frac{5}{9}$

22. $3\frac{6}{7}$

23. $4\frac{1}{8}$

24. $5\frac{9}{10}$

CHAPTER 5
THE MEANING OF FRACTIONS

UNIT 1: INTRODUCTION TO FRACTIONS

THE MEANING OF A FRACTION

A fraction is an indicated division problem. The fraction $\frac{1}{5}$ means $1 \div 5$ or 1(one) divided into 5 (five) equal parts. The number 1 is above the fraction bar. It is called the **numerator** of the fraction. The number below the fraction bar is called the **denominator**.

1 ✍

Every fraction is an indicated division problem.
$\frac{1}{3}$ means $1 \div 3$. $\frac{1}{4}$ means ______. $1 \div 4$

2
A fraction is an indicated division problem.
$\frac{5}{6}$ means $5 \div 6$. $\frac{3}{4}$ means ______. $3 \div 4$

3
Write $\frac{2}{3}$ as a division problem. ______ $2 \div 3$

4
$7 \div 10$ can be written as a fraction $\frac{7}{10}$.
Write $6 \div 11$ as a fraction. ______ $\frac{6}{11}$

5
Write $13 \div 8$ as a fraction. ______ $\frac{13}{8}$

6
Write $7 \div 9$ as a fraction. ______ $\frac{7}{9}$

7
The numerator of $\frac{5}{9}$ is 5 and the denominator is 9.

The numerator of $\frac{2}{3}$ is 2 and the denominator is ______. 3

8
The numerator of $\frac{4}{7}$ is ______ and the denominator is 7. 4

9
The numerator of $\frac{7}{11}$ is ______. 7

10
The denominator of $\frac{1}{4}$ is ______. 4

VISUALIZING FRACTIONS

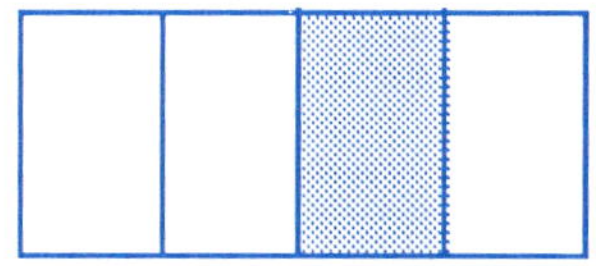

In the figure at the right, the large rectangle has been divided into four smaller rectangles. One of those smaller rectangles has been shaded.

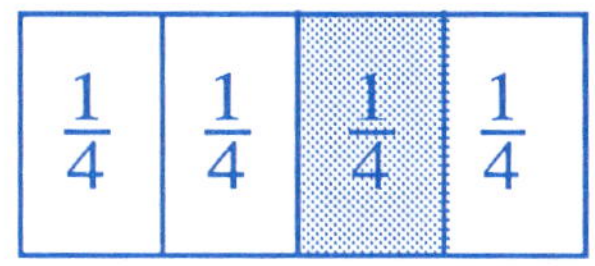

Because the large rectangle has been divided into four equal parts, each of the smaller rectangles represents $\frac{1}{4}$ of the large rectangle.

The shaded part of the figure is $\frac{1}{4}$ of the large rectangle.

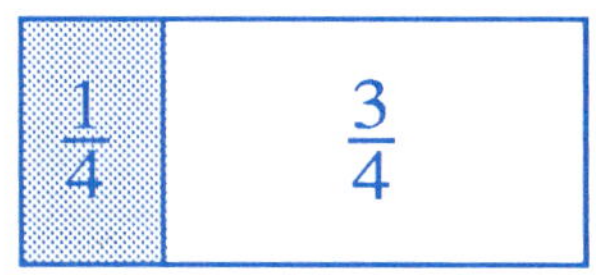

The unshaded part of the figure is $\frac{3}{4}$ of the large rectangle.

11
In the figure at the right, the large rectangle has been divided into ______ equal parts.

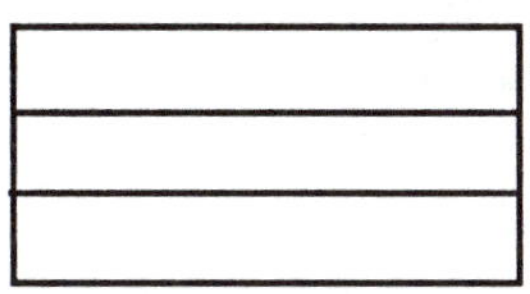

3

12
In the figure of the preceding frame, each of the smaller rectangles represents $\frac{1}{3}$ of the large rectangle. $\frac{1}{3}$ means $1 \div$ ______.

3

13
In the figure at the right, some of the smaller rectangles are shaded. What fraction of the large rectangle is shaded? ______

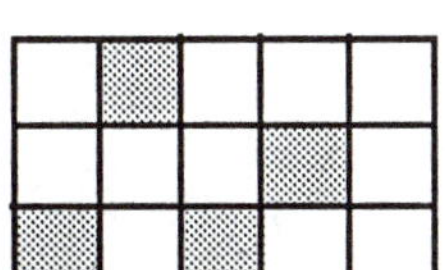

$\frac{4}{15}$

14

In the figure of the preceding frame, the shaded part represents $\frac{4}{15}$ of the large rectangle. What fraction is represented by the unshaded part of the figure? ______ $\frac{11}{15}$

15

At the right a number of circles are shown. What fraction of the circles is shaded? ______

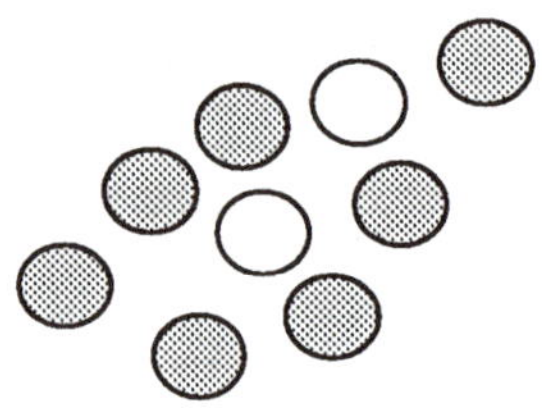

$\frac{7}{9}$

16

In the figure of the preceding frame, what fraction of the circles is unshaded? ______ $\frac{2}{9}$

17

The circle at the right has been divided into six equal parts. Each pie-shaped part represents ______ of the circle.

$\frac{1}{6}$

18

In the figure above, each pie-shaped part represents $\frac{1}{6}$ of the circle. How many pie-shaped parts are needed to represent $\frac{5}{6}$ of the circle? ______ 5

19

In the figure at the right, the square has been divided into four equal parts. What fraction of the square is shaded? ______

$\frac{3}{4}$

20
In the figure in frame 19, what fraction of the square is unshaded? ______

$\frac{1}{4}$

21
In the figure at the right, the rectangle has been divided into four smaller rectangles. What fraction of the large rectangle is represented by each smaller rectangle? ______

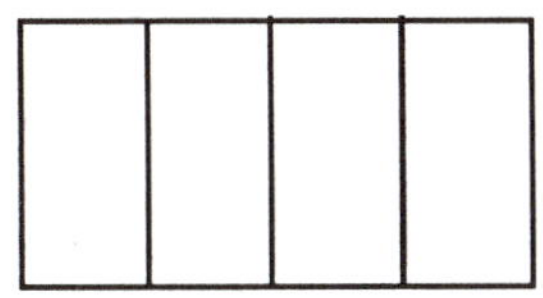

$\frac{1}{4}$

22
The figure at the right is similar to the figure in frame 21, but now each smaller rectangle is divided into two triangles. How many triangles are in the large rectangle? ______

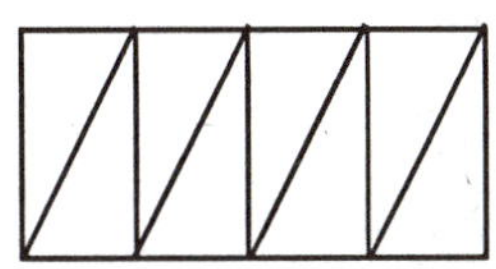

8

23
In the figure of frame 23, each small rectangle represents $\frac{1}{4}$ of the large rectangle. Each triangle represents ______ of the large rectangle.

$\frac{1}{8}$

24
In the figure at the right, one of the smaller rectangles is shaded which means $\frac{1}{4}$ of the figure is shaded. Also, two of the triangles are shaded which means ______ of the figure is shaded.

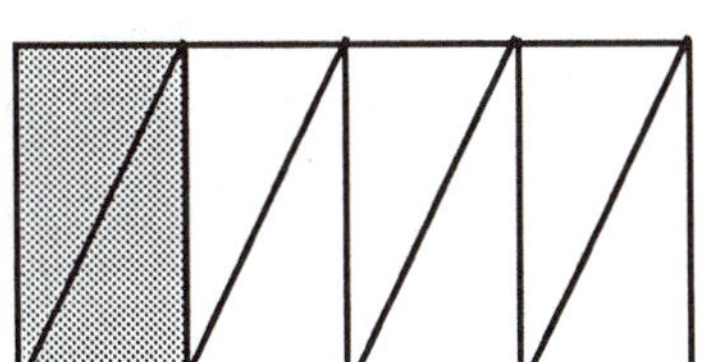

$\frac{2}{8}$

25
The figure of frame 24 shows that $\frac{1}{4}$ of the large rectangle is the same as $\frac{2}{8}$ of the figure. Is $\frac{1}{4}$ equal to $\frac{2}{8}$? ______ yes

26
Using the figure at the right, the unshaded small rectangles represent $\frac{3}{4}$ of the large rectangle. The unshaded triangles represent ______ of the large rectangle.

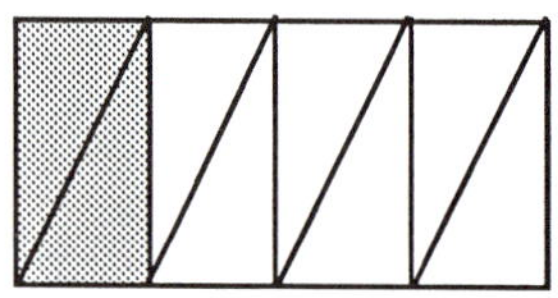

$\frac{6}{8}$

27
Using the figure in frame 26, are $\frac{3}{4}$ and $\frac{6}{8}$ equal? ______ Yes

VISUALIZING FRACTIONS ON A NUMBER RAY

In the figure below, the line segment has its endpoints labeled 0 and 1. Four points labeled A,B,C,D separate the line segment into five equal parts. The point A marks the position of $\frac{1}{5}$; B marks $\frac{2}{5}$; C marks $\frac{3}{5}$; and D marks $\frac{4}{5}$.

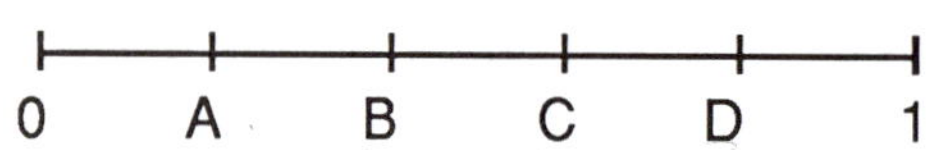

28
The line segment below has been separated into two equal parts. What fraction represents Point A? ______ $\frac{1}{2}$

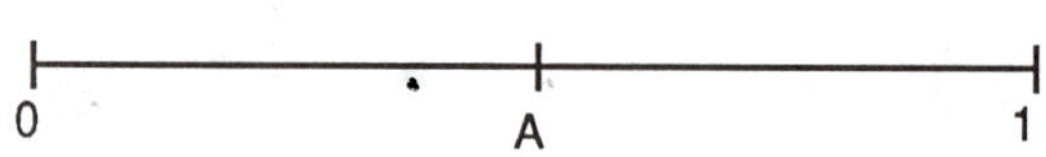

29
The line segment below has been separated into 12 equal parts. Put a capital A at the point that is halfway from 0 to 1.

30
The line segment below has been separated into 6 equal parts. Put a capital A at the point representing $\frac{1}{2}$.

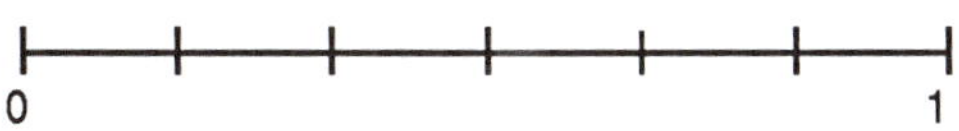

31
The line segment below has been separated into 3 equal parts by the Points A and B. Which point represents $\frac{1}{3}$? ______

A; B is $\frac{2}{3}$

32
The line segment below has been separated into 6 equal parts. Place a capital letter A at the point representing $\frac{1}{3}$.

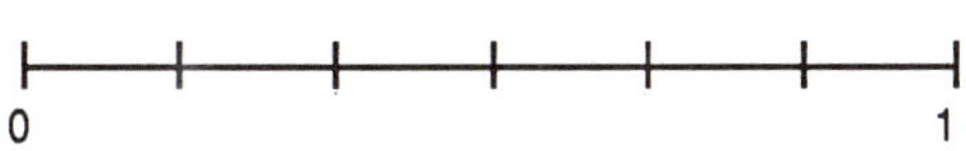

33
In the figure below, the line segment has been separated into 12 equal parts. Place a capital letter A at the point representing $\frac{1}{3}$.

34

The line segment below has been separated into 4 equal parts by the points A, B, and C. Which point represents $\frac{1}{4}$? ______

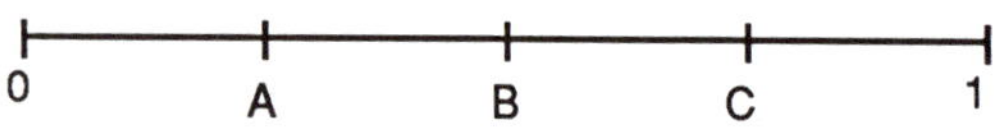

A
B is $\frac{2}{4}$; C is $\frac{3}{4}$

35

The line segment below has been separated into 8 equal parts. Which point represents $\frac{1}{4}$? ______

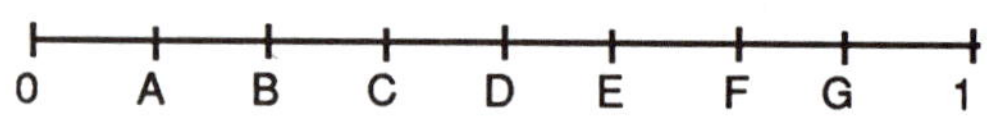

B

36

The line below has been separated into 12 equal parts. Which point represents $\frac{1}{4}$? ______

C

37

The line segment below has been separated into 5 equal parts by the points labeled A, B, C, and D. Which point represents $\frac{1}{5}$? ______

A
B is $\frac{2}{5}$; C is $\frac{3}{5}$; D is $\frac{4}{5}$

38

The line segment below has been divided into 10 equal parts. Mark the point representing $\frac{1}{5}$ with a capital A.

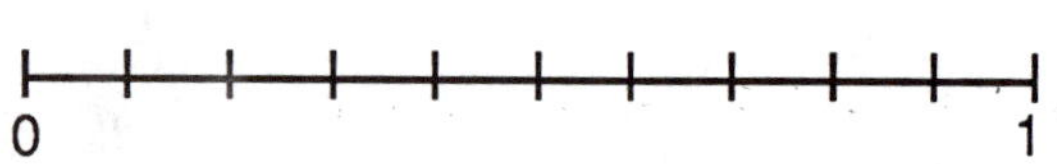

39

The line segment below has been divided into 6 equal parts by the points A, B, C, D, and E. What point represents $\frac{1}{6}$? ______

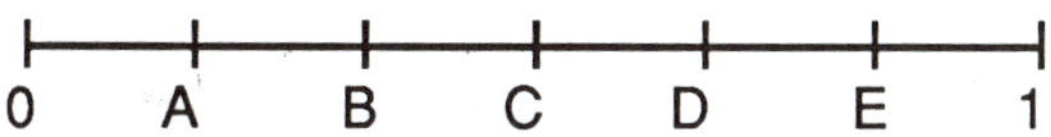

A

B is $\frac{2}{6}$; C is $\frac{3}{6}$

D is $\frac{4}{6}$; E is $\frac{5}{6}$

40

The line segment below has been divided into 12 equal parts. Mark with a capital A the point representing $\frac{5}{6}$.

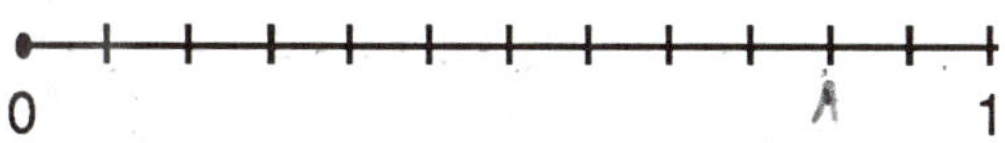

41

Below are shown two line segments; one is divided into two parts and the other into three parts. $\frac{1}{2}$ is greater than $\frac{1}{3}$ because $\frac{1}{2}$ is further to the right. Is $\frac{1}{2}$ greater than $\frac{2}{3}$? ______

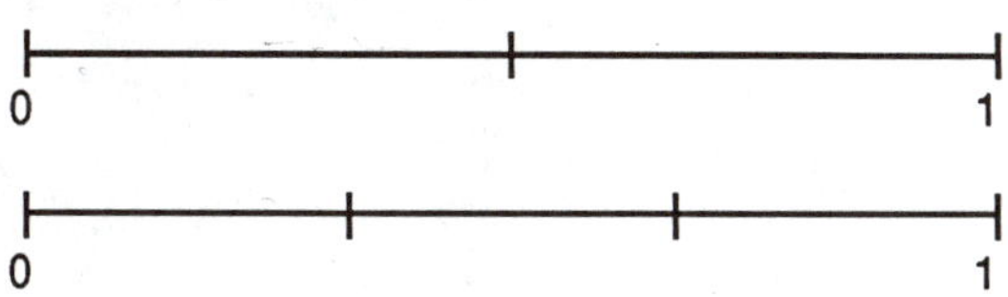

No

42

Below are shown two line segments. $\frac{1}{3}$ is greater than $\frac{1}{4}$. Which is greater, $\frac{2}{3}$ or $\frac{3}{4}$? ______

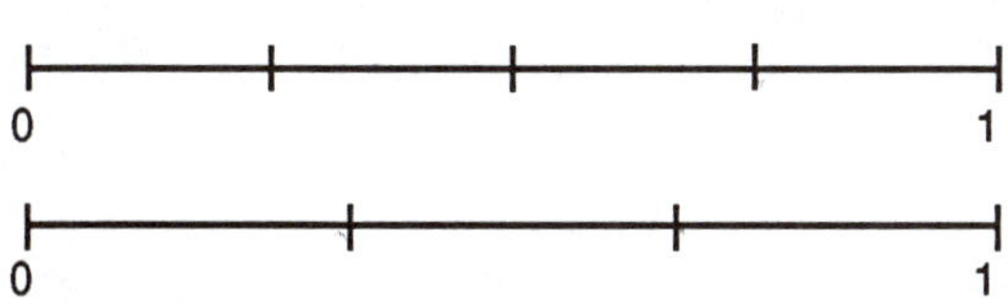

$\frac{3}{4}$

43

Below are shown two line segments. $\frac{1}{4}$ is greater than $\frac{1}{5}$. Is $\frac{3}{4}$ greater than $\frac{4}{5}$? ______ No

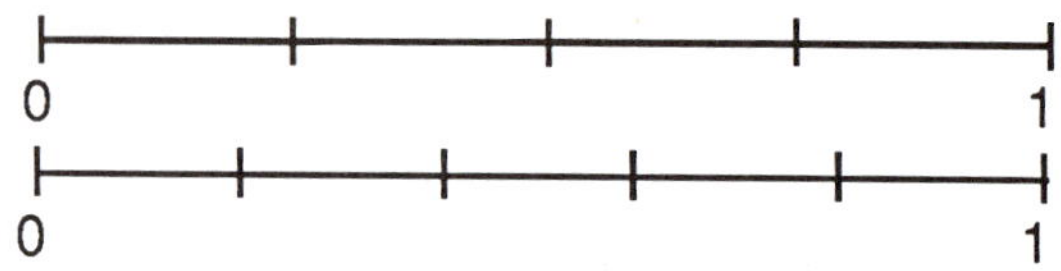

44

Below are shown two line segments.
Is $\frac{2}{4}$ greater than $\frac{1}{2}$? ______ No, they are equal

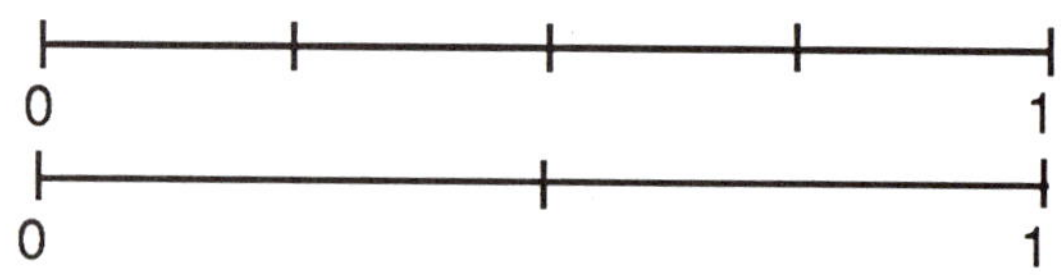

45

Below are shown two line segments.
Is $\frac{1}{3}$ greater than $\frac{2}{6}$? ______ No, they are equal

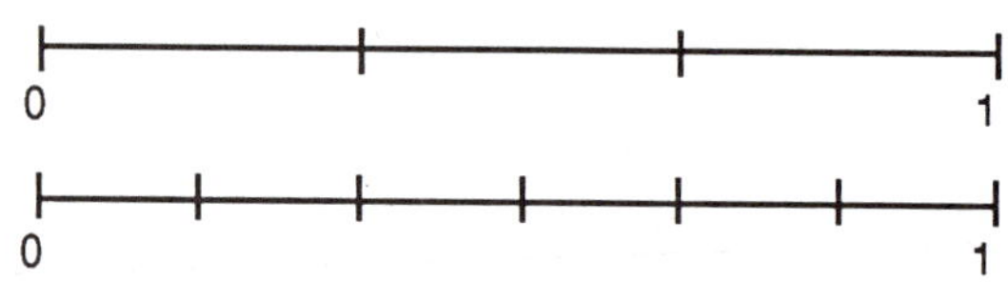

46

If a hospital has 250 beds and 233 patients, the fraction $\frac{233}{250}$ indicates the portion of the hospital beds that are occupied. Write a fraction indicating a hospital's occupancy if it has 140 beds and 73 patients. ______ $\frac{73}{140}$

47

A manufacturing plant with 392 employees has 13 absent employees. What fraction of the employees are absent? ______ $\frac{13}{392}$

48
A classroom has 35 seats and 24 students. Write a fraction indicating the number of occupied seats. ____

$\frac{24}{35}$

FEEDBACK UNIT 1

This quiz reviews the preceding unit. Answers are at the back of the book. Begin the next unit only after all problems of this unit are understood.

1. Write $5 \div 7$ as a fraction.
2. What is the numerator of $\frac{5}{9}$?
3. What is the denominator of $\frac{7}{10}$?
4. Use the figure at the right and determine the fraction of the circles that are shaded.

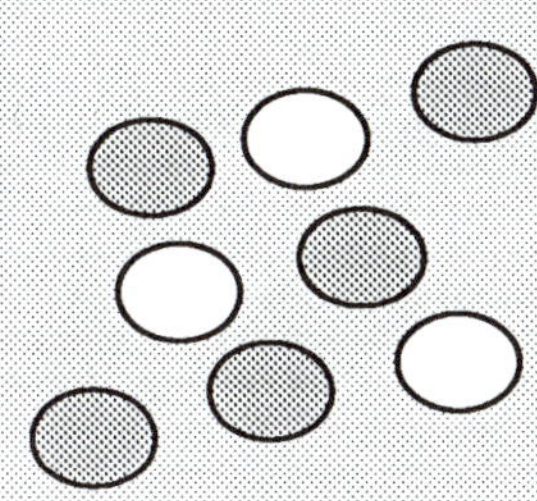

5. Use the two number lines at the right. Which of the following fractions is the largest?

   $\frac{3}{4}$ or $\frac{5}{6}$

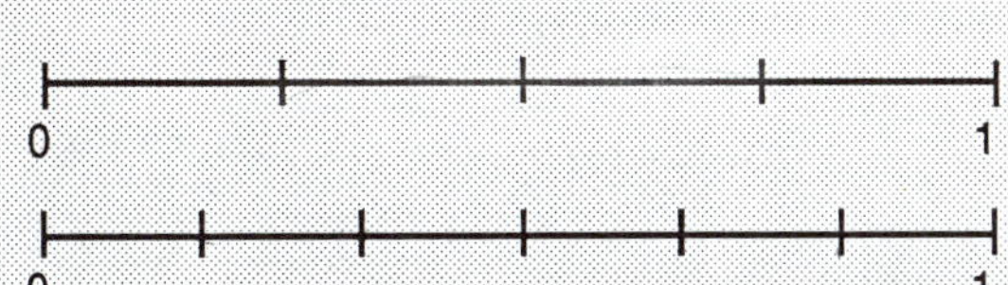

6. A hospital staff has 8 doctors and 45 nurses. Write a fraction indicating the relative number of doctors to nurses.

UNIT 2: FACTORS (DIVISORS) OF A WHOLE NUMBER

FINDING COMMON FACTORS (DIVISORS)

To simplify (reduce) a fraction like $\frac{8}{12}$ it is necessary to find a number that is a divisor of both 8 and 12.

In this unit the skills used in finding factors (divisors) of a counting number are learned.

The factors (divisors) of 20 are: 1, 2, 4, 5, 10, 20.
Each of the factors divides 20 evenly (no remainder).

1
5 is a factor (divisor) of 20 because when 20 is divided by 5 there is a remainder of 0. Is 4 a factor (divisor) of 20? ______

Yes,
$20 \div 4 = 5$ R0

2
6 is a factor (divisor) of 18. When 18 is divided by 6 the remainder is 0. Is 9 a factor of 18? ______

Yes, $18 \div 9 = 2$ R0

3
6 is not a factor of 20. When 20 is divided by 6 the remainder is not zero. Is 9 a factor of 24? ______

No,
$24 \div 9 = 2$ R6

4
Is 8 a factor of 24? ______

Yes, $24 \div 8 = 3$

5
Is 9 a factor of 45? ______

Yes, $45 \div 9 = 5$

6
Is 2 a factor of 24? ______

Yes

7
Is 7 a factor of 32? ______

No, $\frac{32}{7} = 4$ R4

8
A possible factor of a number can always be tested by long division.
Is 11 a factor of 132? ______

Yes

9
To determine whether 7 is a factor of 812, ______ (add, subtract, divide).

divide

10
Is 7 a factor of 546? ______

Yes, $546 \div 7 = 78$
$R = 0$

11
Is 8 a factor of 912? ______

Yes, $912 \div 8 = 114$
$R = 0$

12
Is 2 a factor of 436? ______

Yes

13
Is 3 a factor of 436? ______

No

14
Is 13 a factor of 52? ______

Yes

15
Is 6 a factor of 436? ______

No

FEEDBACK UNIT 2

This quiz reviews the preceding unit. Answers are at the back of the book. Begin the next unit only after all problems of this unit are understood.

1. Is 2 a factor of 604?
2. Is 3 a factor of 814?
3. Is 4 a factor of 538?
4. Is 5 a factor of 436?
5. Is 6 a factor of 525?
6. Is 7 a factor of 475?
7. Is 8 a factor of 864?
8. Is 9 a factor of 618?
9. Is 10 a factor of 903?
10. Is 11 a factor of 454?

UNIT 3: SIMPLIFYING FRACTIONS

DIVIDING NUMERATOR AND DENOMINATOR BY A COMMON FACTOR

The fraction $\frac{8}{12}$ is simplified (reduced) by dividing both the numerator 8 and the denominator 12 by their common factor, 4.

$\frac{8}{12} = \frac{8 \div 4}{12 \div 4} = \frac{2}{3}$ Notice that $\frac{8}{12}$ is equal to $\frac{2}{3}$ in the figure below.

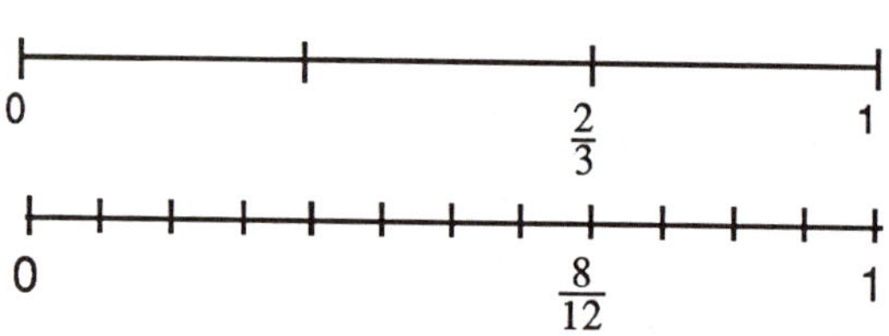

1

Simplify $\frac{14}{16}$ by dividing both 14 and 16 by 2.

$$\frac{14}{16} = \frac{14 \div 2}{16 \div 2} = \_\_\_\_\_\_$$

$\frac{7}{8}$

2

Simplify $\frac{36}{42}$ by dividing both 36 and 42 by 6.

$$\frac{36}{42} = \frac{36 \div 6}{42 \div 6} = \_\_\_\_\_\_$$

$\frac{6}{7}$

3

Simplify $\frac{20}{30}$ by dividing both 20 and 30 by 10.

$$\frac{20}{30} = \frac{20 \div 10}{30 \div 10} = \_\_\_\_\_\_$$

$\frac{2}{3}$

4

Simplify $\frac{48}{88}$ by dividing both 48 and 88 by 8. ______ $\frac{6}{11}$

5

Simplify $\frac{16}{20}$ by dividing both 16 and 20 by 4. ______ $\frac{4}{5}$

6

Simplify $\frac{18}{27}$ by dividing both 18 and 27 by 9. ______ $\frac{2}{3}$

7

Simplify $\frac{6}{15}$ by dividing both 6 and 15 by 3. ______ $\frac{2}{5}$

8

Simplify $\frac{15}{25}$ by dividing both 15 and 25 by 5. ______ $\frac{3}{5}$

9

Simplify $\frac{6}{8}$. ______ $\frac{3}{4}$

10

Simplify $\frac{16}{18}$. ______ $\frac{8}{9}$

11
To simplify $\frac{10}{12}$ it is necessary to find a factor (divisor) for both 10 and ______. 12

12
The factors (divisors) of 10 are 1, 2, 5, and 10. The factors of 12 are 1, 2, 3, 4, 6, and ______. 12

13
Is 2 a factor of both 10 and 12? ______ Yes

14 ✍
2 is a **common factor** of 10 and 12 because 2 is a factor of both 10 and 12. Is 4 a common factor of 10 and 12? ______ No, 4 is not a factor of 10

15 ✍
2 is the **highest common factor** (hcf) of 10 and 12 because 2 is the largest number that is a factor of both 10 and ______. 12

16
2 is the **highest common factor** (hcf) of 10 and 12. Simplify $\frac{10}{12}$ using the hcf 2. ______ $\frac{5}{6}$

17
Both 2 and 4 are common factors of 8 and 12. The **highest common factor** of 8 and 12 is ______ because it is the largest common factor. 4

18
To completely simplify $\frac{8}{12}$ both numerator and denominator are divided by the **hcf** of 8 and 12 which is ______. 4

19
Simplify $\frac{8}{12}$ by dividing both 8 and 12 by their **hcf** which is 4. ______

$\frac{2}{3}$

20
One (1) is a factor of every number. The factors of 6 are 1, 2, 3, and 6. The factors of 11 are ______ and ______.

1,11

21
The fraction $\frac{6}{11}$ cannot be simplified because the **hcf** of 6 and 11 is ______.

1

22
The fraction $\frac{5}{9}$ cannot be simplified because the **hcf** of 5 and 9 is ______.

1

23
Can $\frac{10}{15}$ be simplified? ______

Yes, hcf is 5

24
Simplify $\frac{10}{15}$ ______

$\frac{2}{3}$

25
Can $\frac{12}{17}$ be simplified? ______

No, hcf is 1

26
Can $\frac{15}{33}$ be simplified? ______

Yes, hcf is 3

27
Simplify $\frac{15}{33}$ ______

$\frac{5}{11}$

28
Can $\frac{50}{120}$ be simplified? ______

Yes, hcf is 10

29
Simplify $\frac{50}{120}$ ______ $\frac{5}{12}$

30
Can $\frac{38}{44}$ be simplified? ______ Yes, hcf is 2

31
Simplify $\frac{38}{44}$ ______ $\frac{19}{22}$

32
Can $\frac{24}{35}$ be simplified? ______ No, hcf is 1

33
Can $\frac{35}{56}$ be simplified? ______ Yes, hcf is 7

34
Simplify $\frac{35}{56}$ ______ $\frac{5}{8}$

35
Simplify $\frac{28}{35}$ ______ $\frac{4}{5}$, hcf is 7

36
Simplify $\frac{45}{63}$ ______ $\frac{5}{7}$, hcf is 9

37
Simplify $\frac{30}{50}$ ______ $\frac{3}{5}$, hcf is 10

38
Simplify $\frac{35}{48}$ ______ $\frac{35}{48}$, hcf is 1

39
Simplify $\frac{14}{35}$ ______ $\frac{2}{5}$

40
Simplify $\frac{32}{40}$ ______ $\frac{4}{5}$

AFTER SIMPLIFYING A FRACTION, TRY AGAIN

1, 2, 4, and 8 are all common factors of 32 and 40. If the hcf is not used in simplifying $\frac{32}{40}$ the resulting fraction must be reduced again.

$$\frac{32}{40} = \frac{32 \div 2}{40 \div 2} = \frac{16}{20} \qquad \frac{32}{40} = \frac{32 \div 4}{40 \div 4} = \frac{8}{10} \qquad \frac{32}{40} = \frac{32 \div 8}{40 \div 8} = \frac{4}{5}$$

Both $\frac{16}{20}$ and $\frac{8}{10}$ are not completely simplified. Before ending any simplification, always check to determine that the **hcf** of the numerator and denominator is 1.

41
Simplify $\frac{20}{60}$ ______ — $\frac{1}{3}$, hcf is 20

42
Simplify $\frac{32}{48}$ ______ — $\frac{2}{3}$, hcf is 16

43
Simplify $\frac{12}{36}$ ______ — $\frac{1}{3}$, hcf is 12

44
2 is a factor of 26 and $26 \div 2 = 13$.
Is 13 also a factor of 26? ______ — Yes

45
To simplify $\frac{26}{39}$ first find all the factors of 26.
The factors of 26 are ___, ___, ___, and ___. — 1,2,13,26

46
1, 2, 13, and 26 are the factors of 26. Try each of the factors and find the hcf of 26 and 39. ______ — 13

47
Simplify $\frac{26}{39}$. ______ — $\frac{2}{3}$

48
Is 3 a factor of 51? ______ Yes

49
$51 \div 3 = 17$. Is 17 a factor of 51? ______ Yes

50
List all the factors of 51. ___, ___, ___, ___ 1,3,17,51

51
1, 3, 17, and 51 are factors of 51. Try each factor and find the hcf of 51 and 68. ______ 17

52
Simplify $\frac{51}{68}$. ______ $\frac{3}{4}$

53
Simplify $\frac{38}{57}$. ______
Hint: Find all the factors of 38 and then test each one as a possible factor of 57. $\frac{2}{3}$, hcf is 19

54
Simplify $\frac{33}{88}$. ______ $\frac{3}{8}$, hcf is 11

55
Simplify $\frac{39}{65}$. ______
Hint: First, find all factors of 39. $\frac{3}{5}$, hcf is 13

56
Simplify $\frac{41}{49}$. ______ $\frac{41}{49}$, hcf is 1

57
Simplify $\frac{63}{66}$. ______ $\frac{21}{22}$, hcf is 3

58
Simplify $\frac{16}{24}$. ______

$\frac{2}{3}$, hcf is 8

59
Simplify $\frac{24}{30}$. ______

$\frac{4}{5}$, hcf is 6

60
Simplify $\frac{18}{45}$. ______

$\frac{2}{5}$, hcf is 9

FEEDBACK UNIT 3

This quiz reviews the preceding unit. Answers are at the back of the book. Begin the next unit only after all problems of this unit are understood.

Simplify.

1. $\frac{15}{40}$
2. $\frac{18}{30}$
3. $\frac{7}{35}$
4. $\frac{26}{65}$
5. $\frac{13}{15}$
6. $\frac{40}{70}$
7. $\frac{24}{44}$
8. $\frac{45}{81}$
9. $\frac{57}{76}$
10. $\frac{42}{56}$

UNIT 4: EQUIVALENT FRACTIONS

FINDING EQUIVALENT FRACTIONS WITH LARGER DENOMINATORS

A fraction can be written with a greater denominator by multiplying both numerator and denominator by the same number.

$$\frac{5}{6} = \frac{5 \times 4}{6 \times 4} = \frac{20}{24}$$

1
Multiply both the numerator and denominator of $\frac{7}{8}$ by 3. What is the result? ______

$\frac{7 \times 3}{8 \times 3} = \frac{21}{24}$

2
The fractions $\frac{7}{8}$ and $\frac{21}{24}$ are equal. To change $\frac{7}{8}$ to $\frac{21}{24}$, both numerator and denominator of $\frac{7}{8}$ are multiplied by ______.

3

3
Multiply both numerator and denominator of $\frac{3}{4}$ by 9. What is the result? ______

$\frac{3 \times 9}{4 \times 9} = \frac{27}{36}$

4
Are $\frac{3}{4}$ and $\frac{27}{36}$ equal?______

Yes

5
To change $\frac{3}{4}$ to $\frac{27}{36}$, both numerator and denominator of $\frac{3}{4}$ are multiplied by ______.

9

6
Multiply both numerator and denominator of $\frac{1}{2}$ by 5. What is the result? ______

$\frac{1 \times 5}{2 \times 5} = \frac{5}{10}$

7
Are $\frac{1}{2}$ and $\frac{5}{10}$ equal fractions? ______

Yes

8
To change $\frac{1}{2}$ to $\frac{3}{6}$, both the numerator and denominator of $\frac{1}{2}$ must be multiplied by ______.

3

9
What number must be multiplied by both the numerator and denominator of $\frac{5}{6}$ to give a fraction with denominator 12?______

2

10
Find the numerator of the fraction below.

$$\frac{1}{4} = \frac{?}{12}$$

$$\frac{1 \times 3}{4 \times 3} = \frac{3}{12}$$

Find the numerator of the fractions below.

11

$$\frac{5}{6} = \frac{?}{24}$$

$$\frac{5 \times 4}{6 \times 4} = \frac{20}{24}$$

12

$$\frac{2}{3} = \frac{?}{12}$$

$$\frac{2}{3} = \frac{8}{12}$$

13

$$\frac{5}{8} = \frac{?}{16}$$

$$\frac{5}{8} = \frac{10}{16}$$

14

$$\frac{4}{7} = \frac{?}{21}$$

$$\frac{4}{7} = \frac{12}{21}$$

15
Write $\frac{3}{4}$ with 20 as the denominator. ______

$$\frac{3}{4} = \frac{15}{20}$$

16
Write $\frac{5}{6}$ with 30 as the denominator. ______

$$\frac{5}{6} = \frac{25}{30}$$

LEAST COMMON MULTIPLES (lcm)

Multiples of 4 are: 4, 8, 12, 16, 20, 24, 28, 32, 36, 40, etc.

Multiples of 6 are: 6, 12, 18, 24, 30, 36, 42, etc.

The numbers 12, 24, and 36 appear in both lists of multiples and are **common multiples** of 4 and 6.

The smallest number that appears in both lists of multiples is 12 and 12 is the **least common multiple** (lcm) of 4 and 6.

17
Multiples of 6 are:
6, 12, 18, 24, 30, 36, 42, 48, 54, etc.
List nine multiples of 9.____________

9,18,27,36,45, 54,63,72,81

18 ✍
The numbers 18 and 36 are **common multiples** of 6 and 9. Is 6 x 9 or 54 also a common multiple of 6 and 9? ______

Yes

19
The numbers 18, 36, and 54 are **common multiples** of 6 and 9. Is 24 a common multiple of 6 and 9? ______

No, 24 is not a multiple of 9

20
The product of two whole numbers is always a common multiple for them. Is 45 (9 x 5) a common multiple for 9 and 5? ______

Yes

21
Is 48 (6 x 8) a common multiple for 6 and 8?
______

Yes

22
Is 44 (4 x 11) a common multiple for 4 and 11?
______

Yes

23
The product of two whole numbers ______ (is, isn't) always a common multiple for them.

is

24
A common multiple of 6 and 9 is 54. Is 54 the smallest number that is a common multiple for 6 and 9? ______

No, 18 is the smallest common multiple.

25
A common multiple of 6 and 5 is 30 (6 x 5). Is 30 the smallest number that is a common multiple for 6 and 5? _____

Yes

26
The product of two whole numbers is always a common multiple for them. The product is ______ (always, sometimes) the smallest common multiple.

sometimes

27
List the multiples of 2 and 3. Find the smallest common multiple for them. ______

6

28
List the multiples of 6 and 10 until you find a number in both lists. What is the smallest number in both lists? _____

30

29 ✍
The number 30 is called the **least common multiple** of 6 and 10 because it is the smallest whole number that is a multiple of both 6 and _____.

10

30 ✍
Find the **least common multiple** (lcm) of 4 and 5 by finding the smallest number that is a multiple of both of them. ______

20

31
List the multiples of 12 and 8 until you find their **least common multiple** (lcm). ______ 24

32
List the multiples of 3 and 5 and then select their lcm. ______ 15

33
List the multiples of 3 and 6 and then find their lcm. ______ 6

34
List the multiples of 7 and 5. What is the lcm of 7 and 5? ______ 35

35
Find the lcm of 5 and 9. ______ 45

36
Find the lcm of 12 and 6. ______ 12

WRITING TWO FRACTIONS WITH A COMMON DENOMINATOR

A major use of the least common multiple of two numbers is the finding of a **common denominator**.

To write the fractions $\frac{3}{8}$ and $\frac{5}{6}$ with a common denominator, the following steps are used:

1. Find the lcm of the denominators of $\frac{3}{8}$ and $\frac{5}{6}$.

   The lcm of 8 and 6 is 24.

2. Write both fractions with the lcm, 24, as denominator.

   $\frac{3}{8} = \frac{3 \times 3}{8 \times 3} = \frac{9}{24}$ $\qquad$ $\frac{5}{6} = \frac{5 \times 4}{6 \times 4} = \frac{20}{24}$

37
To write $\frac{3}{4}$ and $\frac{5}{6}$ with a common denominator, the lcm of the denominators is found. What is the lcm of 4 and 6? ______

12

38
$$\frac{3}{4} = \frac{3 \times 3}{4 \times 3} = \frac{9}{12} \quad \text{and} \quad \frac{5}{6} = \frac{?}{12}$$

$\frac{5 \times 2}{6 \times 2} = \frac{10}{12}$

COMPARING SIZES OF TWO FRACTIONS

To compare the sizes of $\frac{5}{8}$ and $\frac{2}{3}$, both fractions are written with a common denominator.

$$\frac{5}{8} = \frac{5 \times 3}{8 \times 3} = \frac{15}{24} \quad \text{and} \quad \frac{2}{3} = \frac{2 \times 8}{3 \times 8} = \frac{16}{24}$$

$\frac{5}{8}$ is **less than** $\frac{2}{3}$ because $\frac{15}{24}$ is **less than** $\frac{16}{24}$.

39
The fraction $\frac{8}{11}$ is less than $\frac{9}{11}$ because the denominators are the same and 8 is less than 9. Is $\frac{15}{17}$ less than $\frac{13}{17}$? ______

No, 15 is greater than 13

40
When fractions have the same denominator, compare the numerators. Is $\frac{14}{20}$ less than $\frac{17}{20}$? ______

Yes, 14 is less than 17

41
If two fractions have the same denominator, the smaller fraction has the ______ (smaller, larger) numerator.

smaller

42
To compare the size of two fractions,

A. Write both fractions with a common denominator.
B. The smaller fraction will have the ______ (smaller, larger) numerator.

smaller

43

$$\frac{5}{8} = \frac{5 \times 7}{8 \times 7} = \frac{35}{56} \quad \text{and} \quad \frac{4}{7} = \frac{4 \times 8}{7 \times 8} = \frac{32}{56}$$

Is $\frac{5}{8}$ less than $\frac{4}{7}$? ______

No, $\frac{35}{56}$ is greater than $\frac{32}{56}$

44

$$\frac{7}{9} = \frac{7 \times 2}{9 \times 2} = \frac{14}{18} \quad \text{and} \quad \frac{5}{6} = \frac{5 \times 3}{6 \times 3} = \frac{15}{18}$$

Is $\frac{7}{9}$ less than $\frac{5}{6}$? ______

Yes, $\frac{14}{18}$ is less than $\frac{15}{18}$

45

Write $\frac{2}{7}$ and $\frac{1}{4}$ with a common denominator.

Is $\frac{2}{7}$ less than $\frac{1}{4}$? ______

No, $\frac{2}{7} = \frac{8}{28}, \frac{1}{4} = \frac{7}{28}$

46

Write $\frac{3}{5}$ and $\frac{7}{10}$ with a common denominator.

Is $\frac{3}{5}$ less than $\frac{7}{10}$? ______

Yes, $\frac{3}{5} = \frac{6}{10}$

47

Write $\frac{5}{11}$ and $\frac{4}{9}$ with a common denominator.

Is $\frac{5}{11}$ less than $\frac{4}{9}$? ______

No, $\frac{45}{99}$ is greater than $\frac{44}{99}$

48

Is $\frac{5}{7}$ less than $\frac{11}{16}$? ______

No, $\frac{80}{112}$ is greater than $\frac{77}{112}$

49

Is $\frac{1}{3}$ less than $\frac{2}{9}$? ______

No, $\frac{3}{9}$ is greater than $\frac{2}{9}$

50

Is $\frac{3}{5}$ less than $\frac{2}{3}$? ______

Yes, $\frac{9}{15}$ is less than $\frac{10}{15}$

51

Is $\frac{8}{13}$ less than $\frac{2}{3}$? ______

Yes, $\frac{24}{39}$ is less than $\frac{26}{39}$

FEEDBACK UNIT 4

This quiz reviews the preceding unit. Answers are at the back of the book. Begin the next unit only after all problems of this unit are understood.

In problems 1-4 write each fraction with the indicated denominator.

1. $\frac{4}{7} = \frac{?}{35}$
2. $\frac{2}{3} = \frac{?}{12}$
3. $\frac{8}{9} = \frac{?}{36}$
4. $\frac{1}{4} = \frac{?}{44}$

In problems 5-8, find the lcm for each of the pairs of numbers.

5. 7, 9
6. 9, 6
7. 12, 4
8. 10, 15

In problems 9-12, select the smaller fraction from each pair.

9. $\frac{7}{8}, \frac{9}{11}$
10. $\frac{3}{4}, \frac{8}{13}$
11. $\frac{3}{10}, \frac{4}{15}$
12. $\frac{7}{9}, \frac{5}{6}$

UNIT 5: CHANGING IMPROPER FRACTIONS TO WHOLE OR MIXED NUMBERS

SIMPLIFYING FRACTIONS TO WHOLE OR MIXED NUMBERS

Fractions like $\frac{5}{7}$, $\frac{11}{12}$, and $\frac{14}{17}$ are all less than 1 because their denominators are greater than their numerators. Such fractions are **proper** fractions.

Fractions like $\frac{4}{4}$, $\frac{11}{11}$, and $\frac{47}{47}$ are equal to 1 because their denominators are equal to their numerators.

Fractions like $\frac{7}{5}$, $\frac{48}{7}$, and $\frac{32}{3}$ are all greater than 1 because their denominators are less than their numerators. Such fractions are **improper** fractions and can be written as **mixed numbers**.

1
Whenever the numerator and denominator of a fraction are equal the fraction can be simplified to 1.

$\frac{4}{4} = 1$ and $\frac{6}{6} =$ ______ — 1

2
Simplify $\frac{37}{37}$. ______ — 1; $37 \div 37 = 1$

3
The fraction $\frac{15}{5}$ is indicated division, $15 \div 5$.
$15 \div 5 =$ ______. — 3

4
Is $\frac{18}{7}$ greater than 1? ______
Hint: The fraction means $18 \div 7$. — Yes, $18 \div 7 = 2$ R4

5
Is $\frac{5}{6}$ greater than 1? ______

No, $5 \div 6 = 0$ R5

6 ✍

The fraction $\frac{37}{5}$ is an **improper** fraction because it is greater than 1. $37 \div 5$ is greater than 1.
Is $\frac{14}{3}$ an improper fraction? ______

Yes, $14 \div 3$ is greater than 1.

7
Is $\frac{95}{8}$ an improper fraction? ______

Yes

8 ✍

The fraction $\frac{3}{8}$ is a **proper** fraction because it is less than one. Is $\frac{5}{19}$ a proper fraction? ______

Yes

9
The fraction $\frac{24}{6}$ is a whole number greater than 1, because $24 \div 6 = 4$. Is $\frac{24}{6}$ an improper fraction?
______

Yes, it is greater than 1

10
$\frac{2}{3}$ is a proper fraction less than 1.
$\frac{15}{7}$ is an improper fraction greater than 1.
Is $\frac{14}{2}$ an improper fraction?

Yes, it is equal to 7

CHANGING AN IMPROPER FRACTION TO A MIXED NUMBER

The improper fraction $\frac{18}{7}$ is equal to the mixed number $2\frac{4}{7}$.

To change $\frac{18}{7}$ to its mixed number equivalent,

1. Divide 18 by 7 $\quad 7\overline{)18}$ quotient 2 R4

2. The quotient, 2 R4, is used to write the mixed number.
   2 R4 becomes $2\frac{4}{7}$

11
Change $\frac{10}{7}$ to a mixed number by dividing 10 by 7. ______

$10 \div 7 = 1$ R3
$\frac{10}{7} = 1\frac{3}{7}$

12
Change $\frac{11}{4}$ to a mixed number by dividing 11 by 4. ______

$11 \div 4 = 2$ R3
$\frac{11}{4} = 2\frac{3}{4}$

13
Change $\frac{15}{7}$ to a mixed number. ______

2 R1 gives $2\frac{1}{7}$

14
Write $\frac{32}{5}$ as a mixed number. ______

6 R2 gives $6\frac{2}{5}$

15
Write $\frac{47}{8}$ as a mixed number. ______

$5\frac{7}{8}$

16
Write $\frac{19}{4}$ as a mixed number. ______

$4\frac{3}{4}$

FEEDBACK UNIT 5

This quiz reviews the preceding unit. Answers are at the back of the book. Begin the next unit only after all problems of this unit are understood.

In problems 1-4, label each fraction as a proper or improper.

1. $\frac{14}{19}$
2. $\frac{15}{13}$
3. $\frac{87}{93}$
4. $\frac{57}{48}$

In problems 5-12, change each fraction to a whole number or mixed number.

5. $\frac{28}{9}$
6. $\frac{35}{6}$
7. $\frac{41}{8}$
8. $\frac{65}{12}$
9. $\frac{42}{7}$
10. $\frac{39}{4}$
11. $\frac{47}{5}$
12. $\frac{8}{8}$

UNIT 6: WRITING MIXED NUMBERS AS IMPROPER FRACTIONS

CONVERTING MIXED NUMBERS TO IMPROPER FRACTIONS

To change $2\frac{5}{7}$ to its improper fraction, the following process is used:

1. Find the numerator of the improper fraction by multiplying the denominator, 7, by the whole number, 2. Then add the numerator of the mixed number fraction to this product.
$$(7 \times 2) + 5 = 19$$

2. Find the denominator of the improper fraction by using the denominator of the mixed number fraction.
$$2\frac{5}{7} = \frac{19}{7}$$

1

In the mixed number $2\frac{5}{7}$, the whole number 2 represents $\frac{14}{7}$ because $14 \div 7 = 2$. In the mixed number $3\frac{1}{8}$, the whole number 3 represents ______ because $24 \div 8 = 3$. $\frac{24}{8}$

2

The whole number 5 is equal to $\frac{10}{2}, \frac{15}{3}, \frac{20}{4}$, etc. To make a fraction equal to the whole number 5 the numerator must be ______ times the denominator. 5

3

The whole number 7 is equal to $\frac{14}{2}, \frac{21}{3}, \frac{28}{4}$, etc. To make a fraction equal to the whole number 7 the numerator must be ______ times the denominator. 7

4

In the mixed number $7\frac{1}{4}$, the whole number 7 equals $\frac{28}{4}$. In the mixed number $5\frac{2}{3}$, the whole number 5 equals $\frac{?}{3}$. ______ $5 = \frac{15}{3}$

5

$$7\frac{1}{4} = 7 + \frac{1}{4} = \frac{28}{4} + \frac{1}{4} = \frac{29}{4}$$

$$5\frac{2}{3} = 5 + \frac{2}{3} = \frac{15}{3} + \frac{2}{3} = \_\_\_\_\_\_$$

$\frac{17}{3}$

6

$$2\frac{5}{8} = 2 + \frac{5}{8} = \frac{16}{8} + \frac{5}{8} = \_\_\_\_\_\_$$

$\frac{21}{8}$

A SHORTCUT FOR FINDING THE NUMERATOR

To find the numerator of a mixed number's improper fraction,

1. Multiply the denominator and the whole number. $4 = \frac{(5 \times 4)}{5}$ or $4 = \frac{20}{5}$

2. Add the numerator of the proper fraction to the product found in step 1.

$$4\tfrac{3}{5} = \frac{(5 \times 4) + 3}{5} = \frac{23}{5}$$

7

The numerator of the improper fraction for $3\frac{5}{9}$ is $(9 \times 3) + 5 = 32$. The denominator is still 9. Write the improper fraction for $3\frac{5}{9}$. ______

$\frac{32}{9}$

8

Find the numerator of the improper fraction for $2\frac{1}{7}$ by multiplying 7 times 2 and adding 1. ______

$(7 \times 2) + 1 = 15$

9

Find the numerator of the improper fraction for $3\frac{2}{7}$ by multiplying 7 times 3 and adding 2. ______

$(7 \times 3) + 2 = 23$

10

$2\frac{1}{5} = \frac{11}{5}$ and $3\frac{2}{7} = \frac{23}{7}$

Find the improper fraction for $2\frac{5}{8}$. ______

$\frac{21}{8}$

11

Change $4\frac{3}{8}$ to an improper fraction. ______

$\frac{35}{8}$

12

Change $6\frac{5}{6}$ to an improper fraction. ______

$\frac{41}{6}$

13
Change $8\frac{3}{5}$ to an improper fraction. ______ $\frac{43}{5}$

14
Change $7\frac{3}{8}$ to an improper fraction. ______ $\frac{59}{8}$

15
Change $9\frac{1}{4}$ to an improper fraction. ______ $\frac{37}{4}$

16
Change $5\frac{7}{11}$ to an improper fraction. ______ $\frac{62}{11}$

17
Change $9\frac{3}{8}$ to an improper fraction. ______ $\frac{75}{8}$

FEEDBACK UNIT 6

This quiz reviews the preceding unit. Answers are at the back of the book.
Begin the Chapter Mastery Test only after all problems of this unit are understood.

Change to improper fractions.

1. $2\frac{5}{8}$
2. $5\frac{9}{11}$
3. $4\frac{3}{4}$
4. $8\frac{1}{5}$
5. $3\frac{2}{9}$
6. $6\frac{3}{7}$
7. $5\frac{4}{5}$
8. $1\frac{3}{8}$
9. $2\frac{4}{7}$
10. $1\frac{1}{3}$

CHAPTER 5 MASTERY TEST

This test provides an opportunity for assuring that **all** of the objectives of Chapter 5 have been learned. Each problem is accompanied by the symbol (U) which indicates the Unit in which that type of problem was studied. If any problems are missed, they need to be corrected. Answers are at the back of the book.

1. Which of the following is the smallest number? $\frac{7}{10}$ or $\frac{2}{3}$ (4)

2. In the fraction equal to $5 \div 6$, the numerator is ______. (1)

3. In the fraction $\frac{10}{9}$, the numerator is ______. (1)

4. What fraction of the figure below is shaded? (1)

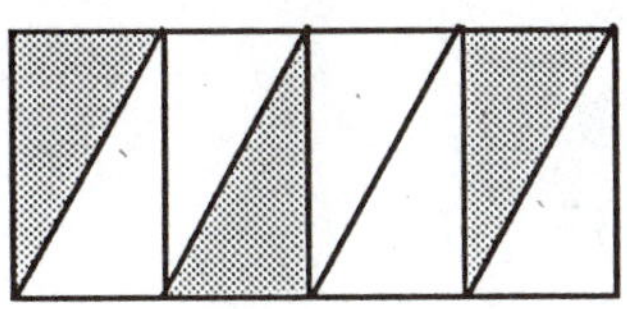

5. Is 2 a divisor of 2221? (2)

6. Is 4 a divisor of 4420? (2)

7. Is 3 a divisor of 412? (2)

8. Is 5 a divisor of 655? (2)

Simplify each fraction in problems 9-12. (3)

9. $\frac{8}{10}$

10. $\frac{12}{36}$

11. $\frac{45}{54}$

12. $\frac{48}{64}$

Write each fraction in problems 13-16 as a mixed number or a whole number. (5)

13. $\frac{9}{7}$

14. $\frac{12}{5}$

15. $\frac{75}{8}$

16. $\frac{40}{7}$

Write each mixed number in problems 17-20 as an improper fraction. (6)

17. $5\frac{7}{8}$

18. $3\frac{10}{13}$

19. $8\frac{4}{5}$

20. $7\frac{1}{9}$

Chapter 6 Objectives

The following problems describe, by example, the skills and knowledge studied in this chapter. Each problem is accompanied by a symbol, (U), which indicates the Unit where the problem is studied. Answers are at the back of the book.

In problems 1-4, find least common multiples (lcm) for each pair of numbers. (3)

1. 4, 5

2. 12, 10

3. 6, 7

4. 6, 12

5. Write $\frac{5}{8}$ with denominator 32. (2)

6. Write $\frac{3}{7}$ with denominator 63. (2)

In problems 7-18, add or subtract as indicated.

7. $\frac{1}{3} + \frac{4}{3}$ (1)

8. $\frac{14}{13} + \frac{5}{13}$ (4)

9. $\frac{2}{3} + \frac{1}{2}$ (1)

10. $\frac{3}{4} + \frac{5}{8}$ (4)

11. $\frac{1}{5} + \frac{5}{6}$ (4)

12. $2\frac{5}{8} + \frac{3}{4}$ (5)

13. $\frac{9}{16} - \frac{3}{16}$ (6)

14. $9\frac{7}{10} - 3\frac{3}{10}$ (7)

15. $\frac{6}{7} - \frac{2}{3}$ (6)

16. $\frac{5}{6} - \frac{5}{12}$ (6)

17. $9\frac{1}{5} - 3\frac{4}{5}$ (7)

18. $7\frac{1}{4} - 5\frac{1}{3}$ (7)

19. Two blocks, one weighing $5\frac{3}{10}$ kilograms and the other weighing $1\frac{1}{4}$ kilograms, were placed on a scale. What was their total weight? (8)

20. A piece of string $6\frac{4}{5}$ centimeters long was cut from a piece $11\frac{1}{2}$ centimeters long. How much was left? (8)

21. How much more is $6\frac{3}{4}$ cups than $2\frac{1}{4}$ cups? (8)

Chapter 6
Addtion And Subtraction Of Fractions

Unit 1: Adding Fractions With The Same Denominator

Common Labels — A Necessity For Addition

It is possible to add 5 apples and 3 apples because the numbers have the same label — apples.

It makes no sense to add 4 apples and 7 oranges because the numbers have different labels — apples and oranges.

Addition and subtraction can only be performed when the numbers have the same label.

1 ✍
It is possible to add 3 chairs and 5 chairs because the numbers have the same label — chairs. What is the sum of 3 chairs and 5 chairs? ______

8 chairs

2 ✍
It makes no sense to add 3 chairs and 2 desks because the numbers have different labels – – chairs and desks. Does it make sense to add 4 pills and 7 beds? ______

No

3
Numbers can only be added when they have the same labels. Can 4 fifths be added to 7 thirds? ______

No, different labels

4 ✍
The denominator of a fraction is its label. The label of $\frac{2}{3}$ is thirds. The label of $\frac{3}{4}$ is ______.

fourths

5
The label of $\frac{1}{9}$ is ______.

ninths

6
Do the fractions $\frac{3}{8}$ and $\frac{5}{8}$ have the same label? ______

Yes, eighths

7
Do the fractions $\frac{3}{10}$ and $\frac{1}{10}$ have the same label?______

Yes, tenths

ADDING FRACTIONS WITH THE SAME DENOMINATOR

When two fractions have the same denominator (label), they can be added by:

1. Adding the numerators. $\frac{5}{13} + \frac{2}{13} = \frac{7}{13}$
2. Keeping the same denominator (label). $\frac{4}{11} + \frac{1}{11} = \frac{5}{11}$

Two examples of adding fractions with the same denominator are shown above.

8
Do the fractions $\frac{3}{17}$ and $\frac{5}{17}$ have the same denominator? ____ Yes

9
Add $\frac{3}{17}$ and $\frac{5}{17}$ by adding their numerators and keeping the same denominator. ______ $\frac{8}{17}$

10
Do the fractions $\frac{1}{5}$ and $\frac{3}{5}$ have the same denominator? ____ Yes

11
Add $\frac{1}{5} + \frac{3}{5}$ by adding the numerators and keeping the same denominator. ______ $\frac{4}{5}$

12
Do the fractions $\frac{4}{7}$ and $\frac{3}{4}$ have the same denominator? _____ No

13
Can $\frac{4}{7} + \frac{3}{4}$ be added by adding the numerators, 4 and 3? ______ No

14

Can $\frac{5}{11}$ and $\frac{1}{11}$ be added by adding the numerators, 5 and 1? ______ Yes

15

$\frac{5}{11} + \frac{1}{11} =$ ______ $\frac{6}{11}$

16

$\frac{4}{7} + \frac{2}{7} =$ ______ $\frac{6}{7}$

17

$\frac{1}{3} + \frac{1}{3} =$ ______ $\frac{2}{3}$

18

$\frac{5}{17} + \frac{8}{17} =$ ______ $\frac{13}{17}$

19

Sometimes the answer to an addition problem is a fraction that should be simplified. Add $\frac{3}{10} + \frac{1}{10}$ and simplify the result. ______ $\frac{4}{10} = \frac{2}{5}$

20

Add and simplify. $\frac{2}{9} + \frac{1}{9} =$ ______ $\frac{3}{9} = \frac{1}{3}$

21

Add and simplify. $\frac{5}{6} + \frac{1}{6} =$ ______ $\frac{6}{6} = 1$

22

$\frac{1}{8} + \frac{3}{8} =$ ______ $\frac{4}{8} = \frac{1}{2}$

23

$\frac{2}{7} + \frac{3}{7} =$ ______ $\frac{5}{7}$

24

$\frac{5}{8} + \frac{1}{8} =$ ______ $\frac{6}{8} = \frac{3}{4}$

25

$\frac{9}{20} + \frac{3}{20} =$ ______

$\frac{12}{20} = \frac{3}{5}$

26

$\frac{9}{14} + \frac{3}{14} =$ ______

$\frac{12}{14} = \frac{6}{7}$

27

Sometimes the answer to an addition problem is greater than one (1) and needs to be changed to a mixed number.

Add and simplify. $\frac{4}{5} + \frac{2}{5} =$ ______

$\frac{6}{5} = 1\frac{1}{5}$

28

Add and simplify. $\frac{4}{7} + \frac{5}{7} =$ ______

$\frac{9}{7} = 1\frac{2}{7}$

29

Add and simplify. $\frac{4}{5} + \frac{4}{5} =$ ______

$\frac{8}{5} = 1\frac{3}{5}$

30

$\frac{5}{7} + \frac{6}{7} =$ ______

$\frac{11}{7} = 1\frac{4}{7}$

31

$\frac{10}{13} + \frac{7}{13} =$ ______

$\frac{17}{13} = 1\frac{4}{13}$

32

$\frac{7}{16} + \frac{3}{16} =$ ______

$\frac{10}{16} = \frac{5}{8}$

33

$\frac{5}{8} + \frac{7}{8} =$ ______

$\frac{12}{8} = 1\frac{4}{8} = 1\frac{1}{2}$

34

$\frac{11}{15} + \frac{7}{15} =$ ______

$\frac{18}{15} = 1\frac{1}{5}$

35

$\frac{5}{6} + \frac{5}{6} =$ ______

$1\frac{2}{3}$

36 $\frac{7}{10} + \frac{9}{10} =$ ______ $1\frac{3}{5}$

37 $\frac{7}{12} + \frac{11}{12} =$ ______ $1\frac{1}{2}$

38 $\frac{9}{13} + \frac{6}{13} =$ ______ $1\frac{2}{13}$

39 $\frac{5}{7} + \frac{1}{7} =$ ______ $\frac{6}{7}$

40 $\frac{8}{15} + \frac{4}{15} =$ ______ $\frac{4}{5}$

41 $\frac{3}{8} + \frac{7}{8} =$ ______ $1\frac{1}{4}$

42 $\frac{5}{9} + \frac{7}{9} =$ ______ $1\frac{1}{3}$

43 $\frac{4}{15} + \frac{7}{15} =$ ______ $\frac{11}{15}$

44 $\frac{14}{15} + \frac{4}{15} =$ ______ $1\frac{1}{5}$

FEEDBACK UNIT 1

This quiz reviews the preceding unit. Answers are at the back of the book. Begin the next unit only after all problems of this unit are understood.

Add.

1. $\frac{3}{10} + \frac{4}{10}$
2. $\frac{5}{13} + \frac{2}{13}$
3. $\frac{3}{5} + \frac{4}{5}$
4. $\frac{8}{9} + \frac{5}{9}$
5. $\frac{5}{8} + \frac{7}{8}$
6. $\frac{9}{16} + \frac{11}{16}$
7. $\frac{6}{13} + \frac{7}{13}$
8. $\frac{5}{6} + \frac{5}{6}$
9. $\frac{8}{15} + \frac{11}{15}$
10. $\frac{9}{20} + \frac{13}{20}$
11. $\frac{7}{18} + \frac{5}{18}$
12. $\frac{8}{9} + \frac{8}{9}$

UNIT 2: WRITING TWO FRACTIONS WITH THE SAME DENOMINATOR

FINDING EQUIVALENT FRACTIONS WITH LARGER DENOMINATORS

A fraction can be written with a greater denominator by multiplying both numerator and denominator by the same number.

$$\frac{5}{6} = \frac{5 \times 4}{6 \times 4} = \frac{20}{24}$$

1
Multiply both the numerator and denominator of $\frac{5}{6}$ by 3. ______ $\frac{5}{6} = \frac{5 \times 3}{6 \times 3} = \frac{15}{18}$

2
The fractions $\frac{5}{6}$ and $\frac{15}{18}$ are equal. To change $\frac{5}{6}$ to $\frac{15}{18}$, both numerator and denominator of $\frac{5}{6}$ are multiplied by ______. 3

3
Multiply both numerator and denominator of $\frac{3}{8}$ by 9. ______ $\frac{3}{8} = \frac{3 \times 9}{8 \times 9} = \frac{27}{72}$

4
Are $\frac{3}{8}$ and $\frac{27}{72}$ equal?______ Yes

5
To change $\frac{3}{8}$ to $\frac{27}{72}$, both numerator and denominator of $\frac{3}{8}$ are multiplied by ______. 9

6
Multiply both numerator and denominator of $\frac{1}{3}$ by 5. ______ $\frac{1}{3} = \frac{1 \times 5}{3 \times 5} = \frac{5}{15}$

7
Are $\frac{1}{3}$ and $\frac{5}{15}$ equal fractions? ______ Yes

8
To change $\frac{1}{2}$ to $\frac{4}{8}$, both the numerator and denominator of $\frac{1}{2}$ must be multiplied by ______. 4

9
What number must be multiplied by both the numerator and denominator of $\frac{3}{4}$ to give a fraction with denominator 12?_____

3

10
Find the numerator of the fraction below. ______

$\frac{1}{4} = \frac{?}{12}$

$\frac{1 \times 3}{4 \times 3} = \frac{3}{12}$

11
Find the numerator of the fraction below. ______

$\frac{5}{6} = \frac{?}{24}$

$\frac{5 \times 4}{6 \times 4} = \frac{20}{24}$

12 $\frac{2}{3} = \frac{?}{12}$

$\frac{2 \times 4}{3 \times 4} = \frac{8}{12}$

13 $\frac{5}{8} = \frac{?}{16}$

$\frac{5 \times 2}{8 \times 2} = \frac{10}{16}$

14 $\frac{4}{7} = \frac{?}{21}$

$\frac{4 \times 3}{7 \times 3} = \frac{12}{21}$

15
Write $\frac{3}{4}$ with 20 as the denominator. ______

$\frac{3}{4} = \frac{15}{20}$

16
Write $\frac{5}{6}$ with 30 as the denominator. ______

$\frac{5}{6} = \frac{25}{30}$

17
Write both $\frac{2}{3}$ and $\frac{5}{8}$ with denominator 24. ____ ____

$\frac{2}{3} = \frac{16}{24}$, $\frac{5}{8} = \frac{15}{24}$

18
Write both $\frac{1}{4}$ and $\frac{5}{6}$ with denominator 12. ____ ____

$\frac{1}{4} = \frac{3}{12}$, $\frac{5}{6} = \frac{10}{12}$

19
Write both $\frac{5}{6}$ and $\frac{2}{9}$ with denominator 18. ____ ____

$\frac{5}{6} = \frac{15}{18}, \frac{2}{9} = \frac{4}{18}$

20
Write both $\frac{3}{10}$ and $\frac{3}{4}$ with denominator 20. ____ ____

$\frac{3}{10} = \frac{6}{20}, \frac{3}{4} = \frac{15}{20}$

21
Write both $\frac{1}{3}$ and $\frac{5}{7}$ with denominator 21. ____ ____

$\frac{1}{3} = \frac{7}{21}, \frac{5}{7} = \frac{15}{21}$

22
Write $\frac{1}{3}$ and $\frac{7}{9}$ with denominator 9.____ ____

$\frac{1}{3} = \frac{3}{9}, \frac{7}{9} = \frac{7}{9}$

23
Write $\frac{1}{4}$ and $\frac{2}{3}$ with denominator 12. ____ ____

$\frac{1}{4} = \frac{3}{12}, \frac{2}{3} = \frac{8}{12}$

FEEDBACK UNIT 2

This quiz reviews the preceding unit. Answers are at the back of the book. Begin the next unit only after all problems of this unit are understood.

1. $\frac{3}{4} = \frac{?}{20}$
2. $\frac{5}{7} = \frac{?}{14}$
3. $\frac{2}{3} = \frac{?}{21}$
4. $\frac{7}{8} = \frac{?}{32}$
5. $\frac{2}{9} = \frac{?}{18}$
6. $\frac{1}{6} = \frac{?}{24}$

In problems 7-10 write both fractions with the stated denominator.

7. $\frac{3}{5}$ and $\frac{5}{8}$ with denominator 40.
8. $\frac{7}{8}$ and $\frac{5}{6}$ with denominator 24.
9. $\frac{2}{9}$ and $\frac{1}{6}$ with denominator 18.
10. $\frac{1}{2}$ and $\frac{7}{10}$ with denominator 10.

UNIT 3: USING lcm's AS COMMON DENOMINATORS

LEAST COMMON MULTIPLES (lcm's)

Multiples of 8 are: 8, 16, 24, 32, 40, 48, 56, etc.

Multiples of 6 are: 6, 12, 18, 24, 30, 36, 42, 48, 54, etc.

The numbers 24 and 48 appear in both lists of multiples and are common multiples of 8 and 6.

The smallest number that appears in both lists of multiples is 24 and 24 is the least common multiple (lcm) of 8 and 6.

1
Multiples of 3 are 3, 6, 9, 12, 15, etc.
List the first seven multiples of 5. ___________ 5,10,15,20,25,30,35

2
15, 30, and 45 are **common multiples** of 3 and 5. Is 25 a common multiple of 3 and 5?
______ No, 25 is not a multiple of 3

3
Is 21 (3 x 7) a **common multiple** of 3 and 7?
______ Yes

4
Is 24 (4 x 6) a common multiple of 4 and 6?
______ Yes

5
Is the product of two whole numbers always a common multiple for them? ______ Yes

6
The product of 6 and 8 (48) is a common multiple of 6 and 8. Is 48 the smallest common multiple for 6 and 8? ______ No, 24

7
The smallest number that is a common multiple for 6 and 8 is 24. 24 is the least common multiple for 6 and 8. What is the smallest common multiple of 4 and 6? ______ 12

8 ✍
The **least common multiple** (lcm) of 4 and 6 is 12. Find the least common multiple (lcm) of 4 and 3. ______ 12

9
The product of two whole numbers is ______ (always, sometimes) a common multiple for them and ______ (always, sometimes) the least common multiple for them.

always

sometimes

10
Find all the common multiples of 6 and 10 up to 6 x 10 (60). What is the least common multiple (lcm) of 6 and 10? ______

30, 60

30

11
Find all the common multiples of 8 and 4 up to 8 x 4 (32). What is the least common multiple (lcm) of 8 and 4? ______

8, 16, 24, 32

8

12
Find all the common multiples of 5 and 9 up to 5 x 9. What is the least common multiple (lcm) of 5 and 9? ______ 45

13
Find the least common multiple (lcm) of 12 and 6. ______ 12

14
Find the lcm of 11 and 6. ______ 66

15
Find the lcm of 8 and 2. ______

8

16
Find the lcm of 12 and 9. ______

36

17
Find the lcm of the denominators of $\frac{1}{4}$ and $\frac{7}{10}$. ______

20

18
Write both $\frac{1}{4}$ and $\frac{7}{10}$ with 20 as the denominator. ______ ______

$\frac{1}{4} = \frac{5}{20}$
$\frac{7}{10} = \frac{14}{20}$

19
Write both $\frac{1}{2}$ and $\frac{5}{6}$ with the same denominator. Use the lcm of the denominators. ______ ______

$\frac{1}{2} = \frac{3}{6}$
$\frac{5}{6} = \frac{5}{6}$

FEEDBACK UNIT 3

This quiz reviews the preceding unit. Answers are at the back of the book. Begin the next unit only after all problems of this unit are understood.

In problems 1-5, find the lcm for each number pair.

1. 3, 8
2. 5, 10
3. 4, 6
4. 9, 6
5. 4, 10

In problems 6-10, write both fractions with the lcm of the denominators.

6. $\frac{4}{5}, \frac{1}{6}$
7. $\frac{5}{8}, \frac{3}{16}$
8. $\frac{4}{9}, \frac{2}{3}$
9. $\frac{1}{2}, \frac{5}{7}$
10. $\frac{5}{6}, \frac{7}{10}$

UNIT 4: FINDING COMMON DENOMINATORS AND ADDING FRACTIONS

ADDING FRACTIONS WITH DIFFERENT DENOMINATORS

The fractions $\frac{3}{4}$ and $\frac{1}{6}$ cannot be added in their present forms because they have different denominators.

The first step in adding $\frac{3}{4}$ and $\frac{1}{6}$ is to write each fraction with the same denominator. After the denominators are made equal, the addition is completed by adding the new numerators.

To add $\frac{3}{4}$ and $\frac{1}{6}$,

1. Find the lcm of the denominators 4 and 6. — The lcm of 4 and 6 is 12
2. Write both fractions with denominator 12.
3. Add the numerators of $\frac{9}{12}$ and $\frac{2}{12}$.

$$\begin{array}{r} \frac{3}{4} = \frac{9}{12} \\ +\frac{1}{6} = \frac{2}{12} \\ \hline \frac{11}{12} \end{array}$$

1
Can the fraction problem $\frac{3}{7} + \frac{2}{7}$ be completed by adding the numerators, 3 and 2? ______

Yes, $\frac{3}{7} + \frac{2}{7} = \frac{5}{7}$

2
Can the fraction problem $\frac{4}{11} + \frac{2}{13}$ be completed by adding the numerators, 4 and 2? ______

No, the denominators are different

3

To add the problem shown at the right, it is necessary to first find a common denominator. What is the lcm of 7 and 9? ______

$$\frac{3}{7} + \frac{4}{9}$$

63

4

Write each of the fractions in the problem shown at the right with 63 as its denominator.

______ ______

$$\frac{3}{7} + \frac{4}{9}$$

$\frac{3}{7} = \frac{27}{63}$, $\frac{4}{9} = \frac{28}{63}$

5

Complete the addition of $\frac{3}{7}$ and $\frac{4}{9}$ shown at the right by adding the new numerators, 27 and 28.

$$\frac{3}{7} = \frac{27}{63} \qquad + \frac{4}{9} = \frac{28}{63}$$

$\frac{55}{63}$

6 ✍

Which of the following is the correct first step in adding $\frac{1}{4}$ and $\frac{2}{5}$? ______

a. Add the numerators 1 and 2.

b. Write each fraction with the same denominator.

b

7

For the problem shown at the right, write both fractions with the same denominator. ______ ______

$$\frac{1}{4} + \frac{2}{5}$$

$\frac{1}{4} = \frac{5}{20}$, $\frac{2}{5} = \frac{8}{20}$

8

Complete the addition shown at the right. ______

$$\frac{1}{4} = \frac{5}{20} \qquad + \frac{2}{5} = \frac{8}{20}$$

$\frac{13}{20}$

9
Which of the following is the correct first step in adding $\frac{3}{8}$ and $\frac{1}{6}$? ______
a) Add the numerators, 3 and 1.
b) Write each fraction with the same denominator. b

10
For the problem shown at the right, write both fractions with the same denominator. ______ ______

$$\begin{array}{r} \frac{3}{8} \\ + \frac{1}{6} \\ \hline \end{array}$$

$\frac{3}{8} = \frac{9}{24}, \frac{1}{6} = \frac{4}{24}$

11
Complete the addition shown at the right. ______

$$\begin{array}{r} \frac{3}{8} = \frac{9}{24} \\ + \frac{1}{6} = \frac{4}{24} \\ \hline \end{array}$$

$\frac{13}{24}$

12
Which of the following is the correct first step in adding $\frac{7}{11}$ and $\frac{2}{11}$? ______
a) Add the numerators, 7 and 2.
b) Write the fractions with a common denominator. a

13
$\frac{7}{11} + \frac{2}{11} =$ ______ $\frac{9}{11}$

14
Which of the following is the correct first step in adding $\frac{7}{10}$ and $\frac{1}{4}$? ______
a) Add the numerators, 7 and 1.
b) Write the fractions with a common denominator. b

15

For the problem shown at the right, write both fractions with a common denominator. ____ ____

$$\begin{array}{r} \frac{7}{10} \\ + \frac{1}{4} \\ \hline \end{array}$$

$\frac{7}{10} = \frac{14}{20}, \frac{1}{4} = \frac{5}{20}$

16

Complete the addition shown at the right. ______

$$\begin{array}{rcl} \frac{7}{10} & = & \frac{14}{20} \\ + \frac{1}{4} & = & \frac{5}{20} \\ \hline \end{array}$$

$\frac{19}{20}$

17

Add $\frac{3}{8} + \frac{5}{12}$ by first finding a common denominator. ______

$$\begin{array}{rcl} \frac{3}{8} & = & \frac{9}{24} \\ + \frac{5}{12} & = & \frac{10}{24} \\ \hline & & \frac{19}{24} \end{array}$$

18

$\frac{3}{10} + \frac{3}{8} =$ ______

$\frac{12}{40} + \frac{15}{40} = \frac{27}{40}$

19

$\frac{2}{3} + \frac{1}{7} =$ ______

$\frac{14}{21} + \frac{3}{21} = \frac{17}{21}$

20

$\frac{5}{9} + \frac{1}{3} =$ ______

$\frac{5}{9} + \frac{3}{9} = \frac{8}{9}$

21

$\frac{6}{11} + \frac{1}{4} =$ ______

$\frac{24}{44} + \frac{11}{44} = \frac{35}{44}$

22

$\frac{9}{13} + \frac{1}{5} =$ ______

$\frac{45}{65} + \frac{13}{65} = \frac{58}{65}$

23 ✍

$\frac{7}{11} + \frac{10}{11} = \frac{17}{11} = 1\frac{6}{11}$

In the addition problem above, the addition resulted in an improper fraction, $\frac{17}{11}$, which was then changed to a mixed number, $1\frac{6}{11}$. Follow the same process to add $\frac{5}{7} + \frac{4}{7} =$ ______

$\frac{9}{7} = 1\frac{2}{7}$

24 $\frac{7}{8} + \frac{1}{2} =$ ______

$\frac{7}{8} + \frac{4}{8} = \frac{11}{8} = 1\frac{3}{8}$

25 $\frac{5}{8} + \frac{3}{4} =$ ______

$\frac{5}{8} + \frac{6}{8} = \frac{11}{8} = 1\frac{3}{8}$

26 $\frac{5}{9} + \frac{4}{5} =$ ______

$\frac{25}{45} + \frac{36}{45} = \frac{61}{45} = 1\frac{16}{45}$

27 $\frac{3}{8} + \frac{11}{12} =$ ______

$\frac{9}{24} + \frac{22}{24} = \frac{31}{24} = 1\frac{7}{24}$

28 $\frac{5}{6} + \frac{1}{4} =$ ______

$\frac{10}{12} + \frac{3}{12} = \frac{13}{12} = 1\frac{1}{12}$

29 $\frac{2}{5} + \frac{7}{15} =$ ______

$\frac{6}{15} + \frac{7}{15} = \frac{13}{15}$

30 $\frac{7}{10} + \frac{1}{3} =$ ______

$\frac{21}{30} + \frac{10}{30} = \frac{31}{30} = 1\frac{1}{30}$

31 $\frac{6}{7} + \frac{1}{2} =$ ______

$\frac{12}{14} + \frac{7}{14} = \frac{19}{14} = 1\frac{5}{14}$

32 $\frac{1}{3} + \frac{1}{4} =$ ______

$\frac{4}{12} + \frac{3}{12} = \frac{7}{12}$

FEEDBACK UNIT 4

This quiz reviews the preceding unit. Answers are at the back of the book. Begin the next unit only after all problems of this unit are understood.

Add.

1. $\frac{5}{8} + \frac{1}{4}$
2. $\frac{3}{7} + \frac{2}{5}$
3. $\frac{4}{9} + \frac{5}{6}$
4. $\frac{8}{11} + \frac{3}{4}$
5. $\frac{2}{9} + \frac{2}{3}$
6. $\frac{3}{4} + \frac{3}{8}$
7. $\frac{5}{9} + \frac{7}{10}$
8. $\frac{5}{8} + \frac{3}{7}$
9. $\frac{4}{15} + \frac{1}{4}$
10. $\frac{7}{8} + \frac{5}{6}$
11. $\frac{3}{5} + \frac{8}{35}$
12. $\frac{15}{16} + \frac{19}{20}$

UNIT 5: THE ADDITION OF MIXED NUMBERS

ADDING MIXED NUMBERS

To add two mixed numbers, $3\frac{1}{5}$ and $5\frac{2}{3}$,

1. Add the two proper fractions.
2. Add the two whole numbers.

$$\begin{array}{r} 3\frac{1}{5} = 3\ \frac{3}{15} \\ +5\frac{2}{3} = 5\ \frac{10}{15} \\ \hline 8\ \frac{13}{15} \end{array}$$

1 ✍

The mixed number $4\frac{3}{8}$ is the sum of the whole number, 4, and the fraction, $\frac{3}{8}$. $4\frac{3}{8} = 4 + \frac{3}{8}$
$9\frac{4}{5}$ is the sum of the whole number ___ and the fraction ____.

$9, \frac{4}{5}$

2

$4\frac{3}{7} = 4 + \frac{3}{7}$ and $2\frac{5}{12} =$ ______

$2 + \frac{5}{12}$

3

To add $4\frac{1}{5}$ and $3\frac{2}{5}$,

1. Add the fractions, $\frac{1}{5}$ and $\frac{2}{5}$
2. Add the whole numbers, ______ and ______

4, 3

4

Complete the addition shown at the right by adding the whole numbers. _______

$$\begin{array}{r} 4\ \frac{1}{5} \\ +\ 3\ \frac{2}{5} \\ \hline \frac{3}{5} \end{array}$$

7

5 ✍

Add the problem at the right by

1. Adding the fractions.
2. Adding the whole numbers.

$$\begin{array}{r} 7\,\frac{3}{11} \\ +\ 2\,\frac{4}{11} \\ \hline \end{array}$$

$9\frac{7}{11}$

6

Add $2\frac{1}{7} + 4\frac{4}{7}$ by separately adding the proper fractions and the whole numbers. ______

$6\frac{5}{7}$

7

Add $5\frac{3}{17} + 8\frac{10}{17}$ by separately adding the proper fractions and the whole numbers. ______

$13\frac{13}{17}$

8

$7\frac{2}{13} + 12\frac{5}{13} =$ ______

$19\frac{7}{13}$

9

$6\frac{3}{11} + \frac{5}{11} =$ ______

$6\frac{8}{11}$

10

Always check the sum of two fractions to see if it can be simplified.

$4\frac{3}{8} + 9\frac{1}{8} =$ ______

$13\frac{4}{8} = 13\frac{1}{2}$

11

$3\frac{3}{16} + 6\frac{9}{16} =$ ______

$9\frac{12}{16} = 9\frac{3}{4}$

12

$26\frac{5}{8} + 9\frac{1}{8} =$ ______

$35\frac{6}{8} = 35\frac{3}{4}$

13

$$9 + \frac{5}{3} = 9 + 1\frac{2}{3} = 10\frac{2}{3}$$

The example above shows how to add a whole number, 9, and an improper fraction, $\frac{5}{3}$. Use the example to complete the following problem.

$6\frac{8}{9} + 4\frac{5}{9} =$ ______ $10\frac{13}{9} = 11\frac{4}{9}$

14

$2\frac{5}{7} + 8\frac{3}{7} =$ ______ $10\frac{8}{7} = 11\frac{1}{7}$

15

$4\frac{5}{11} + 1\frac{8}{11} =$ ______ $5\frac{13}{11} = 6\frac{2}{11}$

16

$6\frac{3}{5} + 7\frac{4}{5} =$ ______ $13\frac{7}{5} = 14\frac{2}{5}$

17

$5\frac{3}{8} + 9\frac{7}{8} =$ ______ $14\frac{10}{8} = 15\frac{2}{8} = 15\frac{1}{4}$

18

$4\frac{7}{10} + 1\frac{3}{10} =$ ______ $5\frac{10}{10} = 6$

19

To add $5\frac{3}{8}$ and $12\frac{1}{6}$ a common denominator must be found for the fraction addition. What will be the common denominator? ______ 24

20

The example at the right shows the addition of two mixed numbers where the denominators of the fractions are different. Complete the problem. ______

$$\begin{array}{rcr} 5\frac{3}{8} & = & 5\frac{9}{24} \\ +\ 12\frac{1}{6} & = & 12\frac{4}{24} \\ \hline \end{array}$$

$17\frac{13}{24}$

21

Add $3\frac{2}{5} + 11\frac{1}{2}$ by first finding a common denominator for the fractions. ______

$3\frac{4}{10} + 11\frac{5}{10} = 14\frac{9}{10}$

22

Add $3\frac{1}{3} + 5\frac{1}{9}$ by first finding a common denominator for the fractions. ______

$3\frac{3}{9} + 5\frac{1}{9} = 8\frac{4}{9}$

23

Add $1\frac{5}{8} + 4\frac{1}{6}$ by first finding a common denominator for the fractions. ______

$1\frac{15}{24} + 4\frac{4}{24} = 5\frac{19}{24}$

24

$6\frac{7}{12} + 5\frac{1}{3} =$ ______

$6\frac{7}{12} + 5\frac{4}{12} = 11\frac{11}{12}$

25

$2\frac{5}{8} + 1\frac{1}{3} =$ ______

$2\frac{15}{24} + 1\frac{8}{24} = 3\frac{23}{24}$

26

$3\frac{5}{6} + 1\frac{2}{3} =$ ______

$4\frac{9}{6} = 5\frac{1}{2}$

27

$7\frac{3}{4} + 2\frac{5}{8} =$ ______

$9\frac{11}{8} = 10\frac{3}{8}$

28

$4\frac{6}{7} + 3\frac{1}{2} =$ ______

$7\frac{19}{14} = 8\frac{5}{14}$

FEEDBACK UNIT 5

This quiz reviews the preceding unit. Answers are at the back of the book. Begin the next unit only after all problems of this unit are understood.

Add.

1. $4\frac{1}{3} + 3\frac{1}{3}$
2. $6\frac{5}{8} + 7$
3. $\frac{4}{7} + 3\frac{2}{7}$
4. $2\frac{7}{11} + 8\frac{8}{11}$
5. $2\frac{5}{16} + 5\frac{3}{16}$
6. $6\frac{3}{8} + 4\frac{3}{4}$
7. $8\frac{9}{10} + 2\frac{4}{5}$
8. $9\frac{6}{7} + 3\frac{1}{2}$
9. $1\frac{2}{5} + 2\frac{3}{8}$
10. $6\frac{3}{4} + 5\frac{5}{6}$
11. $89\frac{2}{5} + 3\frac{7}{15}$
12. $9\frac{3}{8} + 74\frac{7}{10}$

UNIT 6: THE SUBTRACTION OF PROPER FRACTIONS

SUBTRACTING FRACTIONS

To subtract fractions they must have the same denominator.

The subtraction of $\frac{3}{8} - \frac{1}{5}$ is completed by:

1. Writing the fractions with the same denominator.
   $$\frac{3}{8} = \frac{15}{40}$$
   $$-\frac{1}{5} = \frac{8}{40}$$
2. Subtracting the numerators.
   $$\frac{7}{40}$$

1
Addition and subtraction only make sense when the numbers have the same label. Does it make sense to subtract 3 doctors from 7 nurses? ______

No, labels are different

2 ✍
The denominator of a fraction is its label. $\frac{4}{5}$ can be read as 4 fifths. $\frac{5}{7}$ can be read as 5 sevenths. Do $\frac{4}{5}$ and $\frac{5}{7}$ have the same label? ______

No

3
Subtract the problem shown at the right by subtracting the numerators.

$$\begin{array}{r} \frac{7}{11} \\ -\ \frac{3}{11} \\ \hline \end{array}$$

$\frac{4}{11}$

4
Subtract, if possible. $\frac{5}{7} - \frac{2}{7} =$ ______

$\frac{3}{7}$

5
Subtract, if possible. $\frac{9}{10} - \frac{1}{3} =$ ______

denominators are different

6
Subtract, if possible. $\frac{11}{13} - \frac{5}{13} =$ ______

$\frac{6}{13}$

7
Subtract, if possible. $\frac{8}{9} - \frac{5}{9} =$ ______

$\frac{3}{9} = \frac{1}{3}$

8
The subtraction problem $2 - 5 =$ ____ has no whole number answer because the subtrahend, 5, is greater than the minuend, 2. Does the problem $8 - 5 =$ ______ have a whole number answer? ______

Yes, 3

9 ✍

The problem $\frac{2}{7} - \frac{5}{7} = X$ has its answer indicated by the capital letter X because $\frac{5}{7}$ is greater than $\frac{2}{7}$.

$\frac{5}{12} - \frac{7}{12} =$ ______ X

10

$\frac{4}{5} - \frac{1}{5} =$ ______ $\frac{3}{5}$

11

$\frac{6}{7} - \frac{3}{7} =$ ______ $\frac{3}{7}$

12

$\frac{6}{13} - \frac{8}{13} =$ ______ X

13

$\frac{8}{23} - \frac{15}{23} =$ ______ X

14

$\frac{8}{15} - \frac{4}{15} =$ ______ $\frac{4}{15}$

SUBTRACTING FRACTIONS WITH DIFFERENT DENOMINATORS

To subtract $\frac{5}{8}$ from $\frac{11}{12}$,

1. Write the problem with the subtrahend and minuend in the correct order. $\frac{11}{12} - \frac{5}{8}$
2. Find a common denominator and write each fraction with it. $\frac{11}{12} = \frac{22}{24}$, $-\frac{5}{8} = \frac{15}{24}$
3. Subtract the numerators of the new fractions. $\frac{7}{24}$

15

For the problem at the right, write each of the fractions with a common denominator. _____

$$\begin{array}{r} \frac{9}{10} = \frac{?}{?} \\ -\ \frac{1}{8} = \frac{?}{?} \\ \hline \end{array}$$

$\frac{9}{10} = \frac{36}{40}, \frac{1}{8} = \frac{5}{40}$

16

Complete the subtraction shown at the right. A common denominator has already been found. _____

$$\begin{array}{r} \frac{9}{10} = \frac{36}{40} \\ -\ \frac{1}{8} = \frac{5}{40} \\ \hline \end{array}$$

$\frac{31}{40}$

17 ✍

Complete the subtraction shown at the right. A common denominator needs to be found. _____

$$\begin{array}{r} \frac{1}{2} = \frac{?}{?} \\ -\ \frac{1}{3} = \frac{?}{?} \\ \hline \end{array}$$

$\frac{3}{6} - \frac{2}{6} = \frac{1}{6}$

18

Complete the subtraction shown at the right.

$$\begin{array}{r} \frac{3}{4} \\ -\ \frac{1}{6} \\ \hline \end{array}$$

$\frac{9}{12} - \frac{2}{12} = \frac{7}{12}$

19 $\frac{1}{3} - \frac{1}{4} =$ _____

$\frac{4}{12} - \frac{3}{12} = \frac{1}{12}$

20 $\frac{9}{10} - \frac{1}{5} =$ _____

$\frac{9}{10} - \frac{2}{10} = \frac{7}{10}$

21 $\frac{3}{4} - \frac{1}{2} =$ _____

$\frac{3}{4} - \frac{2}{4} = \frac{1}{4}$

22 $\frac{3}{8} - \frac{1}{2} =$ _____

$\frac{3}{8} - \frac{4}{8} =$ X

23 $\frac{5}{6} - \frac{7}{10} =$ ____ $\frac{25}{30} - \frac{21}{30} = \frac{4}{30} = \frac{2}{15}$

24 $\frac{5}{8} - \frac{2}{7} =$ ____ $\frac{35}{56} - \frac{16}{56} = \frac{19}{56}$

25 $\frac{9}{10} - \frac{3}{8} =$ ____ $\frac{36}{40} - \frac{15}{40} = \frac{21}{40}$

26 $\frac{7}{15} - \frac{3}{5} =$ ____ $\frac{7}{15} - \frac{9}{15} \doteq$ X

27 $\frac{2}{3} - \frac{1}{4} =$ ____ $\frac{8}{12} - \frac{3}{12} = \frac{5}{12}$

28 $\frac{3}{8} - \frac{2}{3} =$ ____ $\frac{9}{24} - \frac{16}{24} =$ X

29 $\frac{5}{6} - \frac{5}{9} =$ ____ $\frac{15}{18} - \frac{10}{18} = \frac{5}{18}$

FEEDBACK UNIT 6

This quiz reviews the preceding unit. Answers are at the back of the book. Begin the next unit only after all problems of this unit are understood.

Subtract.

1. $\frac{9}{13} - \frac{2}{13}$	5. $\frac{5}{8} - \frac{1}{6}$	9. $\frac{5}{6} - \frac{3}{4}$
2. $\frac{7}{10} - \frac{9}{10}$	6. $\frac{5}{8} - \frac{1}{4}$	10. $\frac{9}{16} - \frac{3}{4}$
3. $\frac{4}{5} - \frac{3}{10}$	7. $\frac{4}{11} - \frac{1}{2}$	11. $\frac{4}{5} - \frac{2}{3}$
4. $\frac{2}{7} - \frac{1}{8}$	8. $\frac{5}{7} - \frac{3}{8}$	12. $\frac{5}{7} - \frac{3}{10}$

UNIT 7: THE SUBTRACTION OF MIXED NUMBERS

SUBTRACTING MIXED NUMBERS

To subtract two mixed numbers,

1. Subtract the fractions, and
2. Subtract the whole numbers.

1

For the subtraction problem at the right, the fractions and whole numbers are subtracted separately. Complete the subtraction by subtracting the whole numbers. ______

$$\begin{array}{r} 4\,\frac{5}{7} \\ -\,3\,\frac{2}{7} \\ \hline \frac{3}{7} \end{array}$$

$1\frac{3}{7}$

2 ✍

Subtract $2\frac{3}{11}$ **from** $7\frac{5}{11}$ by subtracting the fractions and then subtracting the whole numbers. ______

$$\begin{array}{r} 7\,\frac{5}{11} \\ -\,2\,\frac{3}{11} \\ \hline \end{array}$$

$5\frac{2}{11}$

3

Subtract $3\frac{1}{13}$ **from** $10\frac{6}{13}$ by first subtracting the fractions and then subtracting the whole numbers. ______

$$\begin{array}{r} 10\,\frac{6}{13} \\ -\,3\,\frac{1}{13} \\ \hline 7\,\frac{5}{13} \end{array}$$

4

$5\frac{8}{9} - 3\frac{7}{9} =$ ______

$2\frac{1}{9}$

5

$8\frac{2}{3} - 1\frac{1}{3} =$ ______

$7\frac{1}{3}$

6

$9\frac{8}{11} - 5\frac{5}{11} =$ ______

$4\frac{3}{11}$

7

Can the fractions of $6\frac{4}{7} - 2\frac{5}{7}$ be subtracted? ______

No, $\frac{4}{7} - \frac{5}{7} = X$

"BORROWING" FROM A MIXED NUMBER

The "borrowing" process is needed in some subtraction problems.

The example below shows the steps needed for "borrowing" from the mixed number $6\frac{4}{7}$.

1. Take one (1) away from the whole number, 6, and put it with the proper fraction, $\frac{4}{7}$.
   $$6\frac{4}{7} = 5 + 1\frac{4}{7}$$
2. Change $1\frac{4}{7}$ to an improper fraction.
   $$6\frac{4}{7} = 5 + 1\frac{4}{7} = 5 + \frac{11}{7} = 5\frac{11}{7}$$

8

$5\frac{6}{7} = 4 + 1\frac{6}{7} = 4\frac{13}{7}$

Is $5\frac{6}{7}$ equal to $4\frac{13}{7}$? ______

Yes, $1\frac{6}{7} = \frac{13}{7}$

9 ✍

$6\frac{1}{8} = 5 + 1\frac{1}{8} =$ ______

Complete the equality above by changing $1\frac{1}{8}$ to an improper fraction.

$6\frac{1}{8} = 5\frac{9}{8}$

10

Is $6\frac{1}{8}$ equal to $5\frac{9}{8}$? ______

Yes, $1\frac{1}{8} = \frac{9}{8}$

11

$7\frac{3}{5} = 6 + $ ______

Complete the equality above by finding the improper fraction that equals $1\frac{3}{5}$.

$7\frac{3}{5} = 6\frac{8}{5}$

12

To subtract the problem at the right, borrowing is necessary. Why? ______

$$\begin{array}{r} 5\frac{3}{5} \\ -\ 2\frac{4}{5} \\ \hline \end{array}$$

$\frac{3}{5} - \frac{4}{5} = X$

13

Complete the subtraction problem shown at the right.

$$\begin{array}{rcrcr} 7\frac{2}{5} & = & 6 + 1\frac{2}{5} & = & 6\frac{7}{5} \\ -2\frac{4}{5} & = & -2\frac{4}{5} & = & -2\frac{4}{5} \\ \hline \end{array}$$

______

$4\frac{3}{5}$

BORROWING IN THE SUBTRACTION OF MIXED NUMBERS

To subtract the problem shown at the right, the following steps are used: $8\frac{4}{9} - 3\frac{8}{9}$

1. To borrow one (1) from $8\frac{4}{9}$, take one away from 8 and place it with the proper fraction. $8\frac{4}{9} = 7 + 1\frac{4}{9}$
2. Write $1\frac{4}{9}$ as an improper fraction. $7 + 1\frac{4}{9} = 7\frac{13}{9}$
3. Subtract. $8\frac{4}{9} - 3\frac{8}{9} = 7\frac{13}{9} - 3\frac{8}{9} = 4\frac{5}{9}$

14 ✍

The subtraction problem at the right requires borrowing. Show the borrowing and complete the problem.

$$\begin{array}{rcr} 6\frac{4}{9} & = & 5\frac{?}{?} \\ -\ 3\frac{8}{9} & = & 3\frac{8}{9} \\ \hline \end{array}$$

______

$5\frac{13}{9} - 3\frac{8}{9} = 2\frac{5}{9}$

15

The subtraction problem at the right requires borrowing. Show the borrowing and complete the problem. ______

$$\begin{array}{rcl} 4\frac{5}{8} & = & ?\,\frac{?}{?} \\ -\ 3\frac{6}{8} & = & 3\,\frac{6}{8} \\ \hline \end{array}$$

$3\frac{13}{8} - 3\frac{6}{8} = \frac{7}{8}$

16

Subtract $1\frac{4}{5}$ **from** $3\frac{2}{5}$ by first borrowing from $3\frac{2}{5}$. ______

$2\frac{7}{5} - 1\frac{4}{5} = 1\frac{3}{5}$

17

Use the borrowing process to subtract.

$5\frac{3}{11} - 2\frac{7}{11} =$ ______

$4\frac{14}{11} - 2\frac{7}{11} = 2\frac{7}{11}$

18

$7\frac{9}{16} - 2\frac{10}{16} =$ ______

$6\frac{25}{16} - 2\frac{10}{16} = 4\frac{15}{16}$

19

$5\frac{3}{10} - 2\frac{9}{10} =$ ______

$4\frac{13}{10} - 2\frac{9}{10} = 2\frac{2}{5}$

20

$9\frac{1}{8} - 5\frac{7}{8} =$ ______

$8\frac{9}{8} - 5\frac{7}{8} = 3\frac{1}{4}$

21 ✍

The whole number 5 is equal to the mixed number $4\frac{7}{7}$.

$5 - 1\frac{5}{7} =$ ______

$4\frac{7}{7} - 1\frac{5}{7} = 3\frac{2}{7}$

22

$4\frac{2}{8} - 3\frac{7}{8} =$ ______

$3\frac{10}{8} - 3\frac{7}{8} = \frac{3}{8}$

SUBTRACTING MIXED NUMBERS WITH DIFFERENT DENOMINATORS

To subtract two mixed numbers which have different denominators, a common denominator must first be found.

The steps in subtracting $5\frac{3}{5}$ from $8\frac{1}{4}$ are shown at the right.

$$\begin{array}{rcrcr} 8\frac{1}{4} & = & 8\frac{5}{20} & = & 7\frac{25}{20} \\ -5\frac{3}{5} & = & -5\frac{12}{20} & = & -5\frac{12}{20} \\ \hline & & & & 2\frac{13}{20} \end{array}$$

23

Subtract $4\frac{1}{2}$ **from** $6\frac{3}{4}$ by first finding a common denominator. ______

$6\frac{3}{4} - 4\frac{2}{4} = 2\frac{1}{4}$

24

Subtract $5\frac{1}{16}$ from $5\frac{3}{8}$ by first finding a common denominator. ______

$5\frac{6}{16} - 5\frac{1}{16} = \frac{5}{16}$

25

Subtract $2\frac{1}{2}$ from $7\frac{3}{5}$ by first finding a common denominator. ______

$7\frac{6}{10} - 2\frac{5}{10} = 5\frac{1}{10}$

26

Subtract $2\frac{5}{6}$ from $7\frac{3}{4}$ by first finding a common denominator. Borrowing is also needed. ______

$$\begin{array}{rcr} 7\frac{9}{12} & = & 6\frac{21}{12} \\ -2\frac{10}{12} & = & -2\frac{10}{12} \\ \hline & & 4\frac{11}{12} \end{array}$$

27

$6\frac{3}{7} - 4\frac{1}{2} =$ ______

$5\frac{20}{14} - 4\frac{7}{14} = 1\frac{13}{14}$

28

$5\frac{3}{10} - \frac{4}{5} =$ ______

$4\frac{13}{10} - 0\frac{8}{10} = 4\frac{1}{2}$

29
$6\frac{1}{4} - 2\frac{2}{3} =$ ______

$5\frac{15}{12} - 2\frac{8}{12} = 3\frac{7}{12}$

30
$6\frac{3}{8} - 5\frac{6}{7} =$ ______

$5\frac{77}{56} - 5\frac{48}{56} = \frac{29}{56}$

31
The fraction $\frac{0}{5}$ equals zero (0). Use this fact to subtract: $7\frac{3}{5} - 2 =$ ______

$7\frac{3}{5} - 2\frac{0}{5} = 5\frac{3}{5}$

32
$8\frac{7}{9} - 5 =$ ______

$8\frac{7}{9} - 5\frac{0}{9} = 3\frac{7}{9}$

33
$9 - 2\frac{3}{11} =$ ______

$8\frac{11}{11} - 2\frac{3}{11} = 6\frac{8}{11}$

34
$5\frac{1}{5} - 2\frac{1}{8} =$ ______

$5\frac{8}{40} - 2\frac{5}{40} = 3\frac{3}{40}$

35
$3\frac{5}{16} - 1\frac{1}{2} =$ ______

$2\frac{21}{16} - 1\frac{8}{16} = 1\frac{13}{16}$

Feedback Unit 7

This quiz reviews the preceding unit. Answers are at the back of the book. Begin the next unit only after all problems of this unit are understood.

Subtract.

1. $3\frac{5}{8} - 1\frac{1}{4}$
2. $9\frac{3}{16} - 2\frac{1}{8}$
3. $4\frac{3}{7} - 2\frac{3}{8}$
4. $5\frac{9}{10} - 2\frac{1}{2}$
5. $8\frac{1}{3} - 3\frac{3}{5}$
6. $5\frac{1}{7} - 2\frac{1}{9}$
7. $4 - 3\frac{1}{2}$
8. $6\frac{1}{4} - 3\frac{1}{2}$
9. $8\frac{5}{9} - 3\frac{5}{6}$
10. $9\frac{5}{8} - 2\frac{1}{6}$
11. $6\frac{5}{8} - 2$
12. $5\frac{3}{5} - 1\frac{5}{7}$

UNIT 8: WORD PROBLEMS

WORD PROBLEMS REQUIRING ADDITION OR SUBTRACTION

The numbers in an addition problem are called addends. The answer to an addition problem is its sum.

$$\begin{array}{rl} 4\frac{3}{8} & \textbf{addend} \\ +\ 1\frac{1}{8} & \textbf{addend} \\ \hline 5\frac{4}{8} = 5\frac{1}{2} & \textbf{sum} \end{array}$$

The numbers in a subtraction problem are the minuend and subtrahend. The answer to a subtraction problem is the difference.

$$\begin{array}{rl} 6\frac{7}{12} & \textbf{minuend} \\ -\ 5\frac{4}{12} & \textbf{subtrahend} \\ \hline 1\frac{3}{12} = 1\frac{1}{4} & \textbf{difference} \end{array}$$

---

1 ✍

The answer to an addition problem is its **sum**.
Find the sum of $2\frac{1}{5}$ and $4\frac{3}{5}$. _______

$2\frac{1}{5} + 4\frac{3}{5} = 6\frac{4}{5}$

2
The answer to an addition problem is its _______.

sum

3 ✍

The answer to a subtraction problem is its **difference**.
Find the difference between $6\frac{7}{11}$ and $3\frac{2}{11}$ _______

$6\frac{7}{11} - 3\frac{2}{11} = 3\frac{5}{11}$

4
The answer to a subtraction problem is its _______.

difference

5
Find the sum of $3\frac{5}{8}$ and $2\frac{1}{4}$. _______

$3\frac{5}{8} + 2\frac{2}{8} = 5\frac{7}{8}$

6
Find the difference of $5\frac{9}{10}$ and $1\frac{1}{2}$. _______

$5\frac{9}{10} - 1\frac{5}{10} = 4\frac{2}{5}$

7
Find the sum of $6\frac{8}{11}$ and $4\frac{1}{2}$. ______ $11\frac{5}{22}$

8
Find the difference of $9\frac{1}{6}$ and $1\frac{3}{4}$. ______ $7\frac{5}{12}$

9
What is the sum of $\frac{3}{4}$ lbs of sugar and $1\frac{1}{2}$ lbs of sugar? ______ $2\frac{1}{4}$ lbs of sugar

10
Does it make sense to add $4\frac{1}{2}$ lbs of flour and 3 lbs of sugar? ______ No, different labels

11
Does it make sense to add $4\frac{1}{2}$ ounces of flour and $3\frac{1}{4}$ lbs of flour? ______ No, different labels

12
The words "how much more" usually indicate subtraction. How much more is $2\frac{3}{4}$ yards than $1\frac{5}{8}$ yards? ______ $2\frac{6}{8} - 1\frac{5}{8} = 1\frac{1}{8}$ yds

13
How much more is $7\frac{1}{2}$ days than $1\frac{2}{3}$ days? ______ $6\frac{9}{6} - 1\frac{4}{6} = 5\frac{5}{6}$ days

14
One capsule holds $20\frac{3}{10}$ grams while another holds $15\frac{7}{10}$ grams. How much more does the bigger capsule hold? ______ $19\frac{13}{10} - 15\frac{7}{10} = 4\frac{3}{5}$ grams

15
Ben bought two steaks. One weighed $2\frac{1}{4}$ lbs and the other $1\frac{1}{2}$ lbs. How much steak did Ben buy? ______ $2\frac{1}{4} + 1\frac{2}{4} = 3\frac{3}{4}$ lbs

16
One bottle held $4\frac{3}{8}$ liters of alcohol and another held $1\frac{5}{8}$ liters. How much more did the first bottle hold? _______

$3\frac{11}{8} - 1\frac{5}{8} = 2\frac{3}{4}$ liters

17
A specimen of iron ore weighed $4\frac{7}{10}$ kilograms and another weighed $3\frac{1}{10}$ kilograms. What was the sum of the weights of the two specimens? _______

$4\frac{7}{10} + 3\frac{1}{10} = 7\frac{4}{5}$ kgm

18
Mary worked $2\frac{5}{6}$ hours and Sally worked $7\frac{1}{4}$ hours. How much more did Sally work? _______

$6\frac{15}{12} - 2\frac{10}{12} = 4\frac{5}{12}$ hrs

19
Container A held $5\frac{1}{2}$ cups of milk while container B held only $2\frac{3}{4}$ cups. Find the difference in the amounts of milk the two containers can hold. _______

$4\frac{6}{4} - 2\frac{3}{4} = 2\frac{3}{4}$ cups

20
The price of a stock was $\$10\frac{1}{8}$ on Thursday and $\$8\frac{3}{4}$ on Friday. What was the difference in the price those two days? _______

$9\frac{9}{8} - 8\frac{6}{8} = \$1\frac{3}{8}$

21
Kristi ran $1\frac{1}{4}$ hours in the morning and $2\frac{3}{5}$ hours in the afternoon. How many hours did she run altogether? _____

$3\frac{17}{20}$ hours

22
To add $5\frac{1}{4}$ inches to 1 foot, both measurements need to be expressed as inches. 1 foot equals 12 inches. What is the sum of $5\frac{1}{4}$ inches to 1 foot? ______

$17\frac{1}{4}$ inches

23
The **perimeter** of a geometric figure is the sum of the lengths of its sides. To find the perimeter of a triangle, _____ the lengths of its three sides.

add

24
The **perimeter** of the scalene triangle shown at the right is the sum of the lengths of the three sides. Find the perimeter. _____

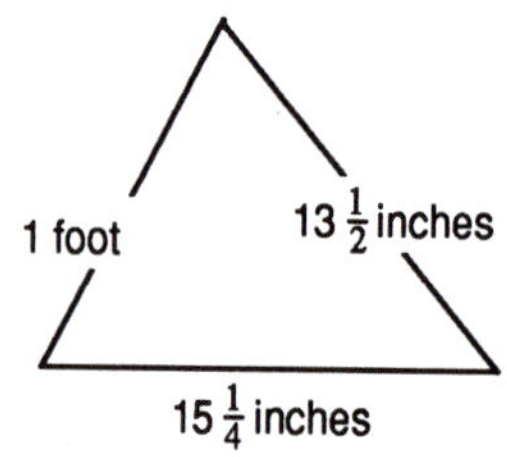

[Note: A **scalene triangle** has three sides of different lengths.]

1 foot = 12 inches

$12 + 13\frac{1}{2} + 15\frac{1}{4}$
$40\frac{3}{4}$ inches is the perimeter

25
The **perimeter** of the **isosceles triangle** shown at the right is the sum of the lengths of the three sides. Find the perimeter. _____

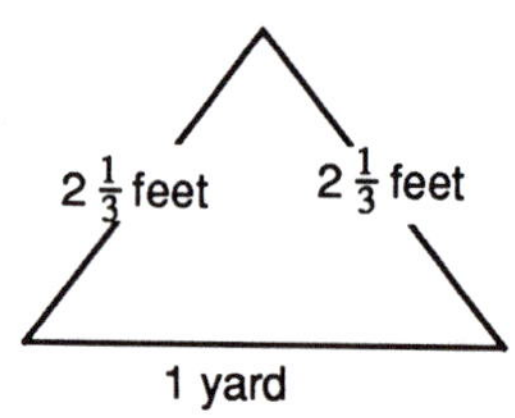

[Note: An **isosceles triangle** has two sides of the same length.]

1 yard = 3 feet

$3 + 2\frac{1}{3} + 2\frac{1}{3}$
$7\frac{2}{3}$ feet is the perimeter

26
Find the perimeter of the **equilateral triangle** shown at the right.

_____

[Note: An **equilateral triangle** has three sides of the same length.]

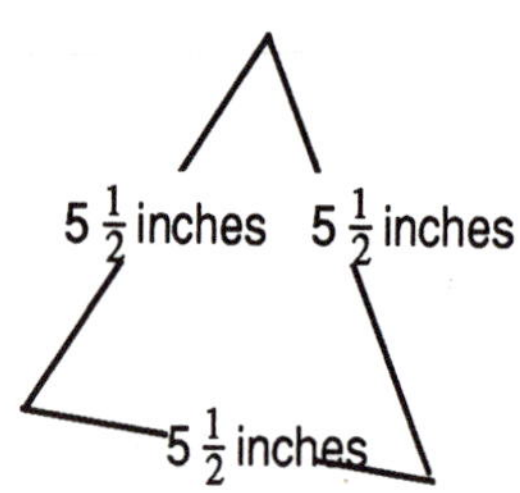

$16\frac{1}{2}$ inches

27
If a triangle has sides of $4\frac{1}{2}$ feet and $5\frac{2}{3}$ feet with a perimeter of $14\frac{1}{4}$ feet, how long is the third side? _____

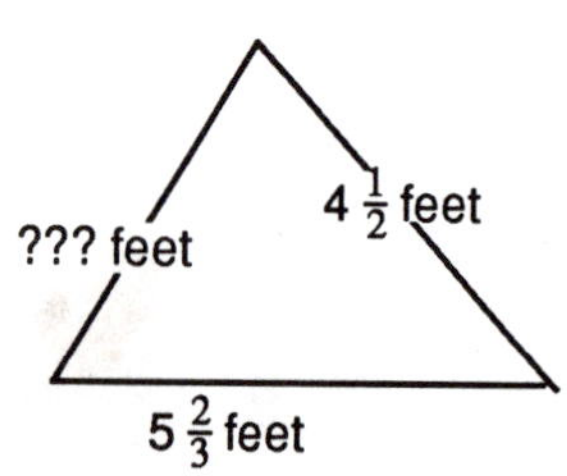

$4\frac{1}{2} + 5\frac{2}{3} = 10\frac{1}{6}$

$14\frac{1}{4} - 10\frac{1}{6} = 4\frac{1}{12}$ feet

FEEDBACK UNIT 8

This quiz reviews the preceding unit. Answers are at the back of the book. Begin the next unit only after all problems of this unit are understood.

1. The answer to a subtraction problem is its ________.

2. The answer to an addition problem is its ________.

3. How much more is $9\frac{1}{4}$ inches than $4\frac{7}{8}$ inches?

4. Sarah bought two pieces of silk. One was $1\frac{2}{3}$ yards long and the other was $1\frac{3}{4}$ yards long. What was the total length of the silk?

5. One stock sells for $\$11\frac{3}{8}$ while another sells for $\$9\frac{1}{4}$. Find the sum of the prices of the stocks.

6. A block of iron weighed $12\frac{7}{10}$ kilograms while a block of gold weighed $2\frac{9}{10}$ kilograms. How much more did the iron weigh than the gold?

7. One capsule weighed $5\frac{1}{10}$ grams. Another weighed $3\frac{3}{5}$ grams. Find the difference in their weights.

8. If $2\frac{73}{100}$ meters of rope is cut from a piece $6\frac{3}{5}$ meters long, how much will be left?

9. Find the perimeter of the isosceles triangle shown below.

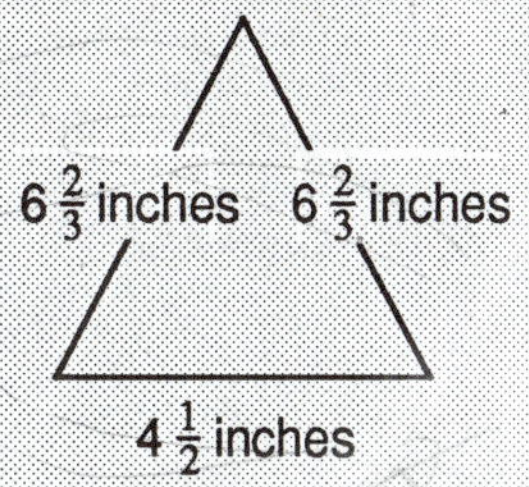

10. A triangle has a perimeter of $20\frac{3}{8}$ inches. If two of its sides are $8\frac{1}{4}$ inches and $5\frac{2}{3}$ inches long, what is the length of the third side?

CHAPTER 6 MASTERY TEST

This test provides an opportunity for assuring that **all** of the objectives of Chapter 6 have been learned. Each problem is accompanied by the symbol (U) which indicates the Unit in which that type of problem was studied. If any problems are missed, they need to be corrected. Answers are at the back of the book.

In problems 1-4, find the least common multiple (lcm) for each pair of numbers. (3)

1. 3, 5
2. 6, 8
3. 12, 15
4. 8, 16
5. Write $\frac{7}{8}$ with denominator 24. (2)
6. Write $\frac{3}{5}$ with denominator 45. (2)

In problems 7-18, add or subtract as indicated.

7. $\frac{1}{2} + 1\frac{1}{2}$ (5)
8. $1\frac{1}{2} + \frac{1}{4}$ (5)
9. $\frac{4}{5} + \frac{1}{5}$ (1)
10. $\frac{3}{10} + \frac{3}{10}$ (1)
11. $\frac{2}{5} + \frac{3}{10}$ (4)
12. $\frac{1}{7} + \frac{2}{3}$ (4)
13. $\frac{4}{7} + 2\frac{3}{4}$ (5)
14. $\frac{8}{10} - 10$ (7)
15. $\frac{5}{6} - \frac{1}{3}$ (6)
16. $2\frac{3}{8} - 1\frac{1}{8}$ (7)
17. $1\frac{1}{8} - \frac{3}{4}$ (7)
18. $1\frac{1}{6} - \frac{1}{12}$ (7)
19. $5\frac{1}{3} - 2\frac{3}{5}$ (7)
20. $7\frac{4}{9} - 1\frac{2}{3}$ (7)
21. Two rocks, one weighing $2\frac{3}{4}$ kilograms and the other weighing $1\frac{1}{2}$ kilograms, were placed on a scale. What was their total weight? (8)
22. A piece of rope $5\frac{3}{4}$ meters long was cut from a rope $10\frac{1}{2}$ meters long. How much was left? (8)
23. How much more is $5\frac{1}{8}$ cups than $3\frac{3}{8}$ cups? (8)
24. Find the perimeter of the scalene triangle shown below. (8)

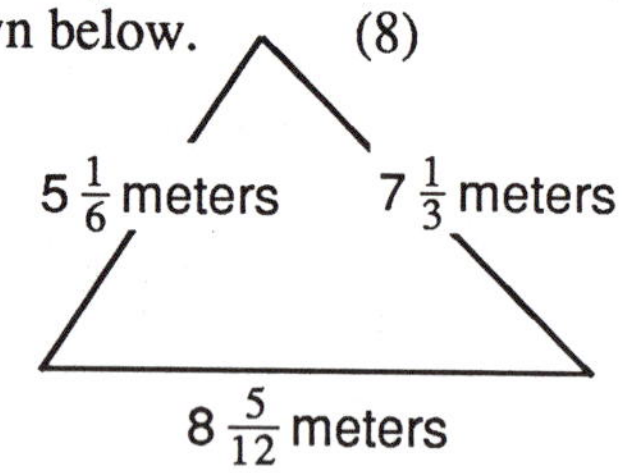

25. A triangle has a perimeter of 2 feet. If two of its sides are $7\frac{1}{3}$ inches and $9\frac{1}{2}$ inches long, what is the length of the third side? (8)

Chapter 7 Objectives

The following problems describe, by example, the skills and knowledge studied in this chapter. Each problem is accompanied by a symbol, (U), which indicates the Unit where the problem is studied. Answers are at the back of the book.

1. $\frac{3}{5} \times \frac{7}{4}$ (1)
2. $7 \times \frac{2}{11}$ (1)
3. $\frac{1}{4} \times \frac{8}{7}$ (2)
4. $9 \times \frac{3}{7}$ (1)
5. $\frac{9}{16} \times 12$ (2)
6. $\frac{10}{7} \times \frac{21}{20}$ (2)
7. $\frac{35}{26} \times \frac{4}{25}$ (2)
8. $2\frac{3}{5} \times 4$ (3)
9. $\frac{5}{22} \times \frac{11}{10}$ (2)
10. $6 \times 1\frac{1}{2}$ (3)
11. $1\frac{2}{3} \times 2\frac{1}{2}$ (3)
12. $1\frac{3}{10} \times 2\frac{1}{3}$ (3)
13. $\frac{1}{2} \times 17$ (1)
14. $\frac{8}{9} \div \frac{10}{21}$ (4)
15. $\frac{3}{7} \div \frac{1}{5}$ (4)
16. $\frac{11}{9} \div \frac{2}{3}$ (4)
17. $1\frac{2}{5} \div 2$ (4)
18. $3\frac{1}{2} \div 1\frac{3}{4}$ (4)
19. $4 \div \frac{1}{7}$ (4)
20. $1\frac{7}{8} \div \frac{3}{4}$ (4)
21. In the problem $\frac{3}{4} \div \frac{9}{8} = \frac{2}{3}$, the divisor is ______, the quotient is ______, and the dividend is ______. (4)
22. Find $\frac{7}{8}$ of 3. (5)
23. A bank has $2\frac{7}{10}$ million dollars. $\frac{2}{3}$ of it was on deposit. How much was on deposit? (5)
24. Spoons that weigh $\frac{1}{10}$ kilogram are to be made from $6\frac{3}{5}$ kilograms of silver alloy. How many spoons can be made? (5)
25. The dealer cost of a car is $\frac{4}{5}$ of the selling price. The sales price is \$3500. Find the dealer cost of the car. (5)

Chapter 7
Multiplication And Division Of Fractions

Unit 1: The Multiplication Of Fractions

Multiplying Fractions

Any two fractions can be multiplied by the following process:

1. Multiply the numerators to find the numerator of the answer (product).

2. Multiply the denominators to find the denominator of the answer (product).

$$\frac{5}{8} \times \frac{3}{4} = \frac{5 \times 3}{8 \times 4} = \frac{15}{32}$$

1
The problem at the right is an example of the multiplication of two fractions. Multiply the second problem shown at the right by separately multiplying the numerators and denominators.

$\frac{3}{4} \times \frac{5}{7} = \frac{15}{28}$

$\frac{6}{7} \times \frac{5}{11} =$ ____ $\frac{30}{77}$

2
In the example multiplication problem shown at the right, the numerator of the answer is 20 and it is the product of 4 and 5. The denominator of the answer is 99 and it is the product of ______ and ______.

$\frac{4}{9} \times \frac{5}{11} = \frac{20}{99}$

9, 11

3 ✍
Multiply $\frac{7}{10} \times \frac{3}{8}$ by separately multiplying the numerators and denominators. ______ $\frac{21}{80}$

4
Multiply $\frac{5}{8} \times \frac{9}{11}$ by separately multiplying the numerators and denominators. ______ $\frac{45}{88}$

5
$\frac{2}{5} \times \frac{2}{7} =$ ______ $\frac{4}{35}$

6
$\frac{6}{7} \times \frac{5}{13} =$ ______ $\frac{30}{91}$

7
$\frac{4}{9} \times \frac{2}{3} =$ ______ $\frac{8}{27}$

8
$\frac{9}{16} \times \frac{1}{2} =$ ______ $\frac{9}{32}$

9

The whole number 6 can be written as $\frac{6}{1}$.

$6 \times \frac{5}{7} =$ _____ 　　 $\frac{6}{1} \times \frac{5}{7} = \frac{30}{7} = 4\frac{2}{7}$

10 $4 \times \frac{3}{7} =$ _____ 　　 $\frac{4}{1} \times \frac{3}{7} = \frac{12}{7} = 1\frac{5}{7}$

11 $\frac{3}{8} \times 3 =$ _____ 　　 $\frac{3}{8} \times \frac{3}{1} = \frac{9}{8} = 1\frac{1}{8}$

12 $\frac{3}{4} \times 7 =$ _____ 　　 $\frac{21}{4} = 5\frac{1}{4}$

13 $\frac{2}{3} \times 5 =$ _____ 　　 $\frac{10}{3} = 3\frac{1}{3}$

14 $\frac{6}{7} \times \frac{2}{5} =$ _____ 　　 $\frac{12}{35}$

15 $\frac{5}{8} \times 3 =$ _____ 　　 $\frac{15}{8} = 1\frac{7}{8}$

16 $\frac{3}{7} \times 9 =$ _____ 　　 $\frac{27}{7} = 3\frac{6}{7}$

17 $4 \times \frac{4}{11} =$ _____ 　　 $\frac{16}{11} = 1\frac{5}{11}$

18 $\frac{4}{7} \times \frac{1}{3} =$ _____ 　　 $\frac{4}{21}$

19 $5 \times \frac{1}{2} =$ _____ 　　 $\frac{5}{2} = 2\frac{1}{2}$

FEEDBACK UNIT 1

This quiz reviews the preceding unit. Answers are at the back of the book. Begin the next unit only after all problems of this unit are understood.

1. $\frac{2}{5} \times \frac{4}{9}$	5. $4 \times \frac{1}{5}$	9. $\frac{8}{11} \times \frac{2}{3}$
2. $\frac{5}{7} \times \frac{3}{11}$	6. $\frac{4}{9} \times \frac{2}{5}$	10. $\frac{8}{9} \times 2$
3. $\frac{4}{5} \times \frac{1}{3}$	7. $\frac{6}{11} \times 3$	11. $7 \times \frac{4}{9}$
4. $\frac{6}{7} \times \frac{3}{5}$	8. $9 \times \frac{3}{5}$	12. $\frac{3}{4} \times 9$

UNIT 2: THE PROCESS OF CANCELLATION

CANCELLING — A VALUABLE SHORTCUT IN MULTIPLYING

One of the most valuable, useful shortcuts in arithmetic is the cancelling that is often possible when multiplying fractions.

Cancelling is possible when a numerator and a denominator have a common divisor greater than one. In the example below, 5 is a common factor. When the 5's are cancelled, the multiplication problem $\frac{3}{5} \times \frac{5}{7}$ becomes $\frac{3}{1} \times \frac{1}{7}$.

$$\frac{3}{5} \times \frac{5}{7} = \frac{3}{\cancel{5}_{1}} \times \frac{\cancel{5}^{1}}{7} = \frac{3}{1} \times \frac{1}{7} = \frac{3}{7}$$

1

In the multiplication problem $\frac{2}{5} \times \frac{1}{2}$, what number is a divisor of both a numerator and a denominator? ______

2

2

The cancelling example shown below is dependent upon the fact that $2 \div 2 =$ ______

1 or $\frac{1}{1}$

$$\frac{2}{5} \times \frac{1}{2} = \frac{\overset{1}{\cancel{2}}}{5} \times \frac{1}{\underset{1}{\cancel{2}}} = \frac{1}{5} \times \frac{1}{1} = \frac{1}{5}$$

3

What number is a divisor of both a numerator and a denominator of $\frac{4}{7} \times \frac{5}{6}$? ______

2 divides 4 and 6

4

The example below shows the cancelling in a multiplication problem. This cancellation is possible because $4 \div 2 = 2$ and $6 \div 2 =$ ______

3

$$\frac{4}{7} \times \frac{5}{6} = \frac{\overset{2}{\cancel{4}}}{7} \times \frac{5}{\underset{3}{\cancel{6}}} = \frac{2}{7} \times \frac{5}{3} = \frac{10}{21}$$

5

For the problem $\frac{14}{15} \times \frac{4}{7}$, what number is a divisor of both a numerator and a denominator? ______

7 divides 14 and 7

6

Complete the multiplication shown below in which cancellation is used to simplify the problem.

$$\frac{14}{15} \times \frac{4}{7} = \frac{\overset{2}{\cancel{14}}}{15} \times \frac{4}{\underset{1}{\cancel{7}}} = \frac{2}{15} \times \frac{4}{1} = \text{\_\_\_\_\_\_}$$

$\frac{8}{15}$

7

For the problem $\frac{3}{8} \times \frac{12}{13}$, what number is a divisor of both a numerator and a denominator? ______

4 divides 8 and 12

8

Complete the multiplication shown below.
Cancelling has been used to simplify the problem.

$$\frac{3}{8} \times \frac{12}{13} = \frac{3}{\cancel{8}_{2}} \times \frac{\cancel{12}^{3}}{13} = \frac{3}{2} \times \frac{3}{13} = \_\_\_\_\_\_$$

$\frac{9}{26}$

9

Use cancellation. Multiply $\frac{3}{4} \times \frac{1}{3}$. ______

$\frac{\cancel{3}^{1}}{4} \times \frac{1}{\cancel{3}_{1}} = \frac{1}{4}$

10

Use cancellation. Multiply $\frac{6}{7} \times \frac{8}{9}$. ______

$\frac{\cancel{6}^{2}}{7} \times \frac{8}{\cancel{9}_{3}} = \frac{16}{21}$

11

Use cancellation. Multiply $\frac{7}{10} \times \frac{5}{9}$. ______

$\frac{7}{\cancel{10}_{2}} \times \frac{\cancel{5}^{1}}{9} = \frac{7}{18}$

12

Use cancellation. Multiply $\frac{6}{11} \times \frac{5}{6}$. ______

$\frac{\cancel{6}^{1}}{11} \times \frac{5}{\cancel{6}_{1}} = \frac{5}{11}$

13

Use cancellation. Multiply $\frac{7}{11} \times \frac{11}{12}$. ______

$\frac{7}{\cancel{11}_{1}} \times \frac{\cancel{11}^{1}}{12} = \frac{7}{12}$

14

$\frac{8}{9} \times \frac{5}{14} = \_\_\_\_\_\_$

$\frac{4}{9} \times \frac{5}{7} = \frac{20}{63}$

15

$\frac{5}{18} \times \frac{12}{17} =$ ______

$\frac{5}{3} \times \frac{2}{17} = \frac{10}{51}$

16

$\frac{18}{23} \times \frac{5}{9} =$ ______

$\frac{2}{23} \times \frac{5}{1} = \frac{10}{23}$

17

$\frac{7}{8} \times \frac{2}{5} =$ ______

$\frac{7}{4} \times \frac{1}{5} = \frac{7}{20}$

18

Sometimes a multiplication problem offers two pairs of cancellation opportunities. Complete the problem shown below.

$$\frac{5}{9} \times \frac{6}{25} = \frac{\overset{1}{\cancel{5}}}{9} \times \frac{6}{\underset{5}{\cancel{25}}} = \frac{1}{\underset{3}{\cancel{9}}} \times \frac{\overset{2}{\cancel{6}}}{5} = \text{\_\_\_\_\_\_}$$

$\frac{1}{3} \times \frac{2}{5} = \frac{2}{15}$

19

The problem below offers two pairs of cancellation opportunities. Complete the problem.

$$\frac{4}{5} \times \frac{15}{16} = \frac{\overset{1}{\cancel{4}}}{5} \times \frac{15}{\underset{4}{\cancel{16}}} = \frac{1}{\underset{1}{\cancel{5}}} \times \frac{\overset{3}{\cancel{15}}}{4} = \text{\_\_\_\_\_\_}$$

$\frac{1}{1} \times \frac{3}{4} = \frac{3}{4}$

20

Cancel wherever possible. $\frac{4}{9} \times \frac{3}{8} =$ ______

$\frac{1}{3} \times \frac{1}{2} = \frac{1}{6}$

21

Cancel wherever possible. $\frac{8}{11} \times \frac{11}{12} =$ ______

$\frac{2}{1} \times \frac{1}{3} = \frac{2}{3}$

22

Cancel wherever possible. $\frac{15}{16} \times \frac{8}{25} =$ ______

$\frac{3}{2} \times \frac{1}{5} = \frac{3}{10}$

23

Cancel wherever possible. $\frac{3}{7} \times \frac{5}{6} =$ ______

$\frac{1}{7} \times \frac{5}{2} = \frac{5}{14}$

24

Cancel wherever possible. $\frac{10}{21} \times \frac{3}{20} =$ ______

$\frac{1}{7} \times \frac{1}{2} = \frac{1}{14}$

25

$\frac{7}{18} \times \frac{27}{28} =$ ______

$\frac{1}{2} \times \frac{3}{4} = \frac{3}{8}$

26

$\frac{5}{21} \times \frac{18}{25} =$ ______

$\frac{1}{7} \times \frac{6}{5} = \frac{6}{35}$

27

$\frac{9}{26} \times \frac{13}{15} =$ ______

$\frac{3}{2} \times \frac{1}{5} = \frac{3}{10}$

28

$\frac{12}{25} \times \frac{15}{16} =$ ______

$\frac{9}{20}$

29

$\frac{35}{38} \times \frac{20}{21} =$ ______

$\frac{50}{57}$

30

$\frac{12}{55} \times \frac{25}{18} =$ ______

$\frac{10}{33}$

31

$\frac{15}{16} \times \frac{12}{35} =$ ______

$\frac{9}{28}$

FEEDBACK UNIT 2

This quiz reviews the preceding unit. Answers are at the back of the book. Begin the next unit only after all problems of this unit are understood.

Multiply, using cancellation wherever possible.

1. $\frac{3}{4} \times \frac{2}{5}$
2. $\frac{4}{9} \times \frac{6}{7}$
3. $\frac{15}{17} \times \frac{17}{19}$
4. $\frac{4}{9} \times \frac{5}{12}$
5. $\frac{5}{22} \times \frac{11}{12}$
6. $\frac{6}{7} \times \frac{14}{15}$
7. $\frac{8}{9} \times \frac{15}{16}$
8. $\frac{12}{13} \times \frac{13}{18}$
9. $\frac{33}{35} \times \frac{7}{22}$
10. $\frac{8}{15} \times \frac{3}{4}$
11. $\frac{5}{8} \times \frac{6}{7}$
12. $\frac{7}{10} \times \frac{3}{14}$

UNIT 3: THE MULTIPLICATION OF MIXED NUMBERS

MULTIPLYING MIXED NUMBERS

The simplest way to multiply two mixed numbers is to change them to improper fractions and then multiply using cancellation wherever possible.

$$2\tfrac{1}{4} \times 5\tfrac{1}{3}$$

$$\frac{9}{4} \times \frac{16}{3} = \frac{3}{1} \times \frac{4}{1} = 12$$

1 ✍

To change $5\frac{2}{3}$ to an improper fraction, multiply the denominator, 3, by the whole number, 5, and then add the numerator, ______. 2

$$5\frac{2}{3} = \frac{(3 \times 5) + 2}{3} = \frac{15 + 2}{3} = \frac{17}{3}$$

2

Complete the process shown below for changing $6\frac{3}{5}$ to an improper fraction.

$$6\frac{3}{5} = \frac{(5 \times 6) + 3}{5} = \frac{30 + 3}{5} = \_\_\_\_\_\_$$

$\frac{33}{5}$

3

Change $2\frac{1}{3}$ to an improper fraction. ______

$\frac{(3 \times 2) + 1}{3} = \frac{7}{3}$

4

Change $4\frac{3}{8}$ to an improper fraction. ______

$\frac{(8 \times 4) + 3}{8} = \frac{35}{8}$

5

Change $5\frac{3}{4}$ to an improper fraction. ______

$\frac{(4 \times 5) + 3}{4} = \frac{23}{4}$

6

To multiply $2\frac{1}{3} \times 2\frac{1}{2}$, first change the mixed numbers to improper fractions.

$$2\frac{1}{3} = \frac{7}{3} \quad \text{and} \quad 2\frac{1}{2} = \_\_\_\_\_\_$$

$\frac{5}{2}$

7 ✍

Complete the multiplication below in which the mixed numbers have been changed to improper fractions.

$$2\frac{1}{3} \times 2\frac{1}{2} = \frac{7}{3} \times \frac{5}{2} = \_\_\_\_\_\_$$

$\frac{35}{6} = 5\frac{5}{6}$

8

To multiply $5\frac{1}{3} \times \frac{4}{5}$, first write $5\frac{1}{3}$ as an improper fraction.

$$5\frac{1}{3} = \_\_\_\_\_\_$$

$\frac{16}{3}$

9

Complete the multiplication below in which the mixed number has been changed to an improper fraction.

$$5\frac{1}{3} \times \frac{4}{5} = \frac{16}{3} \times \frac{4}{5} = \_\_\_\_\_\_$$

$\frac{64}{15} = 4\frac{4}{15}$

10
Multiply $6\frac{1}{2} \times \frac{3}{4}$ by first changing the mixed number to an improper fraction. ______

$\frac{13}{2} \times \frac{3}{4} = \frac{39}{8} = 4\frac{7}{8}$

11
Multiply $3\frac{1}{2} \times \frac{1}{2}$. ______

$\frac{7}{2} \times \frac{1}{2} = \frac{7}{4} = 1\frac{3}{4}$

12
$\frac{3}{5} \times 4\frac{1}{2} =$ ______

$\frac{3}{5} \times \frac{9}{2} = \frac{27}{10} = 2\frac{7}{10}$

13 ✍
Cancellation is a valuable shortcut in multiplying mixed numbers. Use cancellation to complete the problem shown below.

$2\frac{1}{3} \times \frac{2}{7} = \frac{7}{3} \times \frac{2}{7} =$ ______

$\frac{1}{3} \times \frac{2}{1} = \frac{2}{3}$

14
Cancel if possible. $\frac{6}{7} \times 1\frac{1}{2} =$ ______

$\frac{3}{7} \times \frac{3}{1} = \frac{9}{7} = 1\frac{2}{7}$

15
Cancel if possible. $5\frac{1}{3} \times \frac{3}{4} =$ ______

$\frac{4}{1} \times \frac{1}{1} = 4$

16
Cancel if possible. $\frac{2}{3} \times 2\frac{3}{4} =$ ______

$\frac{1}{3} \times \frac{11}{2} = \frac{11}{6} = 1\frac{5}{6}$

17
$\frac{2}{5} \times 3\frac{1}{3} =$ ______

$\frac{2}{1} \times \frac{2}{3} = \frac{4}{3} = 1\frac{1}{3}$

18
$5\frac{1}{2} \times \frac{7}{11} =$ ______

$\frac{1}{2} \times \frac{7}{1} = \frac{7}{2} = 3\frac{1}{2}$

19
$2\frac{1}{2} \times \frac{3}{5} =$ ______

$\frac{3}{2} = 1\frac{1}{2}$

20

$6\frac{3}{4} \times \frac{1}{9} =$ ______ $\frac{3}{4}$

21

$\frac{2}{7} \times 2\frac{4}{5} =$ ______ $\frac{4}{5}$

FEEDBACK UNIT 3

This quiz reviews the preceding unit. Answers are at the back of the book. Begin the next unit only after all problems of this unit are understood.

Cancel wherever possible.

1. $1\frac{2}{7} \times \frac{2}{3}$
2. $1\frac{3}{5} \times \frac{1}{2}$
3. $\frac{4}{5} \times 1\frac{1}{5}$
4. $1\frac{5}{6} \times \frac{6}{11}$
5. $\frac{9}{13} \times 3\frac{1}{4}$
6. $2\frac{2}{3} \times \frac{1}{4}$
7. $3\frac{1}{8} \times \frac{3}{10}$
8. $2\frac{1}{4} \times 1\frac{1}{3}$
9. $5\frac{1}{4} \times \frac{2}{7}$
10. $\frac{1}{5} \times 3\frac{4}{7}$
11. $1\frac{7}{8} \times 1\frac{1}{5}$
12. $1\frac{1}{9} \times \frac{3}{10}$

UNIT 4: THE DIVISION OF FRACTIONS

THE RECIPROCAL OF A FRACTION

The **reciprocal** of a fraction is found by interchanging the numerator and denominator.

The **reciprocal** of $\frac{5}{7}$ is $\frac{7}{5}$.

The **reciprocal** of $\frac{1}{3}$ is $\frac{3}{1}$ or 3.

1

The **reciprocal** of $\frac{3}{4}$ is $\frac{4}{3}$ because the numerator and denominator have been interchanged. Find the reciprocal of $\frac{5}{8}$. ______

$\frac{8}{5}$

2

The reciprocal of $\frac{5}{6}$ is $\frac{6}{5}$. Find the reciprocal of $\frac{9}{10}$. ______

$\frac{10}{9}$

3

Find the reciprocal of $\frac{7}{3}$. ______

$\frac{3}{7}$

4

The reciprocal of $\frac{5}{9}$ is ______ .

$\frac{9}{5}$

5

Multiply $\frac{7}{8}$ by its reciprocal. ______

$\frac{7}{8} \times \frac{8}{7} = 1$

6

Multiply $\frac{17}{5}$ by its reciprocal. ______

$\frac{17}{5} \times \frac{5}{17} = 1$

7

Multiply 5 by its reciprocal. ______

$\frac{5}{1} \times \frac{1}{5} = 1$

DIVIDING FRACTIONS

The symbol ÷ means "**divided by.**" In the problem $\frac{3}{8} \div \frac{2}{7}$, the **dividend** is $\frac{3}{8}$ and the **divisor** is $\frac{2}{7}$.

$\frac{3}{8} \div \frac{2}{7}$
(dividend ÷ divisor)

To divide $\frac{3}{8}$ by $\frac{2}{7}$, the dividend is multiplied by the reciprocal of the divisor.

(dividend x divisor's reciprocal)

$\frac{3}{8} \times \frac{7}{2} = \frac{21}{16} = 1\frac{5}{16}$

$\frac{3}{8} \div \frac{2}{7}$ is the same as $\frac{3}{8} \times \frac{7}{2}$. The dividend is multiplied by the reciprocal of the divisor. The answer, $\frac{21}{16}$ or $1\frac{5}{6}$, is the **quotient**.

8
To divide $\frac{3}{5} \div \frac{4}{7}$, the problem is changed to the multiplication, $\frac{3}{5} \times \frac{7}{4}$. $\frac{7}{4}$ is the reciprocal of ____.

$\frac{4}{7}$

9
To divide $\frac{3}{7} \div \frac{8}{9}$, multiply $\frac{3}{7} \times \frac{9}{8}$ because $\frac{9}{8}$ is the reciprocal of ______ .

$\frac{8}{9}$

10
(dividend ÷ divisor) $\frac{3}{7} \div \frac{8}{9}$
is equivalent to
(dividend x divisor's reciprocal) $\frac{3}{7} \times \frac{9}{8}$
What is the answer for $\frac{3}{7} \times \frac{9}{8}$? ______

$\frac{27}{56}$

11
Write the division problem, $\frac{3}{8} \div \frac{2}{3}$, as a multiplication. ______

$\frac{3}{8} \times \frac{3}{2}$

12
Every division problem can be written as a multiplication of the dividend and the divisor's ____________.

reciprocal

13
Complete the division problem shown below.

$$\frac{4}{9} \div \frac{5}{8} = \frac{4}{9} \times \frac{8}{5} = \_\_\_\_\_\_$$

$\frac{32}{45}$

14
Complete the division problem shown below.

$$2 \div \frac{5}{7} = 2 \times \frac{7}{5} = \_\_\_\_\_\_$$

$\frac{14}{5} = 2\frac{4}{5}$

15
Complete the division problem shown below.

$$\frac{7}{8} \div 5 = \frac{7}{8} \times \frac{1}{5} = \_\_\_\_\_\_$$

$\frac{7}{40}$

16
Use cancellation to complete the problem below.
$\frac{2}{5} \div \frac{3}{5} = \frac{2}{5} \times \frac{5}{3} =$ ______

$\frac{2}{1} \times \frac{1}{3} = \frac{2}{3}$

17
Use cancellation to complete the problem below.
$\frac{8}{9} \div \frac{1}{3} = \frac{8}{9} \times \frac{3}{1} =$ ______

$\frac{8}{3} \times \frac{1}{1} = \frac{8}{3} = 2\frac{2}{3}$

18 $\frac{9}{10} \div \frac{3}{7} =$ ______

$\frac{9}{10} \times \frac{7}{3} = \frac{21}{10} = 2\frac{1}{10}$

19 $\frac{7}{8} \div \frac{7}{5} =$ ______

$\frac{7}{8} \times \frac{5}{7} = \frac{5}{8}$

DIVIDING MIXED NUMBERS

To divide two mixed numbers, first change the mixed numbers to improper fractions. Then complete the problem by multiplying the dividend by the divisor's reciprocal.

$3\frac{1}{5} \div 2\frac{1}{3}$

$\frac{16}{5} \div \frac{7}{3}$

The example at the right shows the steps in dividing two mixed numbers.

$\frac{16}{5} \times \frac{3}{7} = \frac{48}{35} = 1\frac{13}{35}$

20
Divide $3\frac{1}{5} \div \frac{2}{3}$ by first changing $3\frac{1}{5}$ to an improper fraction. ______

$\frac{16}{5} \times \frac{3}{2} = \frac{24}{5} = 4\frac{4}{5}$

21
Divide $\frac{3}{4} \div 2\frac{1}{3}$ by first changing $2\frac{1}{3}$ to an improper fraction. ______

$\frac{3}{4} \times \frac{3}{7} = \frac{9}{28}$

22 $6\frac{1}{4} \div \frac{5}{8} =$ ______

$\frac{25}{4} \times \frac{8}{5} = 10$

23

$3\frac{1}{3} \div \frac{4}{5} =$ ______

$\frac{10}{3} \times \frac{5}{4} = \frac{25}{6} = 4\frac{1}{6}$

24

$2\frac{2}{5} \div \frac{3}{4} =$ ______

$\frac{12}{5} \times \frac{4}{3} = \frac{16}{5} = 3\frac{1}{5}$

25

$\frac{5}{6} \div 1\frac{1}{3} =$ ______

$\frac{5}{6} \times \frac{3}{4} = \frac{5}{8}$

26

$\frac{3}{4} \div 5\frac{3}{4} =$ ______

$\frac{3}{4} \times \frac{4}{23} = \frac{3}{23}$

FEEDBACK UNIT 4

This quiz reviews the preceding unit. Answers are at the back of the book. Begin the next unit only after all problems of this unit are understood.

1. $\frac{3}{8} \div \frac{1}{3}$	5. $\frac{6}{7} \div 2$	9. $2\frac{1}{4} \div 1\frac{1}{8}$
2. $\frac{4}{7} \div \frac{2}{9}$	6. $2\frac{1}{7} \div \frac{5}{8}$	10. $3\frac{2}{5} \div 1\frac{2}{15}$
3. $\frac{4}{5} \div 2$	7. $3\frac{1}{9} \div \frac{7}{9}$	11. $1\frac{2}{5} \div 1\frac{5}{9}$
4. $5 \div \frac{3}{8}$	8. $5\frac{1}{5} \div 1\frac{1}{3}$	12. $5\frac{1}{4} \div 2\frac{1}{3}$

UNIT 5: APPLICATIONS OF MULTIPLICATION OR DIVISION OF FRACTIONS

CHECKING DIVISION

The long division problem 4025 ÷ 7 is checked by multiplying the divisor and the quotient.

$$7\overline{)4025}\quad 575$$

The check shows the division to have been done correctly because the product of the divisor and quotient is the dividend.

Check: $575 \times 7 = 4025$

Division of fractions is checked in exactly the same way. When the product of the divisor and quotient is the dividend, the division checks.

1
$\frac{5}{7} \div \frac{3}{4} = \frac{5}{7} \times \frac{4}{3} = \frac{20}{21}$
Is the product of the divisor, $\frac{3}{4}$, and quotient, $\frac{20}{21}$, equal to the dividend? ______

Yes, $\frac{3}{4} \times \frac{20}{21} = \frac{5}{7}$

2
Show the check for the problem below.
$\frac{7}{9} \div \frac{2}{3} = \frac{7}{9} \times \frac{3}{2} = \frac{7}{6}$ ____________

$\frac{2}{3} \times \frac{7}{6} = \frac{7}{9}$

3
Divide and check. $\frac{8}{9} \div 6 =$ ______

$\frac{4}{27}$ and $6 \times \frac{4}{27} = \frac{8}{9}$

4
Divide and check. $\frac{3}{8} \div \frac{3}{4} =$ ______

$\frac{1}{2}$ and $\frac{3}{4} \times \frac{1}{2} = \frac{3}{8}$

5
Divide and check. $4 \div \frac{3}{5} =$ ______

$\frac{20}{3}$ and $\frac{3}{5} \times \frac{20}{3} = 4$

6
Divide and check. $\frac{4}{9} \div 2\frac{1}{3} =$ ______

$\frac{4}{21}$ and $\frac{7}{3} \times \frac{4}{21} = \frac{4}{9}$

7
Divide and check. $1\frac{1}{4} \div 2\frac{3}{5} =$ ______

$\frac{25}{52}$, $\frac{13}{5} \times \frac{25}{52} = \frac{5}{4} = 1\frac{1}{4}$

A FRACTION "OF" A NUMBER MEANS MULTIPLICATION

The phrase "a fraction **of** a number" always can be interpreted as indicating multiplication of the fraction and the number.

$\frac{3}{4}$ **of** 28 means $\frac{3}{4} \times 28$. $\frac{7}{5}$ **of** 46 means $\frac{7}{5} \times 46$.

8
To find $\frac{3}{5}$ **of** 6, use the fact that "$\frac{3}{5}$ **of** 6" means $\frac{3}{5} \times 6$. To find $\frac{2}{3}$ **of** 8, ______ (multiply, add) the numbers.

multiply

9
To find $\frac{7}{8}$ of 5, multiply $\frac{7}{8} \times 5$. What is $\frac{7}{8}$ of 5?
______

$\frac{7}{8} \times \frac{5}{1} = \frac{35}{8} = 4\frac{3}{8}$

10
To find $\frac{9}{10}$ of 15, ______ (divide, multiply) $\frac{9}{10}$ and 15.

multiply

11
Find $\frac{9}{10}$ of 15. ______

$\frac{9}{10} \times \frac{15}{1} = \frac{27}{2} = 13\frac{1}{2}$

12
Find $\frac{3}{4}$ of 8. ______

$\frac{3}{4} \times \frac{8}{1} = 6$

13
Find $\frac{5}{8}$ of 24. ______

$\frac{5}{8} \times \frac{24}{1} = 15$

14
Suppose $\frac{3}{4}$ of a class of 20 students were absent.
How many students were absent? ______

$\frac{3}{4} \times 20 = 15$ students

15
If a carpenter has finished $\frac{2}{3}$ of a 30 hour job, how much has he finished? ______

$\frac{2}{3} \times 30 = 20$ hours

16
If a student has read $\frac{5}{6}$ of his 120 page reading lesson, how many pages have been read? ______

$\frac{5}{6} \times 120 = 100$ pages

DIVISION IS INDICATED BY THE QUESTION "HOW MANY ARE IN?"

Problem: How many 5 cent pencils can be purchased for 85 cents?

The problem can be rephrased as: How many 5's are in 85?

The question is answered by dividing 85 by 5, (85 ÷ 5).

Whenever a problem asks:
"How many of a first number are in a second number?"
the answer is found by dividing the first number into the second.

17
Problem: How many $\frac{1}{2}$ inches are in 8 inches?
To answer the question, 8 is divided by $\frac{1}{2}$. Which of the following shows 8 divided by $\frac{1}{2}$? ______

a. $\frac{1}{2} \div 8$ b. $8 \div \frac{1}{2}$

b

18
How many $\frac{1}{2}$ inches are in 8 inches? ______

$8 \div \frac{1}{2} = 16$ ($\frac{1}{2}$ inches)

19 ✍

To find how many $\frac{1}{4}$ years there are in 5 years, ______ (multiply, divide).

divide

20

Find how many $\frac{1}{4}$ years there are in 5 years. ______

$5 \div \frac{1}{4} = 20$ ($\frac{1}{4}$ years)

21

Find how many $\frac{1}{3}$ acres there are in 12 acres. ______

$12 \div \frac{1}{3} = 36$ ($\frac{1}{3}$ acres)

22

Find how many $\frac{1}{2}$ ounces there are in 32 ounces. ______

$32 \div \frac{1}{2} = 64$ ($\frac{1}{2}$ ounces)

23

How many $\frac{1}{4}$ inches are in 3 inches? ______

$3 \div \frac{1}{4} = 12$ ($\frac{1}{4}$ inches)

24

How many $\frac{1}{10}$ grams are in 12 grams? ______

$12 \div \frac{1}{10} = 120$ ($\frac{1}{10}$ grams)

25

How many $\frac{4}{5}$ liters are in 7 liters? ______

$7 \div \frac{4}{5} = \frac{35}{4} = 8\frac{3}{4}$ ($\frac{4}{5}$ liters)

26 ✍

Problem: How many $\frac{3}{5}$ gram rings can be made from 10 grams of gold?

To solve the problem rephrase it as:

How many ______ are in ______ .

$\frac{3}{5}$, 10

27

How many $\frac{3}{5}$ gram rings can be made from 10 grams of gold? ______

$10 \div \frac{3}{5} = \frac{50}{3} = 16\frac{2}{3}$

or 16 rings

28

Problem: Find the number of $\frac{7}{10}$ grams pills that can be made from 21 grams of medicine.

To solve the problem rephrase it as:

How many ______ are in ______ .

$\frac{7}{10}$, 21

29

Find the number of $\frac{7}{10}$ grams pills that can be made from 21 grams of medicine. ______

$21 \div \frac{7}{10} = 30$ pills

30

How many $3\frac{4}{5}$ kilogram bricks can be made from 100 kilograms of concrete? ______

$100 \div 3\frac{4}{5} = 26\frac{6}{19}$ or 26 bricks

31

Razor blades weighing $1\frac{3}{10}$ grams each are to be made from 91 grams of steel. How many can be made? ______

$91 \div 1\frac{3}{10} = 70$ blades

32

If capsules containing $3\frac{4}{5}$ milligrams of medicine are to be filled from a supply of 76 mg, how many can be made? ______

$76 \div 3\frac{4}{5} = 20$ capsules

33

How many home mortgages are there if the average interest per year is $\$1\frac{1}{2}$ thousand for each mortgage and the total interest collected per year is $45 thousand? _____

$45 \div 1\frac{1}{2} = 30$ mortages

34
The **area** of the **rectangle** shown at the right is the product of its length and width. Find the area. ______

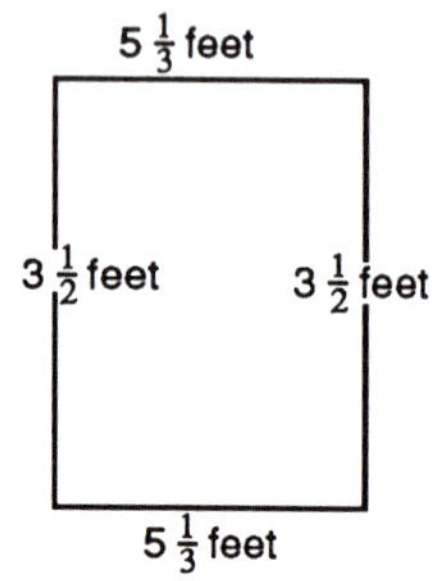

$3\frac{1}{2} \times 5\frac{1}{3} = \frac{56}{3}$
$18\frac{2}{3}$ square feet

35
The formula for finding the area of a **rectangle** is $A = LW$ where L is the length of the rectangle and W is the width. Use the formula to find the area of the rectangle shown at the right. ______

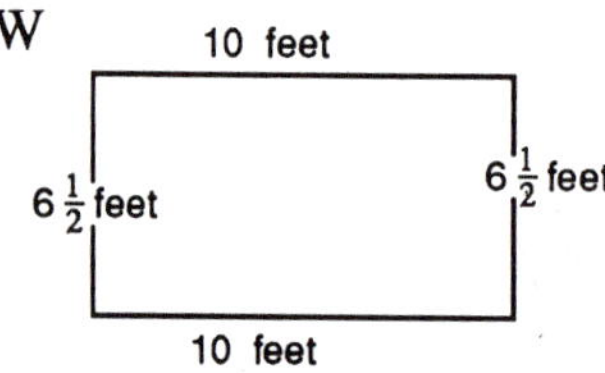

$A = L \times W$
$= 10 \times 6\frac{1}{2}$
65 square feet

36 ✍
Area is always measured in square units such as square inches, square feet, or square centimeters. Find the area of a rectangle with length 12 yards and width $4\frac{1}{3}$ yards.

______

$A = L \times W$
$= 12 \times 4\frac{1}{3}$
52 square yards

37
The formula for finding the area of a **triangle** is $A = \frac{1}{2}BH$ where B is the base of the triangle and H is the height. For the triangle shown at the right, the base is $12\frac{3}{4}$ meters and the height is ______.
(The height is perpendicular to the base.)

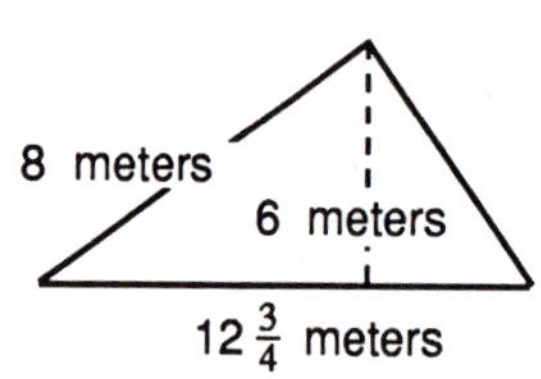

6 meters

8 meters is not the height; it is the length of a side.

38
Use the formula $A = \frac{1}{2}BH$ to find the area of the triangle shown at the right. ______

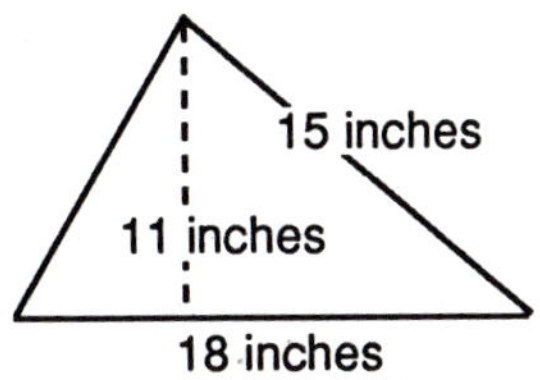

$A = \frac{1}{2} \times B \times H$

$= \frac{1}{2} \times 18 \times 11$

99 square inches

39
The formula for finding the area of a **circle** is $A = \pi r^2$ where π is the number Pi (approximately $3\frac{1}{7}$), r is the radius, and r^2 means r x r. Find the area of the circle shown at the right.

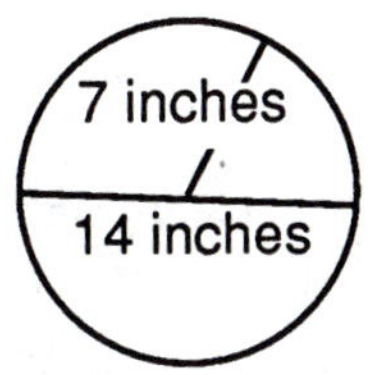

$A = \pi r^2$

$= 3\frac{1}{7} \times 7^2$

$= \frac{22}{7} \times 7 \times 7$

$= 22 \times 1 \times 7$

$=$ 154 square inches

40
Use the formula $A = \pi r^2$ to find the area of the circle shown at the right. ______
[Use $3\frac{1}{7}$ as an approximation for π.]

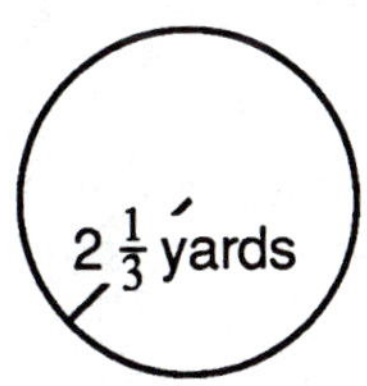

$A = \pi r^2$

$= 3\frac{1}{7} \times (2\frac{1}{3})^2$

$= \frac{22}{7} \times \frac{7}{3} \times \frac{7}{3}$

$= 22 \times \frac{1}{3} \times \frac{7}{3}$

$= \frac{154}{9}$

$= 17\frac{1}{9}$ square yards

FEEDBACK UNIT 5

This quiz reviews the preceding unit. Answers are at the back of the book. Take the Chapter Mastery Test only after all problems of this unit are understood.

1. Divide and check. $\frac{6}{7} \div \frac{3}{5}$
2. Divide and check. $2\frac{1}{4} \div 4\frac{1}{2}$
3. Find $\frac{4}{7}$ of 21.
4. Find how many $\frac{1}{3}$ years there are in 6 years.
5. Find how many $\frac{1}{2}$ quart bottles can be filled with 8 quarts of liquid.
6. A general estimates that $\frac{3}{8}$ of his 32 battalions are well-prepared. How many battalions are well-prepared?
7. How many $\frac{1}{4}$ ounce coins can be made from 25 ounces of silver?
8. How many $\frac{3}{5}$ kilogram blocks of lead can be made from 63 kilograms of lead?
9. Bill has to put $\frac{3}{5}$ of a 7 ton load of sand on a vacant lot. How many tons of sand should he put on the lot?
10. Julie spends \$86 to buy stock that costs $\$5\frac{3}{8}$ per share. How many shares does she buy?
11. Find the area of the rectangle shown below.

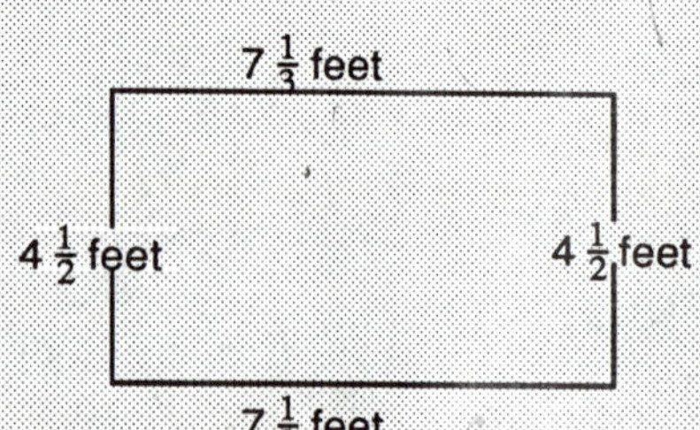

12. Find the area of the triangle shown below.

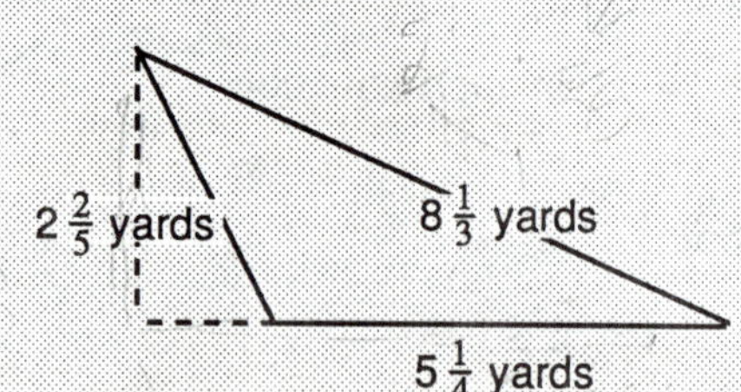

13. Find the area of the circle shown at the right. (Use $3\frac{1}{7}$ as the approximation for Pi.)

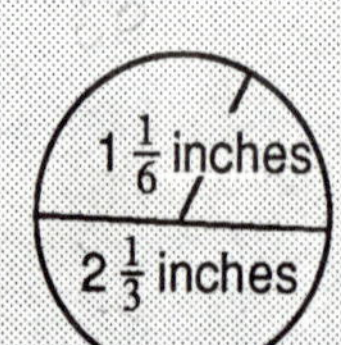

CHAPTER 7 MASTERY TEST

This test provides an opportunity for assuring that **all** of the objectives of Chapter 7 have been learned. Each problem is accompanied by the symbol (U) which indicates the Unit in which that type of problem was studied. If any problems are missed, they need to be corrected. Answers are at the back of the book.

1. $\frac{2}{5} \times \frac{4}{9}$ (1)
2. $\frac{5}{7} \times \frac{2}{9}$ (1)
3. $\frac{3}{5} \times \frac{2}{11}$ (1)
4. $8 \times \frac{2}{5}$ (1)
5. $\frac{7}{8} \times 3$ (1)
6. $\frac{3}{4} \times \frac{2}{5}$ (2)
7. $\frac{4}{9} \times \frac{7}{16}$ (2)
8. $\frac{5}{22} \times \frac{11}{10}$ (2)
9. $\frac{15}{16} \times \frac{32}{35}$ (2)
10. $\frac{9}{10} \times \frac{22}{18}$ (2)
11. $1\frac{5}{8} \times 3$ (3)
12. $1\frac{2}{3} \times \frac{1}{2}$ (3)
13. $2\frac{3}{4} \times 1\frac{1}{3}$ (3)
14. $\frac{10}{21} \div \frac{7}{3}$ (4)
15. $\frac{3}{8} \div \frac{1}{3}$ (4)
16. $\frac{6}{7} \div \frac{2}{3}$ (4)
17. $1\frac{3}{5} \div 3$ (4)
18. $3\frac{1}{2} \div 1\frac{3}{4}$ (4)
19. $8 \div \frac{1}{5}$ (4)
20. $2\frac{3}{4} \div \frac{3}{7}$ (4)
21. In the problem $\frac{2}{3} \div \frac{5}{9} = \frac{6}{5}$, the quotient is ______, the dividend is ______, and the divisor is ______. (4)
22. Find $\frac{4}{5}$ of 25. (5)
23. Beth had $3\frac{2}{3}$ liters of medicine. One-half of it was needed on a wound. How much was needed? (5)
24. Clay cups weighing $\frac{3}{10}$ kilograms are to be made from $5\frac{7}{10}$ kilograms of clay. How many cups can be made? (5)
25. The cost of a motorcycle is $\frac{5}{6}$ of the sales price. The sales price is \$1800. Find the cost of the motorcycle. (5)
26. Find the area of the rectangle shown below. (5)

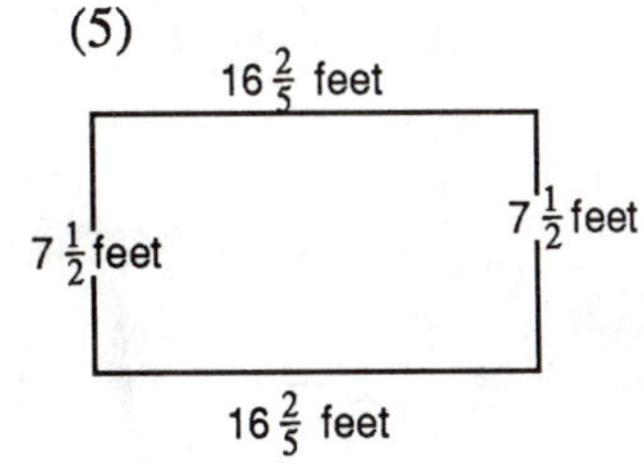

27. Find the area of the triangle shown below. (5)

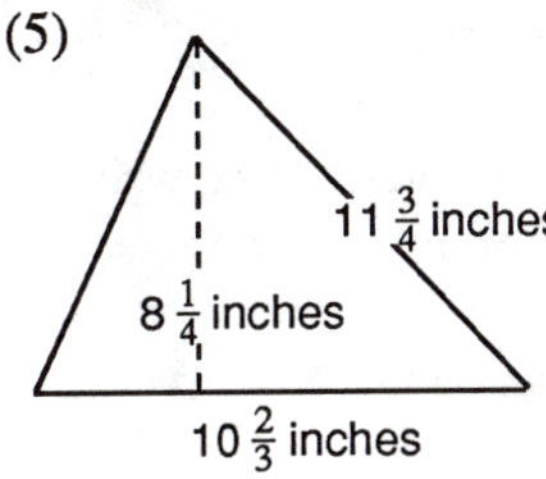

28. Find the area of the circle shown below. (5)

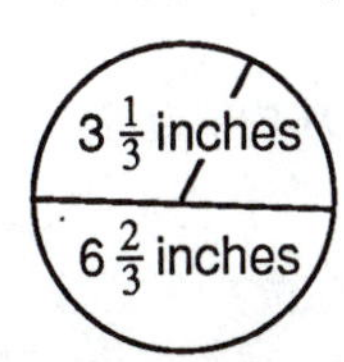

Chapter 8 Objectives

The following problems describe, by example, the skills and knowledge studied in this chapter. Each problem is accompanied by a symbol, (U), which indicates the Unit where the problem is studied. Answers are at the back of the book.

In problems 1-8 fill in the blanks.

	Fraction		Decimal	
1.	$\frac{4}{1000}$	=	______	(1)
2.	______	=	2.8	(1)
3.	$\frac{54}{10}$	=	______	(1)
4.	$\frac{521}{100}$	=	______	(1)
5.	______	=	0.045	(1)
6.	______	=	0.4	(1)
7.	$\frac{10293}{100}$	=	______	(1)
8.	______	=	5.01	(1)

In problems 9-12 write each numeral using scientific notation.

9. 57.814 (2)

10. 0.0007134 (2)

11. 56,700,000 (2)

12. 0.01406 (2)

In problems 13-16 label each statement true or false.

13. 58.0 = 58 (3)

14. 0.07 > 0.1 (3)

15. 0.0038 < 0.019 (3)

16. 30.5 = 30.50 (3)

For problems 17-20 fill in the blanks with =, <, or >.

17. 0.036 ____ 0.3 (3)

18. 1.38 ____ 1.382 (3)

19. 0.0483 ____ 0.2 (3)

20. 0.6 ______ 0.600 (3)

In problems 21-24 round off as indicated.

21. 4.538 to one decimal place (4)

22. 6.103 to the nearest one. (4)

23. 15.509 to the hundredth's place. (4)

24. 59.03928 to two decimal places. (4)

CHAPTER 8
THE MEANING OF DECIMAL FRACTIONS

UNIT 1: FRACTION AND DECIMAL EQUIVALENTS

WRITING FRACTIONS AS DECIMAL FRACTIONS

Decimal notation provides an easy way to write fractions with denominators of 10, 100, 1000, and so forth.

The numbers 10, 100, 1000, etc. are called powers of ten and are sometimes written with exponents, small numbers written to the upper right of 10.

$$10 = 10^1 \qquad 100 = 10^2 \qquad 1000 = 10^3$$

$\frac{1}{10}$ is written as 0.1	$\frac{1}{10} = 0.1$
$\frac{1}{100}$ or $\frac{1}{10^2}$ is written as 0.01	$\frac{1}{100} = \frac{1}{10^2} = 0.01$
$\frac{1}{1000}$ or $\frac{1}{10^3}$ is written as 0.001	$\frac{1}{1000} = \frac{1}{10^3} = 0.001$

1

0.01 is the decimal notation for $\frac{1}{100}$ or $\frac{1}{10^2}$. Notice that 100 has 2 zeros; the exponent of 10^2 is 2; and the "1" in 0.01 is ___ places to the right of the decimal point. 2

2

0.0003 is the decimal notation for $\frac{3}{10000}$ or $\frac{3}{10^4}$. Notice that 10000 has 4 zeros; the exponent of 10^4 is 4; and the "3" in 0.0003 is ___ places to the right of the decimal point. 4

3

0.012 is the decimal notation for $\frac{12}{1000}$ or $\frac{12}{10^3}$. Notice that 1000 has 3 zeros; the exponent of 10^3 is 3; and the "2" in 0.012 is ___ places to the right of the decimal point. 3

4

0.67 is the decimal notation for $\frac{67}{100}$ or $\frac{67}{10^2}$. Write $\frac{372}{100000}$ using a decimal numeral. ______ 0.00372

5

Write $\frac{28}{1000}$ as a decimal fraction. ______ 0.028

6

Write $\frac{3}{1000000}$ as a decimal fraction. ______ 0.000003

7

Write $\frac{1246}{100000}$ as a decimal fraction. ______ 0.01246

8

0.8 is less than one.
1.2 is more than one.
0.92 is ______ (less, greater) than one. less

9

$\frac{4}{10}$ = 0.4 and is less than one.

$\frac{15}{10}$ = 1.5 and is greater than one.

$\frac{17}{10}$ is ______(greater,less) than one. greater

10

True or false? ______ 0.923 is less than one. true

11

True or false? ______ 1.001 is greater than one. true

12

True or false? ______ 0.817 is greater than one. false

13

Which of the following are greater than one?

a) 1.03 c) 0.89278
b) 220.1 d) 2.05

a, b, d

DECIMAL FRACTIONS THAT ARE GREATER THAN ONE

To find the decimal equal to $\frac{258}{100}$,

1) Write the denominator as a power of ten. $\frac{258}{100} = \frac{258}{10^2}$

2) Since the exponent is 2, place the "8" in 258 two places right of the decimal point. 2.58

14

Write $\frac{7533}{10}$ as a decimal using the fact that $10 = 10^1$._____ 753.3

15

Write $\frac{3689}{100}$ as a decimal using the fact that $100 = 10^2$._____ 36.89

16

Write $\frac{4671}{1000}$ as a decimal using the fact that $1000 = 10^3$._____ 4.672

17

Write $\frac{102394}{10000}$ as a decimal using the fact that $10000 = 10^4$._____ 10.2394

18

Find the decimal equal to $\frac{152}{100}$. _____ 1.52

19

Find the decimal equal to $\frac{4205}{100}$. _____ 42.05

20

Find the decimal equal to $\frac{5428}{1000}$. _____ 5.428

21

Find the decimal equal to $\frac{38}{1000}$. _____ 0.038

22

Find the decimal equal to $\frac{45125}{100}$. _____ 451.25

23
Find the decimal equal to $\frac{101}{10}$. ______ 10.1

24
Find the decimal equal to $\frac{37}{10}$. ______ 3.7

25
Find the decimal equal to $\frac{37}{100}$. ______ 0.37

26
Find the decimal equal to $\frac{125}{10}$. ______ 12.5

27
Find the decimal equal to $\frac{7}{10000}$. ______ 0.0007

28
Find the decimal equal to $\frac{23006811}{100000}$. ______ 230.06811

29
Find the decimal equal to $\frac{251}{10}$. ______ 25.1

30
Find the decimal equal to $\frac{48}{100000}$. ______ 0.00048

31
Find the decimal equal to $\frac{1001}{1000}$. ______ 1.001

32
Find the decimal equal to $\frac{23}{10}$. ______ 2.3

33
Find the decimal equal to $\frac{895765}{10}$. ______ 89,576.5

PLACE VALUES OF DECIMAL FRACTIONS

The diagram below shows the names of the place values right of the decimal.

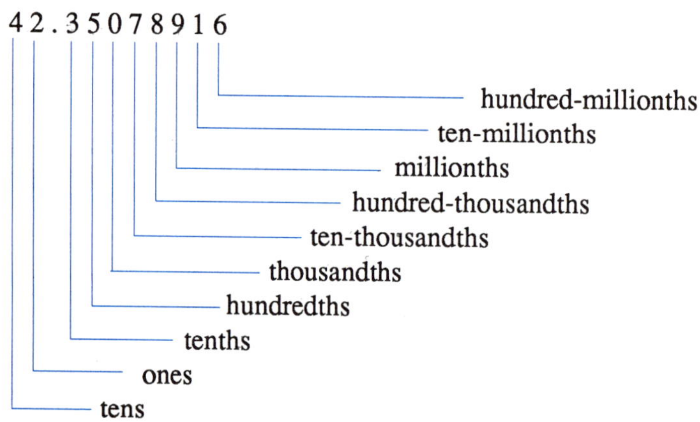

There are several important lessons to be learned about the place value names.

1) Each place is one-tenth as great as the place on the left.

   $\frac{1}{10}$ is one-tenth as great as 1.

   $\frac{1}{100}$ is one-tenth as great as $\frac{1}{10}$.

2) There is no one's place to the right of the decimal.

34
The thousandth's place is three places right of the decimal point. The millionth's place is ______ places right of the decimal point. 6

35
The one's place is one place left of the decimal point. The hundredth's place is ______ places right of the decimal. 2

36
The ten-thousandth's place is ______ places right of the decimal.

4

37
The tenth's place is one place ______ (left, right) of the decimal.

right

38
The hundred's place is three places ______ (left, right) of the decimal.

left

39
The hundredth's place is ______ places ______ of the decimal.

2, right

40
The millionth's place is ______ places ______ of the decimal.

6, right

41
The tenth's place is ______ place/s ______ of the decimal.

1, right

42
The hundred-thousandth's place is ______ places ______ of the decimal.

5, right

43
In 0.5682 the digit in the ten-thousandth's place is 2. In 0.049382 the digit in the thousandth's place is ______.

9

44
In 0.5682 the digit in

a) the tenth's place is ______ .
b) the ten-thousandth's place is ______ .
c) the hundredth's place is ______ .

5
2
6

45
In 42.056387 the digit in the
a) one's place is ______ 2
b) tenth's place is ______ 0
c) ten's place is ______ 4
d) hundred-thousandth's place is ______ 8

46
In 560.738942:
a) 0 is in the ______ place. one's
b) 5 is in the ______ place. hundred's
c) 8 is in the ______ place. thousandth's
d) 2 is in the ______ place. millionth's

47
In 0.00092:
a) 9 is in the ______ place. ten-thousandth's
b) 2 is in the ______ place. hundred-thousandth's

WRITING DECIMAL NUMERALS AS FRACTIONS

To write the decimal numeral 45.07 as a fraction,

1) Determine the place value of the last digit on the right.

   7 is the last digit on the right of 45.07
   and the place value of 7 is hundredths.

2) Write the decimal numeral omitting the decimal point as the numerator and use the place value found in step 1 to write the denominator.

$$45.07 = \frac{4507}{100}$$

48
The digit 9 is in the hundredth's place of 0.09 and the decimal numeral is written as a fraction by:

$$0.09 = \frac{009}{100} = \frac{9}{100}$$

Write 0.7 as a fraction. $\frac{7}{10}$

49

The digit 7 is in the thousandth's place of 0.037 and the decimal numeral is equal to:

$$0.037 = \frac{0037}{1000} = \frac{37}{1000}$$

Write 0.247 as a fraction. ______

$\frac{247}{1000}$

50

The decimal numeral 1.07 equals the fraction $\frac{107}{100}$.
Write 1.925 as a fraction. ______

$\frac{1925}{1000}$

51

Write 0.0008 as a fraction. ______

$\frac{8}{10000}$

52

Write 0.000204 as a fraction. ______

$\frac{204}{1000000}$

53

Write 28.2 as a fraction. ______

$\frac{282}{10}$

54

Write 90.07 as a fraction. ______

$\frac{9007}{100}$

55

Write 0.0900 as a fraction. ______

$\frac{900}{10000} = \frac{9}{100}$

56

Write $\frac{87}{100}$ as a decimal numeral. ______

0.87

57

Write 0.028 as a fraction. ______

$\frac{28}{1000}$

58

Write $\frac{125}{1000}$ as a decimal numeral. ______

0.125

59
Write $\frac{478}{100}$ as a decimal numeral. ______ 4.78

60
Write 10.8 as a fraction. ______ $\frac{108}{10}$

61
Write $\frac{208}{10000}$ as a decimal numeral. ______ 0.0208

FEEDBACK UNIT 1

This quiz reviews the preceding unit. Answers are at the back of the book. Begin the next unit only after all problems of this unit are understood.

Write each fraction in problems 1-8 as a decimal numeral.

1. $\frac{125}{1000}$
2. $\frac{1}{100}$
3. $\frac{37}{10000}$
4. $\frac{3748}{100}$
5. $\frac{98}{10}$
6. $\frac{12}{100}$
7. $\frac{398}{1000}$
8. $\frac{483}{10000}$

9. In a decimal numeral the name of the place value
   a) three places right of the decimal is ________.
   b) one place left of the decimal is ________.
   c) one place right of the decimal is ________.
   d) two places right of the decimal is ________.
   e) six places right of the decimal is ________.

10. In 39.028741 the digit in the
   a) hundredth's place is ______.
   b) ten's place is ______.
   c) ten-thousandth's place is ______.
   d) tenth's place is ______.
   e) millionth's place is ______.

Write each of the decimal numerals in problems 11-14 as fractions.

11. 2.07
12. 0.092
13. 107.52
14. 0.0050

UNIT 2: SCIENTIFIC NOTATION

1

If 4.71 is multiplied by 100 (10^2), the answer is 471.

$$4.71 \times 100 = \frac{471}{100} \times 100 = \frac{471}{\cancel{100}} \times \cancel{100} = 471$$

What is the answer when 65.8 is multiplied by 10? ____

$\frac{658}{\cancel{10}} \times \cancel{10}$

658

2

If .005 is multiplied by 1000 (10^3), the answer is 5.

$$.005 \times 1000 = \frac{5}{1000} \times 1000 = \frac{5}{\cancel{1000}} \times \cancel{1000} = 5$$

What is the answer when 87.3 is multiplied by 100? ____

$\frac{873}{10} \times 100 = 873 \times 10$

8,730

3

If 4.76 is multiplied by 10 (10^1), the answer is 47.6.

$$4.76 \times 10 = \frac{476}{100} \times 10 = \frac{476}{10} = 47.6$$

If 6.12 is multiplied by 1000 (10^3), the product is ______.

$6.12 \times 1000 = 6,120$

or $6.12 \times 10^3 = 6,120$

4 ✍

$4.71 \times 100 = 4.71 \times 10^2 = 471.$

$.005 \times 1000 = .005 \times 10^3 = 5$

$4.76 \times 10 = 4.76 \times 10^1 = 47.6$

When a decimal fraction is multiplied by a power of ten, the exponent shows the number of places the decimal point is to be moved.

$56.873 \times 100 = 56.873 \times 10^2 =$ _____

5,687.3

5

To multiply 100,000 by 4.37 first write 100,000 as 10^5. Then use the exponent to move the decimal point of 4.37 five places to the right.

$4.37 \times 100{,}000 = 4.37 \times 10^5 = 437{,}000.$

$0.29 \times 100{,}000 = 0.29 \times 10^5 =$ ______ 29,000

6

Multiply 61.25 by 1000 using the fact that $1000 = 10^3$. ______ 61,250

7

Multiply .00059 by 1000 using the fact that $1000 = 10^3$. ______ 0.59

8

Multiply 6.31 by 100 using the shortcut for multiplication by a power of ten. ______ $6.31 \times 10^2 = 631$

9

If 6.12 is multiplied by 1000, the product is ______. $612 \times 10^3 = 6{,}120$

10

Use the shortcut for multiplying by a power of ten. $0.0074 \times 10{,}000 =$ ______ $0.0074 \times 10^4 = 74$

11

$0.0971 \times 100 =$ ______ $0.0971 \times 10^2 = 9.71$

12

$46.31 \times 1000 =$ ______ $46.31 \times 10^3 = 46{,}310$

13

$1.293 \times 10{,}000 =$ ______ $1.293 \times 10^4 = 12{,}930$

14

0.039 x 10 = ______ $0.039 \times 10^1 = 0.39$

15

100 x 2.0398 = ______ $10^2 \times 2.0398 = 203.98$

16

100,000 x 0.3 = ______ $10^5 \times 0.3 = 30{,}000$

POWERS OF TEN

10, 100, 1000, etc. are powers of ten which are written using counting numbers as exponents.

Other powers of ten are 0.1, 0.01, 0.001, etc. These powers of ten are written with negative number exponents. A list of many powers of ten is shown below.

One million	1,000,000.	$= 10^6$
One hundred thousand	100,000.	$= 10^5$
Ten thousand	10,000.	$= 10^4$
One thousand	1,000.	$= 10^3$
One hundred	100.	$= 10^2$
Ten	10.	$= 10^1$
One	1.	$= 10^0$
One tenth	.1	$= 10^{-1}$
One hundredth	.01	$= 10^{-2}$
One thousandth	.001	$= 10^{-3}$
One ten thousandth	.0001	$= 10^{-4}$
One hundred thousandth	.00001	$= 10^{-5}$
One millionth	.000001	$= 10^{-6}$

When the exponent is a counting number, the power of ten is greater than 1.

When the exponent is zero, then 10^0 equals 1.

When the exponent is negative, the power of ten is less than 1.

17

$1{,}000 = 10^3$ and $10{,}000 =$ ____ 10^4

18

$1{,}000{,}000 = 10^6$ and $100 =$ ____ 10^2

19

$10 = 10^1$ and $100{,}000 =$ ____ 10^5

20

$10^{-3} = .001$ and $10^{-4} =$ .0001 .0001

21

$10^{-1} = .1$ and $10^{-2} =$ .01 .01

22

If 4.71 is multiplied by 0.01 (10^{-2}), the answer is .0471

$$4.71 \times .01 = \frac{471}{100} \times \frac{1}{100} = \frac{471}{10000}$$

What is the answer when 65.8 is multiplied by 0.01? ____

$$\frac{658}{10} \times \frac{1}{100} = \frac{658}{1000}$$

.658

23

If .005 is multiplied by 0.001 (10^{-3}), the answer is .000005

$$.005 \times .001 = \frac{5}{1000} \times \frac{1}{1000} = \frac{5}{1000000}$$

What is the answer when 87.3 is multiplied by .01? ____

$$\frac{873}{10} \times \frac{1}{100} = \frac{873}{1000}$$

.873

$$\frac{873}{10} \times \frac{1}{100} = \frac{873}{1000}$$

.873

24

If 4.76 is multiplied by 0.1 (10^{-1}), the answer is .476

$$4.76 \times 10 = \frac{476}{100} \times \frac{1}{10} = \frac{476}{1000}$$

If 6.12 is multiplied by 0.001 (10^{-3}), the product is ______.

.00612

25

$4.71 \times .01 = 4.71 \times 10^{-2} = .0471$

$.005 \times .001 = .005 \times 10^{-3} = .000005$

$4.76 \times .1 = 4.76 \times 10^{-1} = .476$

When a decimal fraction is multiplied by a power of ten, the exponent shows the number of places the decimal point is to be moved.

$56.873 \times .01 = 56.873 \times 10^{-2} =$ ______

.56873

26

To multiply .00001 by 4.37 first write .00001 as 10^{-5}. Then use the exponent to move the decimal point of 4.37 five places to the left.

$4.37 \times .00001 = 4.37 \times 10^{-5} = .0000437$

$0.29 \times .00001 = 0.29 \times 10^{-5} =$ ______

.0000029

27

Multiply 61.25 by .001 using the fact that $.001 = 10^{-3}$. ______

.06125

28

Multiply .00059 by .1 using the fact that $.1 = 10^{-1}$. ______

.000059

29

Multiply 6.31 by .01 using the shortcut for multiplication by a power of ten. ______

$6.31 \times 10^{-2} = .0631$

30
If 6.12 is multiplied by .001, the product is ______.

$6.12 \times 10^{-3} = .00612$

31
Use the shortcut for multiplying by a power of ten.

$0.0074 \times .01 =$ ______

$0.0074 \times 10^{-2} = .000074$

32

$0.0971 \times .1 =$ ______

$0.0971 \times 10^{-1} = .00971$

33

$46.31 \times .001 =$ ______

$46.31 \times 10^{-3} = .04631$

34

$1.293 \times .0001 =$ ______

$1.293 \times 10^{-4} = .0001293$

35

$0.039 \times .1 =$ ______

$0.039 \times 10^{-1} = .0039$

36

$.01 \times 2.0398 =$ ______

$10^{-2} \times 2.0398 = .020398$

37

$.00001 \times 0.3 =$ ______

$10^{-5} \times 0.3 = .000003$

38
When multiplying by 100 (10^2), the exponent is a counting number and the decimal is moved two places to the right.

$100 \times 36.7 =$ ______

$10^2 \times 36.7 = 3{,}670$

39
When multiplying by .001 (10^{-3}), the exponent is negative and the decimal is moved three places to the left.

$.001 \times 4.83 =$ ______

$10^{-3} \times 4.83 = .00483$

40

54.3 x 1000 = ______

$54.3 \times 10^{3} = 54{,}300$

41

8.732 x .001 = ______

$8.732 \times 10^{-3} = .008732$

42

.0067 x 10,000 = ______

$.0067 \times 10^{4} = 67$

43

8,300 x .1 = ______

$8{,}300 \times 10^{-1} = 830$

WRITING NUMERALS IN SCIENTIFIC NOTATION

Every decimal fraction can be written as a number between 1 and 10 multiplied by a power of ten. When a number is written in that way it is called its **scientific notation**.

To write 83,000 in scientific notation:

1. Place a decimal point in 83,000 so that the resultant number is between 1 and 10. — 8.3000
2. Find a power of ten that can be multiplied by 8.3 to give 83,000. — $83{,}000 = 8.3 \times 10^{4}$

The scientific notation numeral for 83,000 is 8.3×10^{4}.

44

Write .0039 in scientific notation in the following steps:

Move the decimal point of .0039 so that the result is a number between 1 and 10. ______

3.9

Find a power of ten to complete the following equality.

.0039 = 3.9 x _____

10^{-3}

45
Write 256.7 in scientific notation in the following steps:

Move the decimal point of 256.7 so that the result is a number between 1 and 10. ______ 2.567

Find a power of ten to complete the following equality.
256.7 = 2.567 x _____ 10^2

46
Write 4,850,000 in scientific notation in the following steps:

Move the decimal point of 4,850,000 so that the result is a number between 1 and 10. ______ 4.85

Find a power of ten to complete the following equality.
4,850,000 = 4.85 x _____ 10^6

47
Complete the process for writing the scientific notation numeral for 57.4 shown below:
57.4 = 5.74 x _____ 10^1

48
The scientific notation for 6.314 is ______. 6.314×10^0

49
The scientific notation for 0.014 is ______. 1.4×10^{-2}

50
The scientific notation for 596 is ______. 5.96×10^2

51
The scientific notation for 0.00062 is ______. 6.2×10^{-4}

52
The scientific notation for 0.00293 is ______. 2.93×10^{-3}

53
The scientific notation for 19.03 is ______. 1.903×10^1

FEEDBACK UNIT 2

This quiz reviews the preceding unit. Answers are at the back of the book. Begin the next unit only after all problems of this unit are understood.

In problems 1-10 find each of the products using powers of ten.

1. 813.4 x 10,000
2. .007 x 100
3. 91.32 x .001
4. 27.07 x .1
5. 8,350 x .001
6. 234.6 x 1000
7. 8.903 x .01
8. .0075 x 1,000,000
9. 3,402 x .00001
10. 4,312 x .001

In problems 11-20 write each numeral in its scientific notation form.

11. 36
12. 70,000
13. .008
14. 780,000
15. 92,000,000,000
16. .09273
17. 468,290
18. .5005
19. .00000000000451
20. 800,000,000,000

UNIT 3: ROUNDING OFF AT TWO DECIMAL PLACES (HUNDREDTHS)

1

0.56 is less than 0.63 because both numbers are hundredths and 56 is less than 63. Is 0.874 less than 0.415? _____ No

2

0.25 is less than 0.26 because 25-hundredths is less than 26-hundredths. Is 0.32 less than 0.25?______ No

3

0.21 is greater than 0.09 because 21-hundredths is greater than 9-hundredths. Which is greater, 0.12 or 0.06? ______ 0.12

4

Which is greater, 0.37 or 0.25? ______ 0.37

5

Which is greater, 0.09 or 0.92? _____ 0.92

COMPARING SIZES OF TWO DECIMAL FRACTIONS

To compare the sizes of two fractions like $\frac{5}{7}$ and $\frac{2}{3}$, a common denominator needs to be found. To compare the sizes of 0.95 and 0.461 a common denominator is also found. Common denominators for decimal numerals are found by writing the numerals with the same number of decimal places.

0.95 has two decimal places and means 95 hundredths.
0.461 has three decimal places and means 461 thousandths.

0.95 is written as 0.950 which is 950 thousandths.

0.950 is greater than 0.461. Therefore, 0.95 is greater than 0.461

6
To compare the sizes of 0.012 and 0.09 first write both numerals with three decimal places by affixing a zero to 0.09 to give:
0.012 and 0.090
Which is greater, 0.012 or 0.090? ______ 0.090

7
To compare the sizes of 0.29 or 0.3 first write both with the same number of decimal places. Write 0.3 with two decimal places by affixing a zero. _____ 0.30

8
Which is greater, 0.29 or 0.3? ______ 0.3

9
To compare the sizes of 0.0099 or 0.043 first write both with the same number of decimal places. Write 0.043 with four decimal places by affixing a zero. _____ 0.0430

10
Which is greater, 0.0099 or 0.043? ______ 0.043

11 ✍
To find the greater of 0.2 or 0.13, arrange them vertically as shown at the right. Then affix a zero to 0.2 so that it has the same number of decimal places as 0.13 and make the comparison. Which is greater, 0.2 or 0.13? ______

0.2
0.13

0.20
0.13

0.2

12
To determine whether 0.45 is greater than 0.169, arrange them vertically as shown at the right. Affix a zero to 0.45. Which is greater, 0.45 or 0.169? ______

0.45
0.169

0.45 or 0.450

13

To determine whether 0.0941 is greater than 0.12, arrange them vertically as shown at the right. Affix two zeros to 0.12. Which is greater, 0.0941 or 0.12? ______

0.0941
0.12

0.12 = 0.1200

14

Find the smaller of 0.375 and 0.47 by first writing 0.47 as a 3-place decimal. Which is smaller, 0.375 or 0.47? ______

0.375 is smaller than 0.470

15

To decide which is smaller, 0.2 or 0.1750, write the numerals with their decimal points aligned vertically as shown at the right. Which is smaller, 0.2 or 0.1750? ______

0.2
0.1750

0.1750 is smaller than 0.2000

16

Which is greater, 0.98 or 0.1? _____

0.98 is greater than 0.10

17

Which is greater, 0.11 or 0.9? _____

0.9 = 0.90

18

Which is smaller, 0.09 or 0.3? _____

0.09

19

True or false? 0.28 is greater than 0.0926. _____

True, 0.28 = 0.2800

20

0.2 is less than 1 because 1 = 1.0

True or false? 0.8 is less than 2. ______

True

21

True or false? 0.5 is greater than 0.38 _____

True

22
True or false? 0.4 is greater than 0.05. ______

True

23
True or false? 0.1 is greater than 1. ______

False

24
The symbol “>” means “greater than.”
0.05 > 0.0009 is true because 0.0500 is greater than 0.0009. True or false? 0.02 > 0.1 ______

False, 0.1 = 0.10

25
True or false? 0.08 > 0.009 ______

True

26
True or false? 0.07 > 0.056 ______

True

27
True or false? 0.00125 > 0.125 ______

False,
0.125 = 0.12500

28
The symbol “<“ means “less than.” 0.098 < 0.0981 is true because 0.098 = 0.0980 and is less than 0.0981. True or false? 0.0923 < 0.092 ______

False, 0.092 = 0.0920

29
True or false? 0.12 < 0.13 ______

True

30
True or false? 0.05 < 0.006 ______

False

31
True or false? 0.07 > 0.004 ______

True

32
True or false? 0.0092 < 0.00925 ______

True

33
True or false? $3.6 > 0.0928$ ______ True

34
True or false? $1.2 = 1.200$ ______ True

35
Fill in the blanks with $<$, $>$, or $=$.

0.3	______	0.4	$<$
0.096	______	0.72	$<$
0.0078	______	0.72	$<$
12.0	______	12	$=$
0.093	______	0.09300	$=$

FEEDBACK UNIT 3

This quiz reviews the preceding unit. Answers are at the back of the book. Begin the next unit only after all problems of this unit are understood.

For problems 1-10 label each statement as true or false.

1. $56 = 56.0$
2. $0.67 > 0.0671$
3. $1.03 > 1.9$
4. $0.02 < 0.2$
5. $0.0127 < 0.0128$
6. $2.6 < 2.60$
7. $0.0029 = 0.00029$
8. $0.007 > 0.00089$
9. $2.08 = 02.08$
10. $0.092 > 0.0920$

For problems 11-16 fill in the blanks with $<$, $>$, or $=$ to make a true statement.

11. 1.2901 ______ 0.12901
12. 0.008 ______ 0.02
13. 0.103 ______ 0.1003
14. 4.00 ______ 4
15. 0.0928 ______ 0.00929
16. 0.046 ______ 0.0058

UNIT 4: ROUNDING OFF DECIMAL NUMERALS

APPROXIMATING DECIMAL FRACTIONS

1.573 is between 1.57 and 1.58 because

$$1.573 > 1.570 \quad \text{and} \quad 1.573 < 1.580$$

To **round off** 1.573 to the nearest hundredth, it is necessary to find the two-place decimal that is closest to 1.573.

1
To round 0.632 to two decimal (hundredths) places, look at the digit in the third place. What digit is in the third decimal (thousandths) place?
______ 2

2
0.632 has the digit "2" in the third decimal (thousandths) place. Because 2 is less than 5, 0.632 rounded to two decimal (hundredths) places is 0.63 which is ______(less, greater) than 0.632. less

3
To round 0.964 to two decimal (hundredths) places, look at the digit in the third decimal (thousandths) place. What digit is in the third place? ______ 4

4
The digit in the third decimal (thousandths) place of 0.964 is "4". Because 4 is less than 5, 0.964 is rounded off to two decimal (hundredths) places as 0.96 which is ______ (less, greater) than 0.964. less

ROUNDING OFF DECIMAL FRACTIONS TO TWO PLACES

The rule for rounding off to two decimal places (nearest hundredth) has two parts.

1. If the digit in the third decimal place (thousandths) is less than 5, drop all the digits to the right of the hundredth's place without changing any other digits.

2. If the thousandth's digit is greater than or equal to 5, add one (1) to the digit in the hundredth's place and drop all digits to its right.

Four examples of rounding to two decimal (hundredths) places are listed below.

Numeral	Thousandth's Digit	Rounded to Two Decimal Places
1.273	3	1.27
0.095	5	0.10
14.271	1	14.27
4.506	6	4.51

5

0.352 rounded off to two decimal places is 0.35 because "2" is in the third decimal place. Round off 0.584 to two decimal places. ______ 0.58

6 ✍

Rounding off 1.283 to two decimal places is the same as rounding off 1.283 to the hundredth's place. Round off 1.283 to the hundredth's place. ______ 1.28

7

Round off 0.326 to the hundredth's place. ______ 0.33

8

Round off 1.892 to two decimal places. ______ 1.89

9

0.395 rounded off to two decimal places is 0.40. Round off 0.598 to two decimal places.

______ 0.60

10

0.975 rounded off to the nearest hundredth is 0.98. Round 0.389 to the nearest hundredth.

______ 0.39

11

0.0939 rounded off to the nearest hundredth is ______ 0.09

12

Round 1.995 to two decimal places. ______ 2.00

ROUNDING OFF ANY DECIMAL FRACTION

Decimal fractions can be rounded off to any number of decimal places or to any place value.

The "round off" rule is:

To round a numeral to a particular place, look at the digit right of the place.

1. If it is greater than or equal to 5, add one to the digit in the place and drop all the digits to the right.
2. If it is less than 5, drop all the digits to the right of the particular place without changing the digit.

1.27 rounded to one decimal (tenths) place is 1.3.

0.0948 rounded to three decimal (thousandths) places is 0.095.

0.10293 rounded to four decimal (ten-thousandths) places is 0.1029.

13 ✍

To round 4.26 to one decimal place, look at the digit in the second decimal place. Since 6 is 5 or greater, 1 is added to the first decimal place and the digits on the right are dropped. 4.26 rounded to one decimal place is 4.3. Round 5.16 to one decimal place. ______ 5.2

14 ✍

To round 0.9263 to the thousandth's place, look at the first digit right of the thousandth's place. Since 3 is less than 5, the digits right of the thousandth's place are dropped. 0.9263 rounded to the thousandth's place is ______. 0.926

15

Round off 13.50293 to the nearest:

one	______	14
tenth	______	13.5
hundredth	______	13.50
thousandth	______	13.503

16

Round off 2.91256 to the:

ten thousandth's place	______	2.9126
thousandth's place	______	2.913
hundredth's place	______	2.91
tenth's place	______	2.9
one's place	______	3

17

Round off 1.09052 to:

1 decimal place	______	1.1
2 decimal places	______	1.09
3 decimal places	______	1.091
4 decimal places	______	1.0905

18

Round off the following to two decimal places.

		Answer
109.325	______	109.33
2.0419	______	2.04
9.89529	______	9.90
0.0064	______	0.01
0.001	______	0.00

19

Round off the following to the thousandth's place.

		Answer
1.2995	______	1.300
0.22549	______	0.225
0.1095	______	0.110
29.0004	______	29.000
392.42509	______	392.425

20

Round off 1.30928 to:

		Answer
1 decimal place	______	1.3
2 decimal places	______	1.31
3 decimal places	______	1.309
4 decimal places	______	1.3093

FEEDBACK UNIT 4

This quiz reviews the preceding unit. Answers are at the back of the book. Begin the next unit only after all problems of this unit are understood.

1. Round 1.894 to two decimal places.

2. Round 13.999 to the hundredth's place.

3. Round 0.455 to the second decimal place.

4. Round 301.993 to the nearest hundredth.

Round the numbers in problems 5-8 to the thousandth's place.

5. 26.2999

6. 0.02709

7. 12.4785

8. 0.04749

9. Round 13.5 to the nearest one.

10. Round 0.49 to one decimal place.

11. Round 243.53 to the nearest tenth.

12. Round 4.04938 to four decimal places.

CHAPTER 8 MASTERY TEST

This test provides an opportunity for assuring that all of the objectives of Chapter 8 have been learned. Each problem is accompanied by the symbol (U) which indicates the Unit in which that type of problem was studied. If any problems are missed, they need to be corrected. Answers are at the back of the book.

For problems 1-8 fill in the blank.

	Fraction		Decimal	
1.	$\frac{5}{1000}$	=	______	(1)
2.	______	=	1.8	(1)
3.	$\frac{42}{10}$	=	______	(1)
4.	$\frac{86}{100}$	=	______	(1)
5.	______	=	0.056	(1)
6.	______	=	0.90	(1)
7.	$\frac{2056}{10}$	=	______	(1)
8.	______	=	4.08	(1)

In problems 9-12 write each numeral using scientific notation.

9. 417.94 (2)

10. 0.006107 (2)

11. 830,000,000 (2)

12. 0.216 (2)

In problems 13-16 label each statement true or false.

13. $56.0 = 56$ (3)

14. $0.06 > 0.5$ (3)

15. $0.0026 < 0.042$ (3)

16. $21.05 = 21.050$ (3)

In problems 17-20 fill each blank with =, >, or < to make a true statement.

17. 0.92 ___ .920 (3)

18. 1.27 ___ 1.271 (3)

19. 0.1 ___ 0.0875 (3)

20. 4.027 ___ 4.028 (3)

In problems 21-24 round off as indicated.

21. 0.535 to one decimal place (4)

22. 1.273 to two decimal places (4)

23. 0.625 to the nearest hundredth (4)

24. 125.6327 to three decimal places (4)

Chapter 9 Objectives

The following problems describe, by example, the skills and knowledge studied in this chapter. Each problem is accompanied by a symbol, (U), which indicates the Unit where the problem is studied. Answers are at the back of the book.

For problems 1-4, add. (1)

1. 0.097 and 0.056

2. 1.2, 0.91, and 1.478

3. 1,278.2 and 4,189.8

4. 0.0285, 0.193, and 4.9824

5. In the statement $4 - 3 = 1$ (1)
   The difference is ______
   the minuend is ______
   the subtrahend is ______

6. The subtraction $361 - 294 = 67$ (1)
   checks when
   ______ + ______ = ______

For problems 7-10, subtract. (2)

7. $0.03 - 0.021$

8. $0.0298 - 0.01184$

9. $1,308.3 - 876.7$

10. $10.003 - 0.844$

11. Take 2.58 from 9.26 (3)

12. Find the difference between 1.903 and 1.295 (3)

13. What number is 1.09 less than 8.27? (3)

14. If one capsule weighs 45.2 gm and another weighs 37.8 gm, how much more does the heavier capsule weigh than the lighter capsule? (3)

15. Bill spends \$10.28, \$5.09, \$105.56 and \$28.97 on a shopping trip. Find the sum of his purchases. (1)

16. One bank charges 0.0095 less interest per month than another bank which charges 0.0153 per month. Find the interest charge of the first bank. (3)

CHAPTER 9
ADDITION AND SUBTRACTION OF DECIMAL FRACTIONS

UNIT 1: ADDING DECIMAL NUMERALS IN COLUMNS

ADDING DECIMAL FRACTIONS

The addition of decimal fractions is accomplished vertically by arranging digits with the same place value in columns.

The addition of 0.5 and 0.2 is shown at the right.

0.5	addend
+ 0.2	addend
0.7	sum

1

To add 0.5 and 0.7 the decimals are arranged in a vertical column.

$$\begin{array}{r} 0.5 \\ +\,0.7 \\ \hline \end{array}$$

Arrange the addends 0.5 and 0.27 for vertical addition.

$$\begin{array}{r} 0.5 \\ +\,0.27 \\ \hline \end{array}$$

2

Write the numerals 1.934 and 0.279 with the digits of the same place value in vertical columns.

$$\begin{array}{r} 1.934 \\ +\,0.279 \\ \hline \end{array}$$

3

Write the numerals 0.05 and 0.008 with the digits of the same place value in vertical columns.

$$\begin{array}{r} 0.05 \\ +\,0.008 \\ \hline \end{array}$$

4

Write 1.58 + 72.6 + 0.6 as a vertical addition problem.

$$\begin{array}{r} 1.58 \\ 72.6 \\ +\,0.6 \\ \hline \end{array}$$

5

Add 0.5 + 0.37 by arranging the addends vertically. The decimal in the sum will be directly below the decimals in the addends.

$$\begin{array}{r} 0.5 \\ +\,0.37 \\ \hline 0.87 \end{array}$$

6

Add 0.007 + 0.002 vertically.

$$\begin{array}{r} 0.007 \\ +\,0.002 \\ \hline 0.009 \end{array}$$

ADDING RAGGED DECIMALS

The addition of 0.2 + 0.35 is aided by affixing 0 to the 0.2 to form 0.20.

Two ways of doing the problem are shown at the right. The example on the left shows the "ragged" decimal addition. In the example on the right the decimals are not "ragged". In either case, the decimal points must be aligned in a vertical column.

0.2	0.20	**addend**
+ 0.35	+ 0.35	**addend**
0.55	0.55	**sum**

7
Add 0.05 + 0.028 affixing a zero to eliminate ragged decimals. _____

$$\begin{array}{r} 0.050 \\ +\,0.028 \\ \hline 0.078 \end{array}$$

8
Add 0.007 + 0.1 + 0.0006 affixing zeros where necessary. _____

$$\begin{array}{r} 0.0070 \\ 0.1000 \\ +\,0.0006 \\ \hline 0.1076 \end{array}$$

9
Add 1.4 + 0.24 vertically. _____

$$\begin{array}{r} 1.40 \\ +\,0.24 \\ \hline 1.64 \end{array}$$

10
Add 0.45 + 0.001 + 0.2 affixing zeros where necessary. _____

$$\begin{array}{r} 0.450 \\ 0.001 \\ +\,0.200 \\ \hline 0.651 \end{array}$$

11

The addition example on the right shows "carrying" because the sum of the hundredth's column is 11. One (1) is added to the tenth's column.

$$\begin{array}{r} 1 \\ 0.0\,9 \\ +\,0.0\,2 \\ \hline 0.1\,1 \end{array}$$

Add 0.8 and 0.9 and show the carrying required.

______

$$\begin{array}{r} 1 \\ 0.8 \\ +\,0.9 \\ \hline 1.7 \end{array}$$

12

Add 1.3 and 9.7 and show the carrying. ______

$$\begin{array}{r} 1 \\ 1.3 \\ +\,9.7 \\ \hline 11.0 \end{array}$$

13

Add 1.05 and 2.37 and show the carrying. ______

$$\begin{array}{r} 1 \\ 1.05 \\ +\,2.37 \\ \hline 3.42 \end{array}$$

14

Add

$$\begin{array}{r} 4.6 \\ +\,5.73 \\ \hline \end{array}$$

$$\begin{array}{r} 4.60 \\ +\,5.73 \\ \hline 10.33 \end{array}$$

15

Add

$$\begin{array}{r} 0.137 \\ 0.5 \\ +\,0.094 \\ \hline \end{array}$$

$$\begin{array}{r} 0.137 \\ 0.500 \\ +\,0.094 \\ \hline 0.731 \end{array}$$

16

Add 13.081, 2.5, 4.62, and 0.4 ______

$$\begin{array}{r} 13.081 \\ 2.500 \\ 4.620 \\ +\,0.400 \\ \hline 20.601 \end{array}$$

17

0.027 + 0.99 + 0.038 = ______

$$\begin{array}{r} 0.027 \\ 0.990 \\ +\,0.038 \\ \hline 1.055 \end{array}$$

18

Add 3.098, 0.409, and 1.047 ______

$$\begin{array}{r} 3.098 \\ 0.409 \\ +\,1.047 \\ \hline 4.554 \end{array}$$

19

Add 0.12, 0.378, 0.9, 0.8073 ______

$$\begin{array}{r} 0.1200 \\ 0.3780 \\ 0.9000 \\ +\,0.8073 \\ \hline 2.2053 \end{array}$$

20 ✍

The word "sum" means the answer to an addition problem.

George has 1.05 kilograms of aspirin and 0.98 kilograms of aspirin. Find the sum of the weights of the drug. ______

$$\begin{array}{r} 1.05 \\ +\,0.98 \\ \hline 2.03 \end{array}$$ kilograms

21

Find the sum of 5.6 milligrams (mg) and 23.92 mg.______

$$\begin{array}{r} 5.60 \\ +\,23.92 \\ \hline 29.52 \end{array}$$ mg

22

What is the sum of 0.45 liters and 1.9 liters? ______

$$\begin{array}{r} 0.45 \\ +\,1.90 \\ \hline 2.35 \end{array}$$ liters

23

1.4 kilograms and 0.5 feet cannot be added because the numbers have different labels. Can the sum of 483.25 grams and 37.099 grams (gm) be found? ______

Yes,
483.250
+ 37.099
520.349 gm

24

Can 57.3 pounds be added to 0.49 feet? ______

No, the labels are not the same.

25

Nurse Reynolds needed to give dosages of 6 minims, 28 minims, and 14 minims. Find the sum of the dosages. ______

6
28
+ 14
48 minims

26

Cotes Sporting Goods store received two shipments costing $1,468.05 and $82,025.36. Find the sum of the costs of the shipments. ______

1,468.05
+ 82,025.36
$83,493.41

27

Find the sum of forces on an object if they are 42.5 kilograms (kgm), 1.07 kgm, 0.78 kgm, and 38 kgm. ______

42.50
1.07
0.78
+ 38.00
82.35 kgm

28

A pharmacy received the following amounts of alcohol: 4.05 liters, 0.561 liters, and 0.26 liters. Find the sum of the amounts of alcohol.

______

4.050
0.561
+ 0.260
4.871 liters

29

Bill ran a mile race with the following quarter mile split times: 58.3 sec, 61.1 sec., 58 sec., 59.28 sec. Find the sum of the split times. ______

58.30
61.10
58.00
+ 59.28
236.68 sec

30
Find the sum of the commission fees for a job when the four fees were 1.936%, 2.05%, 2.19%, and 2.335%. ______

1.936
2.050
2.190
+ 2.335
8.511 %

31
To add 45.7 inches to 5 feet, both measurements need to be expressed as inches. 1 foot equals 12 inches. What is the sum of 45.7 inches and 5 feet?

______

5 feet = 60 inches
105.7 inches

32
The **perimeter** of a geometric figure is the sum of the lengths of its sides. To find the perimeter of a rectangle, _____ the lengths of its four sides.

add

33
The **perimeter** of the quadrilateral shown at the right is the sum of the lengths of the four sides. Find the perimeter. ______

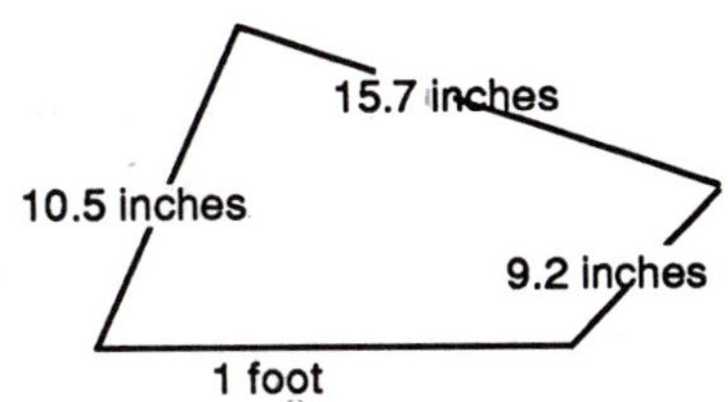

[Note: A **quadrilateral** is a four-sided figure.]

1 foot = 12 inches

12 + 15.7 + 9.2 + 10.5
47.4 inches is the perimeter

34
The **perimeter** of the **trapezoid** shown at the right is the sum of the lengths of the four sides. Find the perimeter. ______

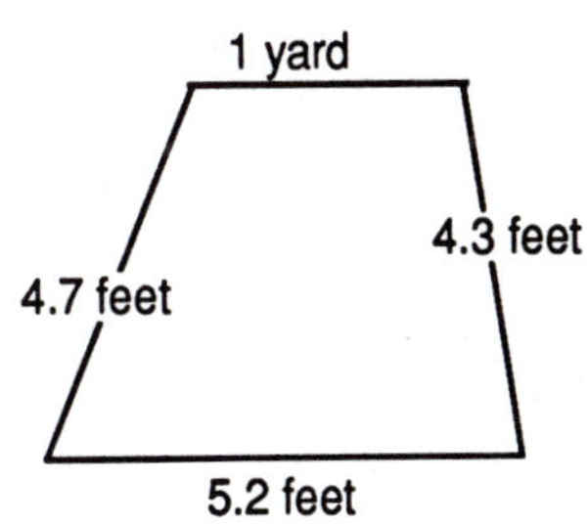

[Note: A **trapezoid** is a quadrilateral with one pair of parallel sides.]

1 yard = 3 feet

3 + 4.3 + 4.7 + 5.2
17.2 feet is the perimeter

35
The perimeter of the **parallelogram** shown at the right is the sum of the lengths of the four sides. Find the perimeter. ______

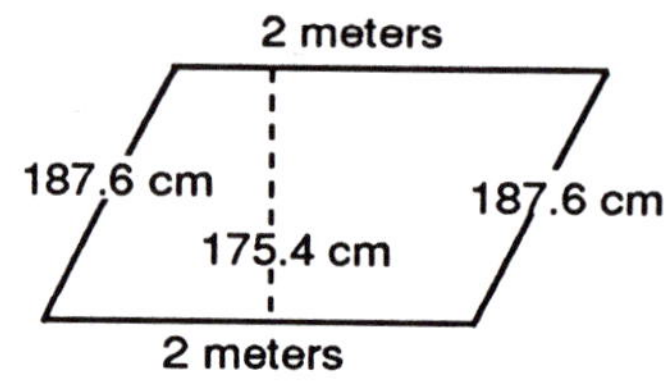

100 cm = 1 meter

200 + 200 + 187.6 + 187.6
775.2 cm is the perimeter

[Note: A **parallelogram** is a quadrilateral with two pairs of parallel sides.]

36
Find the perimeter of the **rectangle** shown at the right .

______

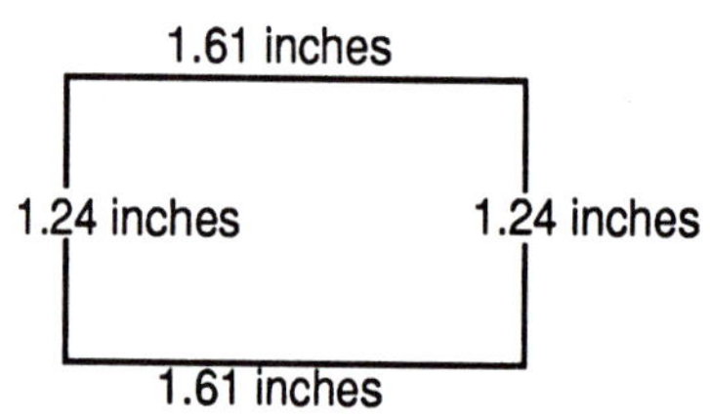

[Note: A **rectangle** is a parallelogram with four right angles.]

1.61 + 1.24 + 1.61 + 1.24
5.7 inches is the perimeter

FEEDBACK UNIT 1

This quiz reviews the preceding unit. Answers are at the back of the book. Begin the next unit only after all problems of this unit are understood.

1. $\begin{array}{r} 0.3 \\ +\,0.05 \\ \hline \end{array}$

2. $\begin{array}{r} 0.08 \\ +\,0.09 \\ \hline \end{array}$

3. $\begin{array}{r} 0.27 \\ +\,0.8 \\ \hline \end{array}$

4. $\begin{array}{r} 1.047 \\ +\,9.128 \\ \hline \end{array}$

5. $\begin{array}{r} 0.037 \\ 0.46 \\ +\,0.1282 \\ \hline \end{array}$

6. $\begin{array}{r} 1.5 \\ 7.05 \\ 2.49 \\ +\,0.21 \\ \hline \end{array}$

7. $\begin{array}{r} 0.006 \\ 0.013 \\ +\,0.49 \\ \hline \end{array}$

8. $\begin{array}{r} 0.0025 \\ 0.1037 \\ +\,0.056 \\ \hline \end{array}$

9. Add 2.3, 1.09, 4.35, and 0.08

10. Add 0.026, 0.055, and 1.9203

11. Add 28.026 and 37.082

12. Add 1,350.5 and 2,259.21

13. An architect's fees were 3.975% and 4.1092%. Find the sum of the fees.

14. The ingredients in a medicine tablet weighed 2.03 mg, 14 mg, 8.785 mg and 0.573 mg. Find the sum of the weights of the ingredients.

15. The amounts of three solutions to be mixed are 4.5 liters, 9.236 liters, and 2 liters. What is the sum of the three amounts?

16. The unit prices of 3 sizes of screws were found to be \$.0397, \$.125, and \$.38275. Find the sum of the prices.

17. Find the perimeter of the parallelogram shown at the right.

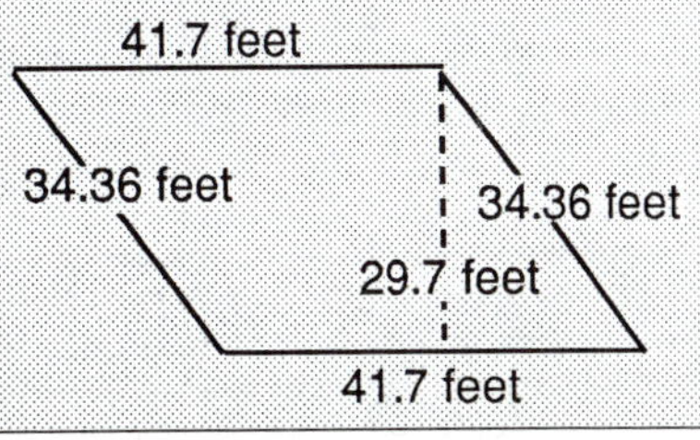

UNIT 2: THE SUBTRACTION OF DECIMAL FRACTIONS

SUBTRACTION OF DECIMAL FRACTIONS

The addition of decimal fractions depends on placing the decimal points in a vertical column so that the digits of the same place value will be in the same column. The subtraction of decimal fractions also depends upon placing the decimal points in a column. Also, both numerals should be written with the same number of decimal places. Sometimes affixing of zeros will be required. The subtraction is completed by subtracting the digits in each column. To subtract 0.03 from 0.09

1) The word "from" indicates the problem 0.09 – 0.03 which is written vertically as shown at the right.

$$\begin{array}{r} 0.09 \\ -\,0.03 \\ \hline \end{array} \quad \begin{array}{l} \textbf{minuend} \\ \textbf{subtrahend} \end{array}$$

2) Complete the subtraction by placing the decimal point in the answer directly below the decimals in the numerals.

$$\begin{array}{r} 0.09 \\ -\,0.03 \\ \hline 0.06 \end{array} \quad \textbf{difference}$$

1

The correct way to write the subtraction 0.09 **from** 1.25 is shown at the right.

$$\begin{array}{r} 1.25 \\ -\,0.09 \\ \hline \end{array}$$

Is the subtraction of 1.28 **from** 3.6 shown correctly at the right? ______

$$\begin{array}{r} 3.6 \\ -\,1.2\,8 \\ \hline \end{array}$$

No, it should be

$$\begin{array}{r} 3.60 \\ -\,1.28 \\ \hline \end{array}$$

2

To subtract decimal fractions vertically, digits of the same place value should be in columns. Which problem below is written correctly? ______

a) $\begin{array}{r} 28.25 \\ -\,0.30 \\ \hline \end{array}$ b) $\begin{array}{r} 28.25 \\ -\,0.300 \\ \hline \end{array}$

a

3 ✍

Every whole number can be written using a decimal point.

5 equals 5.0 or 5.000 or 5.

17 equals 17.0 or 17.0000 or 17.00

Can 28 be written as 28.000? ______ Yes

4

Is 34 equal to 34.0? ______ Yes

5

7 equals 7.0 and 92 equals 92.0

Is 18 equal to 1.8? ______ No, 18.0

6 ✍

To subtract 0.28 **from** 6:

1) Write 6 with a decimal and two zeros so it will have the same number of decimal places as 0.28

   6 = 6.00

2) Place 0.28 in the second row and arrange the decimal points in a vertical column.

   6.00
   – 0.28

Arrange the problem "subtract 1.39 from 17" for vertical subtraction. ______

17.00
– 1.39

7

Write the problem "subtract 0.9 from 1.8" for vertical subtraction. ______

1.8
– 0.9

8

Write the problem "subtract 9 from 39.125" for vertical subtraction. ______

39.125
– 9.000

9
Write the problem "subtract 0.0092 from 0.018" for vertical subtraction. ______

$$\begin{array}{r} 0.0180 \\ -\,0.0092 \\ \hline \end{array}$$

SUBTRACTING VERTICALLY

Subtracting decimal fractions vertically is accomplished as though subtracting whole numbers, but the numerals must first be arranged so that the decimal points are in the same vertical column. The problem at the right shows the subtraction of 0.137 from 0.23

$$\begin{array}{rl} 0.230 & \text{minuend} \\ -\,0.137 & \text{subtrahend} \\ \hline 0.093 & \text{difference} \end{array}$$

Notice that the "borrowing" that was needed in this subtraction was accomplished in the same way as it would have been with whole numbers.

10
Subtract 0.08 from 0.12 and show all your work. ______

$$\begin{array}{r} 0.12 \\ -\,0.08 \\ \hline 0.04 \end{array}$$

11
Subtract 1.5 from 2.5. ______

$$\begin{array}{r} 2.5 \\ -\,1.5 \\ \hline 1.0 \end{array}$$

12
Subtract 0.8 from 4 by first writing 4 with a decimal point. ______

$$\begin{array}{r} 4.0 \\ -\,0.8 \\ \hline 3.2 \end{array}$$

13
Subtract 0.2 from 5. ______

$$\begin{array}{r} 5.0 \\ -\,0.2 \\ \hline 4.8 \end{array}$$

14
Subtract 0.12 from 0.5. ______

$$\begin{array}{r} 0.50 \\ -\,0.12 \\ \hline 0.38 \end{array}$$

15
Subtract 0.025 from 0.8. ______

$$\begin{array}{r} 0.800 \\ -\,0.025 \\ \hline 0.775 \end{array}$$

16
0.039 − 0.018 = ______

$$\begin{array}{r} 0.039 \\ -\,0.018 \\ \hline 0.021 \end{array}$$

17
5 − 0.08 = ______

$$\begin{array}{r} 5.00 \\ -\,0.08 \\ \hline 4.92 \end{array}$$

18
14 − 0.3 = ______

$$\begin{array}{r} 14.0 \\ -\,0.3 \\ \hline 13.7 \end{array}$$

19
0.0427 − 0.0116 = ______

$$\begin{array}{r} 0.0427 \\ -\,0.0116 \\ \hline 0.0311 \end{array}$$

20
15.715 − 0.8 = ______

$$\begin{array}{r} 15.715 \\ -\,0.800 \\ \hline 14.915 \end{array}$$

21
3.65 − 1.025 = ______

$$\begin{array}{r} 3.650 \\ -\,1.025 \\ \hline 2.625 \end{array}$$

22
0.00287 − 0.00092 = ______

$$\begin{array}{r} 0.00287 \\ -\,0.00092 \\ \hline 0.00195 \end{array}$$

FEEDBACK UNIT 2

This quiz reviews the preceding unit. Answers are at the back of the book. Begin the next unit only after all problems of this unit are understood.

Subtract.

1. $\begin{array}{r} 0.48 \\ -0.27 \\ \hline \end{array}$

2. $\begin{array}{r} 0.038 \\ -0.02 \\ \hline \end{array}$

3. $\begin{array}{r} 10.4 \\ -4.079 \\ \hline \end{array}$

4. $\begin{array}{r} 0.6 \\ -0.16 \\ \hline \end{array}$

5. Subtract 1.049 from 9.1052

6. Subtract 0.105 from 0.82

7. 10.4 − 5.7 = _____

8. 0.0493 − 0.0486 = _____

9. 0.1027 − 0.0923 = _____

10. 182.04 − 29.82 = _____

UNIT 3: WORD PROBLEMS REQUIRING SUBTRACTION

CHECKING SUBTRACTION

In the subtraction example below:

0.50 is the **minuend**
0.12 is the **subtrahend**
and 0.38 is the **difference**.

$$\begin{array}{r} 0.50 \\ -0.12 \\ \hline 0.38 \end{array}$$

The subtraction is checked by the addition shown at the right. The sum of the difference and the subtrahend must be the minuend.

$$\begin{array}{rl} 0.38 & \text{difference} \\ +0.12 & \text{subtrahend} \\ \hline 0.50 & \text{minuend} \end{array}$$

1 ✍

In the subtraction example on the right, the minuend is 0.03, the subtrahend is 0.007 and the difference is 0.023. Add the difference and subtrahend. Is the sum equal to the minuend? ______

$$\begin{array}{r} 0.03 \\ -\,0.007 \\ \hline 0.023 \end{array}$$

Yes

$$\begin{array}{r} 0.023 \\ +\,0.007 \\ \hline 0.030 \end{array}$$

2

Does the subtraction problem on the right check? ______

$$\begin{array}{r} 1.0278 \\ -\,0.11 \\ \hline 0.9178 \end{array}$$

Hint: Add the difference and the subtrahend.

Yes

$$\begin{array}{r} 0.9178 \\ +\,0.11 \\ \hline 1.0278 \end{array}$$

3

In the subtraction 0.008 - 0.001 = 0.007,
the minuend is ______,
the subtrahend is ______,
and the difference is ______.

$$\begin{array}{r} 0.008 \\ -\,0.001 \\ \hline 0.007 \end{array}$$

4 ✍

In the subtraction "take 0.02 from 0.08,"
the minuend is ______,
the subtrahend is ______,
and the difference is ______.

$$\begin{array}{r} 0.08 \\ -\,0.02 \\ \hline 0.06 \end{array}$$

5

In the subtraction "take 1.9 from 2.9,"
the minuend is ______,
the subtrahend is ______,
and the difference is ______.

$$\begin{array}{r} 2.9 \\ -\,1.9 \\ \hline 1.0 \end{array}$$

6
If the minuend is 10.4 and the subtrahend is 4.3, what is the difference? ______

$$\begin{array}{r} 10.4 \\ -\,4.3 \\ \hline 6.1 \end{array}$$

7

$$\begin{array}{rl} 0.08 & \text{minuend} \\ -\,0.019 & \text{subtrahend} \\ \hline 0.061 & \text{difference} \end{array}$$

If the sum of the difference and the subtrahend is the minuend, a subtraction problem checks. Does the problem above check? ______

Yes, $\begin{array}{r} 0.061 \\ +\,0.019 \\ \hline 0.080 \end{array}$

8

$$\begin{array}{r} 0.023 \\ -\,0.014 \\ \hline 0.019 \end{array} \qquad \begin{array}{r} 0.019 \\ +\,0.014 \\ \hline 0.033 \end{array}$$

Does the subtraction problem above check? ______

No, $0.023 \neq 0.033$

9

$10.9 - 1.8 = 9.1$

In the problem above,
the minuend is
the subtrahend is
and the difference is

$$\begin{array}{r} 10.9 \\ -\,1.8 \\ \hline 9.1 \end{array}$$

Does the difference check? ______

Yes, $\begin{array}{r} 9.1 \\ +\,1.8 \\ \hline 10.9 \end{array}$

10
Subtract and check. $0.5 - 0.2 =$ ______

$$\begin{array}{r} 0.5 \\ -\,0.2 \\ \hline 0.3 \end{array} \qquad \begin{array}{r} 0.3 \\ +\,0.2 \\ \hline 0.5 \end{array}$$

11
A subtraction problem checks when
______ + ______ = ______

difference plus subtrahend = minuend

12

The word "difference" is the name of the answer to a subtraction problem.

If one book weighs 28.3 kilograms and another book weighs 16.2 kilograms (kgm), what is the difference of the weights? ______

$$\begin{array}{r} 28.3 \\ -\,16.2 \\ \hline 12.1 \end{array}\ \text{kgm}$$

13

The difference cannot be found between 2 dozen eggs and 1.9 grams of butter because the labels are not the same. Can the difference between 48.2 milliliters (ml) and 39.1 milliliters be found? ______

Yes, $$\begin{array}{r} 48.2 \\ -\,39.1 \\ \hline 9.1 \end{array}\ \text{ml}$$

14

The phrase

"how much more"

implies a subtraction situation.

Jose has $1,268.09 in his savings account and Maria has $932.78 in hers. How much more is in Jose's account than in Maria's? ______

$$\begin{array}{r} 1{,}268.09 \\ -\,932.78 \\ \hline \$335.31 \end{array}$$

15

Sheerif has $92.56 and Abdullah has 48 kopeks. How much more money does Sheerif have than Abdullah? ______

Cannot be found.
Labels are different.

16

The phrase

"how much was left"

indicates a subtraction situation.

$15,600 was withdrawn from a bank account of $92,500. How much was left? ______

$$\begin{array}{r} 92{,}500 \\ -\,15{,}600 \\ \hline \$76{,}900 \end{array}$$

17
A supply of 38.5 liters of solvent was decreased by 10.6 liters. How much was left? ______

$$\begin{array}{r} 38.5 \\ -\,10.6 \\ \hline 27.9 \end{array}$$ liters

18 ✍
The word "less" implies subtraction.
4 less than 9 means 9 − 4
Does the sum of 8 and 2 mean 8 − 2 ? ______

No, 8 + 2 is the sum

19
A first tablet weighed 3.9 milligrams (mgm) less than another tablet that weighed 12.1 milligrams. Find the weight of the first tablet. ______

$$\begin{array}{r} 12.1 \\ -\,3.9 \\ \hline 8.2 \end{array}$$ mgm

20
The interest rate on Iris's savings account is 1.42% per quarter. Percy's account pays 1.53% per quarter. How much more is Percy's interest rate than Iris'? ______

$$\begin{array}{r} 1.53\,\% \\ -\,1.42\,\% \\ \hline 0.11\,\% \end{array}$$

21
If Vito's salary is raised to $6.85 per hour and Luigi makes $5.87 per hour, how much more does Vito make each hour than Luigi? ______

$$\begin{array}{r} 6.85 \\ -\,5.87 \\ \hline \$0.98 \end{array}$$

22
Sammy makes $1.25 less per hour than Johnny who makes $8.19 per hour. How much does Sammy make per hour? ______

$$\begin{array}{r} 8.19 \\ -\,1.25 \\ \hline \$6.94 \end{array}$$

23
A virus traveled 0.0068 millimeters (mm) through a human cell which had a thickness of 0.0092 millimeters. How much farther does the virus need to go to travel completely through the cell? ______

$$\begin{array}{r} 0.0092 \\ -\,0.0068 \\ \hline 0.0024 \end{array} \text{ mm}$$

24
A specimen of rock feldspar weighs 106.2 milligrams and a specimen of lignite weighs 28.3 milligrams. Find the difference in their weights. ______

$$\begin{array}{r} 106.2 \\ -\,28.3 \\ \hline 77.9 \end{array} \text{ mgm}$$

25
A supply of 42 cubic centimeters (cc) of morphine was decreased by 9.7 cubic centimeters. How much was left? ______

$$\begin{array}{r} 42.0 \\ -\,9.7 \\ \hline 32.3 \end{array} \text{ cc}$$

26
If a triangle has sides of 4.57 feet and 5.9 feet with a perimeter of 14.25 feet, how long is the third side? ________

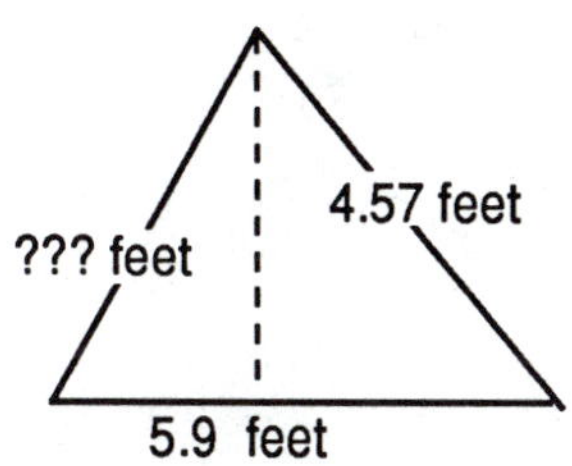

$4.57 + 5.9 = 10.47$

$14.25 - 10.47 = 3.78$ feet

FEEDBACK UNIT 3

This quiz reviews the preceding unit. Answers are at the back of the book. Take the mastery test only after all problems of this unit are understood.

1. $\begin{array}{r} 1.27 \\ -0.16 \\ \hline \end{array}$

2. $\begin{array}{r} 0.0056 \\ -0.0019 \\ \hline \end{array}$

3. $\begin{array}{r} 0.5 \\ -0.13 \\ \hline \end{array}$

4. $\begin{array}{r} 128.57 \\ -75.2 \\ \hline \end{array}$

5. 4.75 − 2.92 = ______

6. 0.406 − 0.01932 = ______

7. 0.0125 − 0.0118 = ______

8. Take 0.7 from 12.2

9. Find the difference between 4.0798 and 3.526.

10. If a subtraction problem has a difference of 12.05 and a subtrahend of 10.39, what is the minuend?

11. Joe's specimen weighed 28.5 grams and Belinda's weighed 37.2 grams. How much more did Belinda's weigh than Joe's?

12. A jar of 0.546 liters of alcohol was drained from a container holding 2.729 liters. How much was left in the container?

13. The perimeter of the triangle shown at the right is 57.3 feet. Find the length of the unknown side.

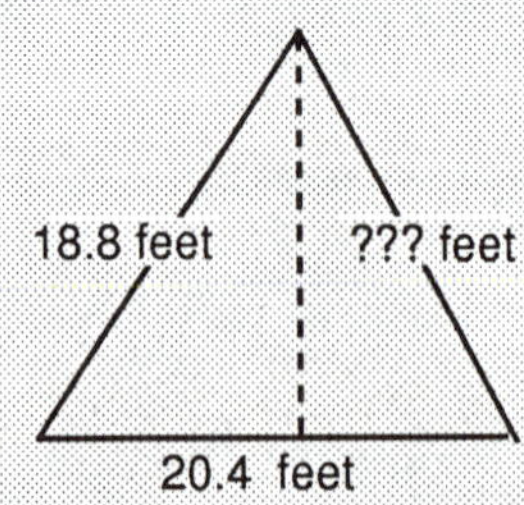

CHAPTER 9 MASTERY TEST

This test provides an opportunity for assuring that **all** of the objectives of Chapter 9 have been learned. Each problem is accompanied by the symbol (U) which indicates the Unit in which that type of problem was studied. If any problems are missed, they need to be corrected. Answers are at the back of the book.

For problems 1-4, add. (1)

1. 0.5 and 0.8
2. 0.07 and 1.2
3. 0.085 and 0.016
4. 0.025, 0.0962, 0.2, and 1.75
5. If the letters A, B, and C represent three different numbers and A − B = C, the minuend is ______, the subtrahend is ______, and the difference is ______. (1)
6. The subtraction problem 378 − 188 = 190 checks when ______ + ______ = ______. (1)

For problems 7-12, subtract. (2)

7. $\begin{array}{r} 0.01 \\ -0.008 \\ \hline \end{array}$
8. $\begin{array}{r} 0.00125 \\ -0.001163 \\ \hline \end{array}$
9. $\begin{array}{r} 1{,}407.2 \\ -926.5 \\ \hline \end{array}$
10. $\begin{array}{r} 100.002 \\ -0.971 \\ \hline \end{array}$
11. 4.027 − 3 = ______
12. 5 − 0.0025 = ______
13. If Tom has $2,049.00 and Sandra has $3,150.75, how much more does Sandra have than Tom? (3)
14. The chemicals in a mixture weigh 0.25 kg, 1.6 kg, 0.097 kg, and 4.97 kg. Find the sum of the weights. (3)
15. A small flask holds 4.51 liters less than a jar holding 8.49 liters. How much does the small flask hold? (3)
16. Robert decreases his interest rate of 8.23% by 1.51%. What is his new interest rate after the decrease? (3)
17. Find the perimeter of the parallelogram below. (1)

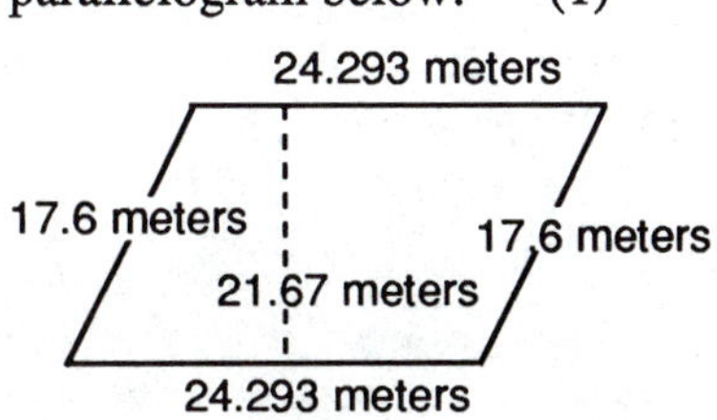

18. The perimeter of the scalene triangle shown at right is 27.6 meters. Find the length of the unknown side. (3)

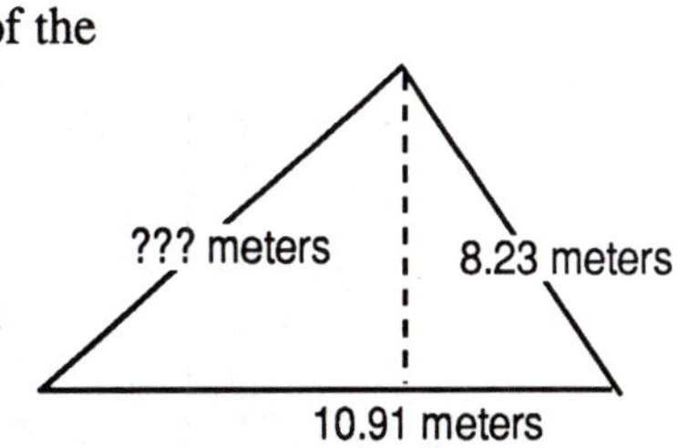

Chapter 10 Objectives

The following problems describe, by example, the skills and knowledge studied in this chapter. Each problem is accompanied by a symbol, (U), which indicates the Unit where the problem is studied. Answers are at the back of the book.

For problems 1-4, multiply. (2)

1. $\begin{array}{r} 1{,}876.2 \\ \times\ 0.08 \\ \hline \end{array}$

2. $\begin{array}{r} 7{,}098 \\ \times\ 0.089 \\ \hline \end{array}$

3. $\begin{array}{r} 2.098 \\ \times\ 100 \\ \hline \end{array}$

4. $\begin{array}{r} 7{,}000 \\ \times\ 35.8 \\ \hline \end{array}$

5. Find 0.37 of 56. (3)

6. If 35 people are to receive dosages of nembutal of 7.48 mg each, how much of the medicine is needed? (4)

7. The First Bank charges $0.109 per year for each dollar it loans. Robert and Linda need a $50,000 loan. How much will the bank charge to loan $50,000 for one year? (4)

8. A force was directed so that 0.37 of the force moved an object along a plane. The force was 60 grams. How much of the force moved the object? (4)

For problems 9 and 10, divide until a zero remainder is obtained. (6)

9. $1.6\overline{)34.4}$

10. $2.8\overline{)0.01652}$

For problems 11 and 12, divide and round off at two decimal places. (7)

11. $57\overline{)91.3}$

12. $0.43\overline{)13.25}$

13. Divide and round off $0.815 \div 2.7$ at three decimal places. (7)

14. Stamps used to cost $.15 each. How many stamps could be bought with $6.45? (8)

15. Sally keeps track of her grocery purchases for four consecutive weeks. She spends $87.52, $102.05, $48.26,and $78.09 in those four weeks. What was her average weekly grocery cost during this period? (8)

Chapter 10
Multiplication And Division Of Decimal Fractions

Unit 1: Placing The Decimal Point In A Product

Multiplying Whole Numbers

The multiplication of decimal fractions is almost identical to the multiplication of whole numbers. The only new task that is required is the placement of the decimal point in the answer (product). If you are uncertain of your ability to correctly multiply whole numbers, it would be wise to retake the Mastery Test for Chapter 3.

The problem at the right offers a quick review of the words and processes used in the multiplication of whole numbers.

36		factor (multiplicand)
x 28		factor (multiplier)
288	= 36 x 8	partial product
72	= 36 x 2	partial product
1008	= 36 x 28	product

1
Multiply and show all your work. _____

```
  406
x 109
```

```
  406
x 109
 3654
 000
406
44254
```

2
Multiply and show all your work. _____

```
  827
x 905
```

```
   827
 x 905
  4135
  000
 7443
748435
```

DECIMAL PLACES IN THE PRODUCT OF DECIMAL FRACTIONS

When multiplying decimal fractions, the location of the decimal point in the product is determined by the total number of digits right of the decimal point in the two factors of the problem.

To determine the number of digits right of the decimal point in the product of 0.12 x 13.005:

1) Count the digits right of the decimal point in each factor.

   0.12 has two digits right of the decimal point.
   13.005 has three digits right of the decimal point.

2) Find the sum of 2 and 3 to find the number of digits right of the decimal point in the product of 0.12 and 13.005.

   2 + 3 = 5 digits right of the decimal point.

3
In the problem 0.3 x 1.27, the factor 0.3 has one digit right of the decimal point. The factor 1.27 has ______ digits right of the decimal.

2

4
The numeral 4.362 has three digits right of the decimal point. How many digits are right of the decimal point in 2.78? ______ 2

5
How many digits are right of the decimal point in 65.814? ______ 3

6
How many digits are right of the decimal point in 8.3? ______ 1

7
How many digits are right of the decimal point in 0.028? ______ 3

8 ✍
0.0028 has four decimal places because the two zeros to the right of the decimal point are counted. How many decimal places are in 0.04? ______ 2

9
In the problem 0.2 x 1.27, the factor 0.2 has one decimal place. How many decimal places in the factor 1.27? ______ 2

10
In the problem 0.2 x 1.27, there is a total of three decimal places.

1 + 2 = 3

In the problem 0.102 x 0.34, there are three decimal places in the factor 0.102 and two decimal places in 0.34 for a total of ______ decimal places. 5

11
46.2 has one decimal place.
0.087 has three decimal places.
In the answer (product) of 46.2 x 0.087 there will be ______ decimal places, which is the sum of 1 and 3. 4

12
How many digits will be right of the decimal point in the product of 3 x 1.343? ______ 3

[Note: There are no decimal places to the right of the decimal point for any whole number.]

13
How many digits will be right of the decimal point in the product of 0.025 x 394? ______ 3

14
How many digits will be right of the decimal point in the product of the problem 10.1 x 0.28387?
______ 6

FEEDBACK UNIT 1

This quiz reviews the preceding unit. Answers are at the back of the book. Begin the next unit only after all problems of this unit are understood.

Give the number of digits right of the decimal point (decimal places) in problems 1 through 8.

1. 4.5
2. 38.2
3. 0.078
4. 1,278.3
5. 10.128
6. 0.00278
7. 0.100009
8. 203

Give the number of digits that should be right of the decimal point in the product of the factors in problems 9 through 12.

9. 0.8 x 0.2
10. 0.03 x 0.9
11. 4 x 0.5
12. 0.28 x 0.03

UNIT 2: MULTIPLYING DECIMAL NUMERALS

PLACING THE DECIMAL POINT IN THE PRODUCT OF DECIMAL FRACTIONS

The number of digits right of the decimal in the product of decimal fractions is:

> The sum of the number of digits right of the decimal points in the two factors which is the total number of decimal places in the two factors.

To multiply 0.24 by 0.02:

1) Multiply the factors as though multiplying whole numbers. Ignore the decimal points.

```
    0.2 4
  x 0.0 2
  -------
      4 8
    0 0 0
  0 0 0
  -------
  0 0 0 4 8
```

2) Since there is a total of four digits right of the decimal points in the two factors, start at the right side of 00048 and count four digits to the left. After the fourth digit, place the decimal point in the product.

0.0 0 4 8 $\qquad$ 0.24 x 0.02 = 0.0048

1
To multiply 1.29 by 0.01:

1) Multiply and ignore the decimal points.

```
    1.2 9
  x 0.0 1
  -------
    1 2 9
  0 0 0
0 0 0
---------
0 0 1 2 9
```

2) Complete the multiplication by correctly placing the decimal point in the product.

______ $\qquad$ 0.0129

2
Multiply 22.3 by 0.3 and place the decimal point in the product. ______

```
  2 2.3
  x 0.3
  6 6 9
0 0 0
0 6.6 9 = 6.69
```

3
The problem at the right requires "carrying." Do it just like a whole number multiplication. Then place the decimal point correctly in the product. ____________

```
   4 3 2 5
   x 0.0 2
```

```
   4 3 2 5
   x 0.0 2
   8 6 5 0
  0 0 0 0
 0 0 0 0
0 0 8 6.5 0 = 86.50
```

4
Multiply the problem shown at the right. Count the total number of decimal places in 0.23 and 0.009; then place the decimal point correctly in the product. ____________

```
   0.2 3
x 0.0 0 9
```

```
     0.2 3
  x 0.0 0 9
     2 0 7
    0 0 0
   0 0 0
  0 0 0
0.0 0 2 0 7
```

5
Multiply. 1.98 x 0.6 ______

```
  1.9 8
  x 0.6
1 1 8 8
0 0 0
1.1 8 8
```

6
Multiply. 0.222 x 0.07 ______

0.01554

7
Multiply. 0.4 x 0.3 ______

0.12

8
Multiply. 2.03 x 0.009 ______

0.01827

9
Multiply. 0.045 x 1.6 ______

0.072

FEEDBACK UNIT 2

This quiz reviews the preceding unit. Answers are at the back of the book. Begin the next unit only after all problems of this unit are understood.

Multiply.

1.	2.	3.	4.
1.02 x 0.3	0.078 x 0.001	1.625 x 0.03	102.037 x 0.008

5. 0.12 x 0.9
6. 5 x 0.01
7. 10.25 x 0.8
8. 7.1 x 7
9. Multiply 3.09 by 0.009.
10. Find the product of 780 and 0.005.

UNIT 3: MORE MULTIPLICATION OF DECIMAL FRACTIONS

MULTIPLYING DECIMAL FRACTIONS WITH TWO DIGITS OR MORE

Multiplying decimal fractions with two or more non-zero digits in the multiplier is done as it was in the preceding unit.

The decimal point is placed according to the total number of digits right of the decimal in the two factors of the problem.

The example at the right shows the multiplication of 8.5 by 0.68.

```
    8.5
 x 0.6 8
  6 8 0
 5 1 0
 0 0
 5.7 8 0
```

Notice that there are three decimal places in the product because the total number of digits right of the decimal points in the multiplier and the multiplicand is three.

$$8.5 \times 0.68 = 5.780 \text{ or } 5.78$$

1
To multiply 2.2 by 0.34:
1) Multiply 22 by 34.
   22 x 34 = 748
2) Count the total number of digits right of the decimal points in the factors 2.2 and 0.34.
   3 digits

Complete the multiplication of 2.2 by 0.34 by placing the decimal point correctly in 748. ______

0.748

2
Multiply. 1.2 x 5.4 ______

```
  1.2
x 5.4
  4 8
 6 0
 6.4 8
```

3
There is a total of five digits right of the decimal points in the problem 1.75 x 0.064. Multiply and place the decimal correctly in the product. ______.

```
      1.7 5
  x 0.0 6 4
      7 0 0
   1 0 5 0
   0 0 0
 0 0 0
 0.1 1 2 0 0 = 0.112
```

4
Multiply. 0.128 x 25 ______

3.2

5
Multiply. 1.032 x 3.5 ______

3.612

6
Multiply. 12.0897 x 68 ______

822.0996

7
Multiply. 3.8 x 12.7 ______

48.26

8
Multiply. 0.483 x 1.11 ______ 0.53613

9
203 x 2.51 = ______ 509.53

10
0.462 x 248 = ______ 114.576

11
92.8 x 1.79 = ______ 166.112

12
921.7 x 856.9 = ______ 789,804.73

MULTIPLYING BY POWERS OF TEN

"Powers of Ten" are numbers such as 10, 100, 1,000, 10,000, etc.

There is a simple shortcut for multiplying any decimal fraction by a power of ten. The shortcut is:

> Move the decimal point in the decimal fraction to the right the same number of places as there are zeros in the power of ten.

5.78 x 1,000 = 5.780. = 5,780 The decimal was moved 3 places right because there are 3 zeros in 1,000.

13 ✍
To multiply 100,000 by 0.281 first count the number of zeros in 100,000 and then move the decimal point right that many places in 0.281.

100,000 x 0.281 = ______ 28,100

14
Multiply 6.31 by 100 using the shortcut for multiplication by a power of ten. ______

631. = 631

15
Use the shortcut for multiplying by a power of ten.
0.0074 x 10,000 = ______

74. = 74

16
0.0971 x 100 = ______

9.71

17
46.31 x 1,000 = ______

46,310

18
1.293 x 10,000 = ______

12,930

19
0.039 x 10 = ______

0.39

20
100 x 2.0398 = ______

203.98

FEEDBACK UNIT 3

This quiz reviews the preceding unit. Answers are at the back of the book. Begin the next unit only after all problems of this unit are understood.

Multiply.

1. 300 x 2.5
2. 4.6 x 7.9
3. 0.97 x 0.08
4. 2,000 x 0.12
5. 1.03 x 7.9
6. 4,028 x 0.906
7. 0.056 x 0.97
8. 0.28 x 1.829
9. 0.28 x 10
10. 0.278 x 100
11. 3.2784 x 1,000
12. 0.00829 x 10,000

UNIT 4: WORD PROBLEMS REQUIRING MULTIPLICATION

THE CLUE WORD "EACH" FOR MULTIPLICATION

In the problem, "What is the cost of 120 books if each book costs $4.95?" the word "each" gives the clue to multiply to find the answer.

120 x $4.95 = $594.00 total cost

1
The word "each" often indicates a multiplication situation. If 250 test tubes hold 102.5 milliliters (ml) each, the total number of milliliters that the test tubes can hold is the product of 250 and 102.5. How many milliliters can the tubes hold? ______

25,625 ml

2
If 4,000 razor blades weigh 0.129 grams each, what is the total weight of the blades? ______

516.000 = 516 gm

3
If each theater ticket costs $3.50, how much do 2,000 cost? ______

$7,000.00

4
If the fee for each dollar value of a travelers check is $0.0075, what is the fee for $300 worth of travelers checks? ______

$2.2500 or $2.25

5
A bank charges $0.11 for each dollar loaned for one year. What will it charge for loaning $12,000 for one year? ______

$1,320.00

6
A nurse has to find the amount of aspirin in 300 tablets that contain 0.05 grams of aspirin each. Find the weight of aspirin in the tablets. ______ 15 gm

7
A test showed that each liter of air in a person's lung contained 0.09 mg of pollutants. If Sue's lungs hold 5 liters of air, find the amount of pollutants in her lungs. ______ 0.45 mg

"OF" AS A CLUE FOR MULTIPLICATION

To find "A fraction of a number," the fraction is always multiplied by the number.

Decimal numerals are fractions and any sentence of the form

"A decimal numeral of a number"

indicates that the decimal numeral should be multiplied by the number.

8 ✍
To find "0.8 of 53," the word "of" indicates that 0.8 should be multiplied by 53.
Find 0.8 of 53. ______ 42.4

9
Find 0.3 of 4.8 by multiplying the numbers. ______ 1.44

10
Find 0.079 of 11,000. ______ 869.000 = 869

11
Find 0.0037 of 4.6. ______ 0.01702

12
In the problem
"John lost .34 of his $300. How much money did he lose?"
the word "of" indicates that 0.34 and 300 should be ____________.

multiplied

13
A general estimates that .35 of his 2,000 rifles are defective. How many rifles are defective? ______

.35 x 2000 = 700

14
An engineer has .42 of his 40 day project completed. How many days are completed? ______

.42 x 40 = 16.8

15
0.05 of a vat of beer is sediment. The vat holds 5,000 liters. The "of" phrase "0.05 of a vat of beer" is equivalent to the phrase
0.05 of 5,000 liters
because the vat holds ______ liters.

5,000

16
1.2 of the cost of a refrigerator is the sale price. The cost is $300. From the information in the two previous sentences the "of" phrase
"1.2 of $300"
can be made. The words "the cost" in the first sentence were replaced by ______.

$300

17
0.6 of the people seeing a doctor have no physical problems. A doctor sees 60 people. The phrase "0.6 of the people" can be replaced by the "of" phrase: 0.6 of ______.

60 people

18
0.875 of the volume of milk is water. A tank truck holds 20,000 liters of milk. The phrase "0.875 of the volume of milk" can be replaced by the "of" phrase ________________________.

0.875 of 20,000 liters

19
Ariane has 300 ml of a medicine. 0.3 of the medicine is a solvent. Using the information in the two sentences above, the phrase "0.3 of the medicine" can be replaced by the "of" phrase ______.

0.3 of 300 ml

20
0.095 of the price of a car goes toward the administrative expense of the company. The price of a car is $5,900. $5,900 can be substituted for "price" in the phrase

0.095 of the price

to obtain the "of" phrase ______.

0.095 of $5,900

21
0.116 of the amount of a mortgage is the interest on the mortgage for one year. The mortgage is for $30,000. How much is the interest? ______

0.116 of $30,000
$3480.00

22
A microwave oven can cook a roast in 0.6 of the time needed by an electric oven. An electric oven cooks a roast in 3 hours. How much time does the microwave oven need to cook the roast? ______

0.6 of 3 hours
1.8 hours

23
0.21 of the atmosphere is oxygen. A closet contains 3 cubic meters of air. How much of the air is oxygen? ______

0.21 of 3 cu. m
0.63 cu. m of oxygen

24
The selling price of a book is $20.50. The printing cost is 0.12 of the selling price. Find the printing cost of the book. ______

0.12 of $20.50
$2.46

25
A diver's tank will hold 203.8 liters of air. 0.79 of the volume of the air is nitrogen. How much of the volume in the tank is nitrogen? ______

161.002 liters

26

0.8 of a bottle of medicine is codeine. The bottle contains 200 drams of medicine. How much codeine is in the bottle? ______

160 drams

27

A loan payment is 1.3 of the principal part of the payment. The principal part of the payment is $80. Find the size of the loan payment. ______

$104.00

28

The **area** of the **rectangle** shown at the right is the product of its length and width. Find the area. ______

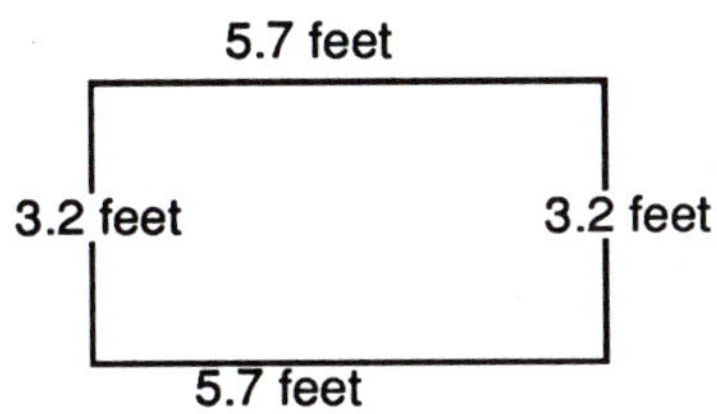

A = L x W
= 3.2 x 5.7
18.24 square feet is the area

29

The formula for finding the area of a **rectangle** is A = LW where L is the length of the rectangle and W is the width. Use the formula to find the area of the rectangle shown at the right. ______

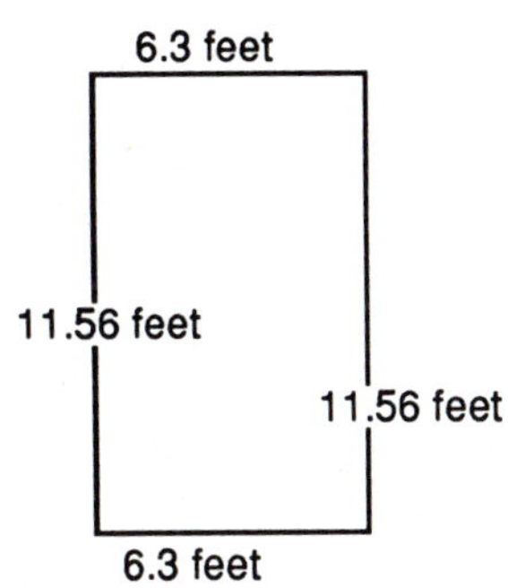

A = L x W
= 11.56 x 6.3
72.828 square feet is the area

30

Area is always measured in square units such as square inches, square feet, or square centimeters. Find the area of a rectangle with length 12.7 yards and width 4.5 yards.

______

A = L x W
= 12.7 x 4.5
57.15 square yards is the area

31
The formula for finding the area of a **triangle** is $A = \frac{1}{2}BH$ where B is the base of the triangle and H is the height. For the triangle shown at the right, the base is 1.7 meters and the height is _______.

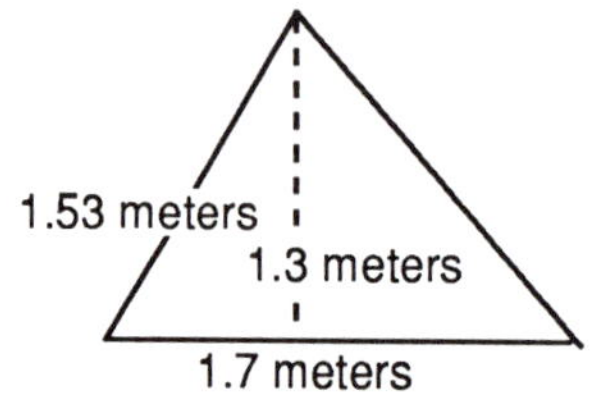

1.3 meters
1.53 meters is not the height; it is the length of a side

32
Use the formula $A = \frac{1}{2}BH$ and the fact that $\frac{1}{2} = .5$ to find the area of the triangle shown at the right. ______

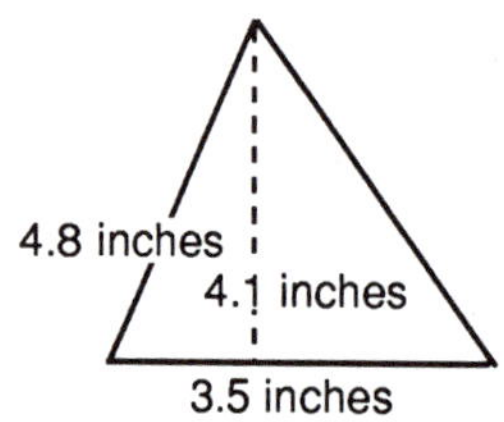

$A = \frac{1}{2} \times 3.5 \times 4.1$
$= .5 \times 3.5 \times 4.1$
7.175 square inches is the area

33
The formula for finding the **circumference** of a circle is $C = \pi D$ where π is the number PI (approximated as 3.14) and D is the diameter. For the circle shown at the right, the circumference is _______.

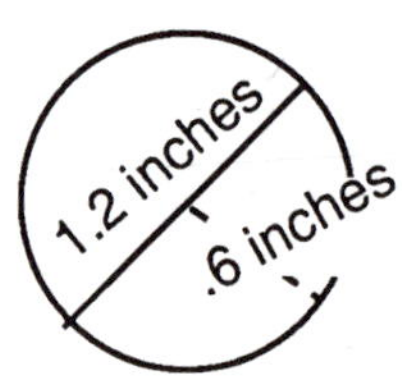

$C = \pi D$
$= 3.14 \times 1.2$
$= 3.768$
3.768 inches is the approximate circumference

34
The formula for finding the **area** of a circle is $A = \pi r^2$ where π is the number PI (approximated as 3.14) and r is the radius. For the circle shown at the right, the area is _______.

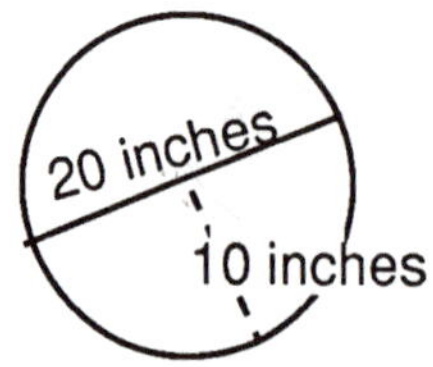

$A = \pi r^2$
$= 3.14 \times 10 \times 10$
$= 3.14 \times 100$
314 square inches is the approximate area

FEEDBACK UNIT 4

This quiz reviews the preceding unit. Answers are at the back of the book. Begin the next unit only after all problems of this unit are understood.

1. If 25 radios cost $9.50 each, what is the cost of all the radios?

2. Each cubic meter of gas weighs 1.56 kgm. A tank holds 40 cubic meters of the gas. Find the weight of the gas in the tank.

3. A department store charges $0.015 a month for each dollar it loans. Janie borrows $300 to buy a set of silver. How much is the interest on the loan for one month?

4. Sam buys 24 peaches. They cost $0.12 each. How much will he pay for the peaches?

5. A processed meat product weighs 2.5 kgm. Each kgm of the product is allowed to have 0.0009 kgm of insect parts. How many kgm of insect parts could be in the product?

6. 0.3 of 8 = ______

7. 1.6 of 500 = ______

8. 0.2 of a tablet is caffeine. The tablet weighs 5 milligrams. How much caffeine is in the tablet?

9. The price of brand X deodorant is 1.13 of the price of Snickle deodorant. Snickle costs $2 per can. What is the price of brand X?

10. In a genetics experiment, 200 brown rats were used. 0.87 of the rats had brown eyes. How many rats had brown eyes?

11. Sam is to receive a raise of 0.15 of his salary for the previous year. Find the size of the raise if his salary last year was $22,000.

12. Find the area of the rectangle shown below.

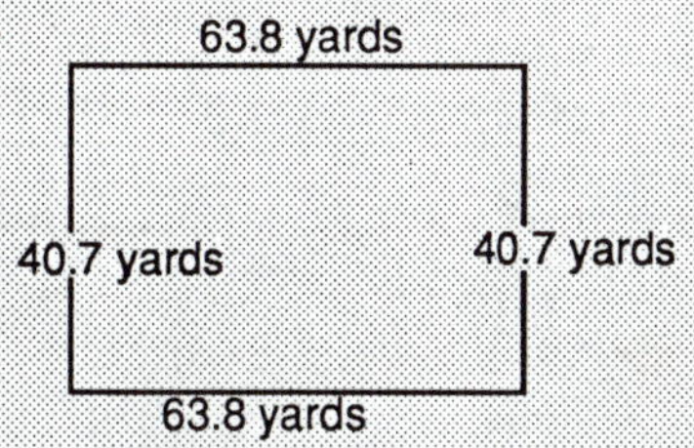

13. Find the area of the triangle shown below.

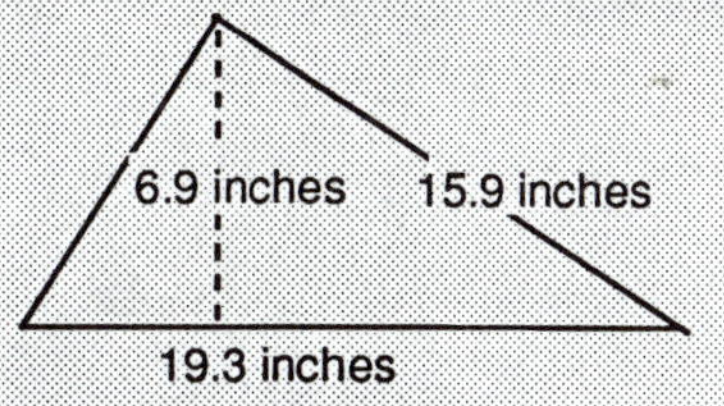

14. Find the circumference of the circle shown at the right. Use 3.14 as an approximation for π.

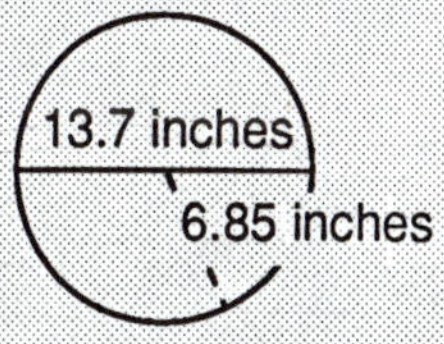

15. Find the area of the circle shown at the right. Use 3.14 as an approximation for π.

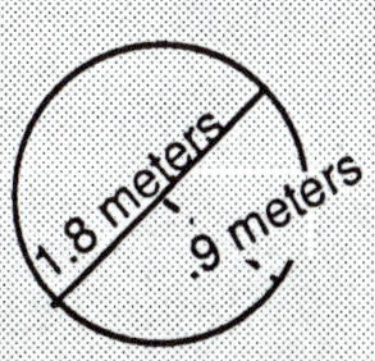

UNIT 5: DIVIDING A WHOLE NUMBER INTO A DECIMAL FRACTION

DIVIDING DECIMAL NUMERALS BY WHOLE NUMBERS

Success in this unit is dependent upon an ability to divide whole numbers. If review is needed, try the Mastery Test for Chapter 4.

The example at the right shows the division of 714 by 17.

In the example,
the divisor is 17
the dividend is 714
the quotient is 42
the remainder is 0.

```
      042
17 )714
   -0
    71
   -68
     34
    -34
      0
```

1
In the problem at the right
the divisor is ______
the dividend is ______
the quotient is ______

```
      0049
53 )2597
   -212
     477
    -477
       0
```

53

2,597

49

2
Divide 3,914 by 19 and show all your work. _____

```
      0206
19 )3914
   - 38
      11
     - 0
     114
    -114
       0
```

3

Divide 58,546 by 73 and show all your work. _____

```
     00802
 73 )58546
    -584
       14
      -0
      146
     -146
        0
```

4

Divide the problem at the right by:

1) Placing the decimal in the quotient directly above its position in the dividend.
2) Dividing as if the numbers were whole numbers.

```
6 )4.734
```

```
   0.7 8 9
6 )4.7 3 4
  -4 2
     5 3
    -4 8
       5 4
      -5 4
         0
```

DIVIDING A DECIMAL FRACTION BY A WHOLE NUMBER

To divide 1.32 (a decimal fraction) by 6 (a whole number):

1) Place the decimal point in the quotient space directly above the decimal point in 1.32.

```
     .
76 )1.32
```

2) Complete the division as if dividing 132 by 6.

```
   0.2 2
6 )1.3 2
  -1 2
     1 2
    -1 2
       0
```

5

In the problem at the right, place the decimal point correctly in the quotient. ______

```
2 )12.6
```

```
      .
2 )12.6
```

6
Complete the division of 12.6 by 2 by dividing as if the problem were 126 divided by 2. ______

$2\overline{)12.6}$ (decimal point placed above)

$06.3 = 6.3$

7
The first step in dividing 1.005 by 5 is to place the decimal point in the quotient. Complete this first step. ______

$5\overline{)1.005}$ (decimal point placed above)

8
Complete the division of 1.005 by 5 by dividing as if the problem were $1005 \div 5$. ______

$5\overline{)1.005}$ (decimal point placed above)

$$\begin{array}{r} .201 \\ 5\overline{)1.005} \end{array}$$

9 ✍
The problem $0.0025 \div 5$ is prepared for division by placing the decimal point in the quotient. Prepare $0.0025 \div 5$ for division. ______

$5\overline{)0.0025}$ (decimal point placed above)

10
Complete the division of 0.0025 by 5 by dividing as if the problem were $00025 \div 5$. ______

$$\begin{array}{r} 0.0005 \\ 5\overline{)0.0025} \end{array}$$

11
Prepare $25.32 \div 4$ for division by placing the decimal point in the quotient. ______

$4\overline{)25.32}$ (decimal point placed above)

12
Complete the division $25.32 \div 4$ by dividing as if the numbers were whole numbers. ______

$$\begin{array}{r} 06.33 \\ 4\overline{)25.32} \end{array} = 6.33$$

13
Divide 37.8 by 2.
First prepare the problem for division. Then divide as if the numbers were whole numbers. ______

$$\begin{array}{r} 18.9 \\ 2\overline{)37.8} \end{array}$$

14
Divide 3.12 by 3.
First prepare the problem for division.
Then divide as if the numbers
were whole numbers. ______

$3\overline{)3.12}$ = 1.04

15
Divide 212.4 by 3. ______

$3\overline{)212.4}$ = 070.8 = 70.8

16
47.24 ÷ 4 = ______

$4\overline{)47.24}$ = 11.81

17
30.1 ÷ 7 = ______

$7\overline{)30.1}$ = 04.3 = 4.3

18
121.5 ÷ 9 = ______

$9\overline{)121.5}$ = 013.5 = 13.5

19
0.008 ÷ 4 = ______

$4\overline{)0.008}$ = 0.002 = .002

20
1.82 ÷ 13 = ______

$13\overline{)1.82}$ = 0.14 = .14

21
47.5 ÷ 19 = ______

$19\overline{)47.5}$ = 02.5 = 2.5

22
1.638 ÷ 39 = ______

$39\overline{)1.638}$ = 0.042 = .042

23

93.6 ÷ 78 = ______

$$78\,\overline{)93.6}\quad 01.2 = 1.2$$

24

2.065 ÷ 59 = ______

$$59\,\overline{)2.065}\quad 0.035 = .035$$

AFFIXING ZEROS TO THE DIVIDEND

The decimal numeral 4.35 is equal to 4.350 or 4.3500. Zeros may be affixed to decimal numerals like 4.35 without changing their value.

In dividing a decimal numeral, zeros may be affixed to the dividend to extend the number of decimal places in the quotient.

25

Divide 0.01 by 5. Affix zeros to 0.01 until a zero remainder is obtained. ______

$$5\,\overline{)0.010}\quad 0.002 = .002$$

26

Divide 1.5 by 6. Affix zeros to 1.5 until a zero remainder is obtained. ______

$$6\,\overline{)1.50}\quad 0.25 = .25$$

27

Divide 1.8 by 16. Affix zeros to 1.8 until a zero remainder is obtained. ______

$$16\,\overline{)1.8000}\quad 0.1125$$

28

Divide 0.14 by 4 until a zero remainder is obtained. ______

$$4\,\overline{)0.140}\quad 0.035$$

29

Divide 0.02 by 125 until a zero remainder is obtained. ______

$$125\,\overline{)0.02000}\quad 0.00016$$

30

Is 426.000 equal to 426? ______

Yes

31
Divide 426 by 80 until a zero remainder is obtained. ______

$$80\,\overline{)426.000}\quad 005.325$$

CHANGING A FRACTION TO A DECIMAL FRACTION

The fraction $\frac{3}{16}$ can be changed to a decimal fraction using long division. The steps in the process are listed below.

1) Write $\frac{3}{16}$ as a long division problem. $\frac{3}{16} = 3 \div 16$. $\quad 16\,\overline{)3}$

2) Affix a decimal point and zeros to the right of the dividend, 3. $\quad 16\,\overline{)3.0000}$

3) Complete the long division. $\quad 16\,\overline{)3.0000}$ with quotient .1875

32
Change $\frac{5}{8}$ to a decimal fraction by dividing 5 by 8 until a zero remainder is obtained. ______

$$8\,\overline{)5.000}\quad 0.625$$

33
Change $\frac{7}{20}$ to a decimal fraction by dividing 7 by 20 until a zero remainder is obtained. ______

$$20\,\overline{)7.00}\quad 0.35$$

34
Change $\frac{13}{4}$ to a decimal fraction. ______

$$4\,\overline{)13.00}\quad 3.25$$

35
Change $\frac{5}{16}$ to a decimal fraction. ______

$$16\,\overline{)5.0000}\quad 0.3125$$

36
Change $\frac{17}{4}$ to a decimal fraction. ______

$$4\,\overline{)17.00}\quad 4.25$$

FEEDBACK UNIT 5

This quiz reviews the preceding unit. Answers are at the back of the book. Begin the next unit only after all problems of this unit are understood.

Divide until a zero remainder is obtained.

1. $0.00765 \div 15$
2. $0.075 \div 3$
3. $0.8 \div 4$
4. $0.0625 \div 25$
5. $80.08 \div 16$
6. $4.07 \div 50$
7. $28.38 \div 48$
8. $24.84 \div 36$

Change each fraction to a decimal fraction.

9. $\frac{3}{8}$
10. $\frac{42}{16}$

UNIT 6: DIVIDING A DECIMAL FRACTION BY A DECIMAL FRACTION

DIVIDING WITH A DECIMAL FRACTION DIVISOR

In the last unit, problems with whole number divisors were studied. In this unit, the divisors are decimal numerals. The method for solving these problems is to change the divisor to a whole number and then divide using the whole number divisor. This is accomplished in the following steps as illustrated in the example shown at the right: 0.048)53.76

1) Move the decimal point in the divisor to the right until the divisor is a whole number. 0048.)53.76

2) Move the decimal point in the dividend to the right the same number of places it was moved in the divisor. This may require the affixing of some zeros. 0048.)53760.

3) Since the divisor is now a whole number, the problem is completed as it was learned in the preceding unit.

$$\begin{array}{r} 1120. \\ 48\,\overline{)53760.} \end{array}$$

1
In the decimal numeral 0.048 there are three decimal places. Move the decimal point in 0.048 three places to the right. What numeral is obtained?
______ 0048.

2
0.048 is not a whole number.
Is 0048. a whole number? ______ Yes, 48

3
Move the decimal point in 53.62 three places to the right (affixing a zero if necessary). ______ 53620.

4
In the problem at the right, change the divisor to a whole number by moving its decimal point two places to the right. Also, move the decimal point in the dividend two places to the right. ______

$0.06\overline{)5.316}$

$6\overline{)531.6}$

5 ✍
Prepare the problem at the right for long division by moving the decimal points in both the divisor and dividend. Place the decimal point in the quotient. ______

$0.4\overline{)2.172}$

$4\overline{)21.72}$

6
Prepare the problem at the right for long division by moving the decimal points in both the divisor and dividend. Place the decimal point in the quotient space. ______

$0.0025\overline{)0.36}$

$25\overline{)03600.}$

7
Divide 0.36 by 0.0025. ______

$\begin{array}{r} 144 \\ 25\overline{)03600.} \end{array}$

8
Divide 1.272 by 5.3. ______

$\begin{array}{r} 00.24 \\ 53\overline{)12.72} \end{array}$

9
Divide 54.4 by 0.008. ______

$\begin{array}{r} 6800. \\ 8\overline{)54400.} \end{array}$

10
Divide 16.054 by 4.6 until a zero remainder is obtained. ______

$\begin{array}{r} 3.49 \\ 46\overline{)160.54} \end{array}$

11
Divide 10.816 by 0.52 until a zero remainder is obtained. ______

$52\overline{)1081.6}$ = 20.8

12
Divide 400 by 0.02 by first changing the divisor to a whole number. ______

$2\overline{)40000.}$ = 20000.

13
Divide 320 by 0.16. ______

$16\overline{)32000.}$ = 2000.

14
Divide 4.256 by 0.2. ______

$2\overline{)42.56}$ = 21.28

15
Divide 5.25 by 0.005. ______

$5\overline{)5250.}$ = 1050.

16
Divide 1.5762 by 0.37. ______

$37\overline{)157.62}$ = 4.26

17
Divide 17.34 by 0.085. ______

$85\overline{)17340.}$ = 204.

FEEDBACK UNIT 6

This quiz reviews the preceding unit. Answers are at the back of the book. Begin the next unit only after all problems of this unit are understood.

Divide until a zero remainder is found.

1. $0.6 \div 0.5$
2. $8.2 \div 0.016$
3. $714 \div 1.25$
4. $4.1 \div 0.0082$
5. $0.04536 \div 7.2$
6. $2.8458 \div 0.93$

UNIT 7: APPROXIMATING QUOTIENTS

ROUNDING OFF DECIMAL DIVISION QUOTIENTS

All previous examples in this chapter had zero remainders when the division was carried to enough decimal places. For some problems the long division of decimals will never arrive at a zero remainder.

Division answers (quotients) are often approximated by rounding off the answer at a particular number of decimal places. The example below illustrates the procedure for rounding off the answer of $1 \div 3$ at two decimal places:

1) Since the quotient is to be rounded off at two places, the division of 1 by 3 is carried to three decimal places.

2) The decimal 0.333 has the digit "3" in its third decimal place and is rounded off to 0.33

```
     .3 3 3
 3 )1.0 0 0
   -9
     1 0
    -9
       1 0
      -9
         1
```

If the digit in the third place is less than 5 then it is dropped and the second digit remains unchanged.

$$0.333 \doteq 0.33$$

The symbol "$\doteq$" means "approximately equal to."

1
Divide 2 by 3 and carry the division to three decimal places. ______

$3\overline{)2.000}$ with quotient 0.666

2
The answer to $2 \div 3$ carried to three decimal places is 0.666. Round it off to two decimal places. Use the rule that if the digit in the third place is 5 or more then one is added to the digit in the second place. ______

$0.666 \doteq 0.67$

3
If 2 is divided by 3 the remainder zero ______ (does, does not) appear after dividing to three decimal places.

does not

4
Divide 8.3 by 7 and carry the division to three decimal places. ______

$7\overline{)8.300}$ = 1.185

5
The answer to $8.3 \div 7$ carried to three decimal places is 1.185. Round off 1.185 to two decimal places. ______

$1.185 \doteq 1.19$

6
If 8.3 is divided by 7 the remainder zero ______ (does, does not) appear after dividing to three decimal places.

does not

7 ✍
The quotient of $2 \div 3$ rounded off to two decimal places is 0.67 and the quotient of $8.3 \div 7$ rounded off to two decimal places is 1.19. To round off a division answer to two decimal places the division should be carried out to ______ (how many) decimal places.

three

8
Round off the quotient of $5 \div 7$ to two decimal places by first dividing 5 by 7 and carrying the division to three decimal places. ______

$7\overline{)5.000}$ = 0.714 $\doteq 0.71$

9
Divide 4.1 by 7 to three decimal places. Then round off the answer to two places. ______

$7\overline{)4.100}$ = 0.585 $\doteq 0.59$

10
Divide 6.2 by 6 to three decimal places. Then round off the answer to two places. ______

$6\overline{)6.200}$ = 1.033 $\doteq 1.03$

11
Divide 8.5 by 4.6 to three decimal places. Then round off the answer to two places. ______

$46.\overline{)85.000}$ = 1.847 ≐ 1.85

12
Divide 0.56 by 0.015 to three decimal places. Then round off the answer to two places. ______

37.333 ≐ 37.33

13
To round off a quotient to two decimal places the division is carried out to ______ (how many) places.

3

14
Round off the quotient of 13.2 ÷ 5.1 to two decimal places. ______

2.588 ≐ 2.59

15
Round off the quotient of 4.28 ÷ 0.75 to two decimal places. ______

5.706 ≐ 5.71

16
Approximate $\frac{9}{11}$ with a two-place decimal by dividing 9 by 11. ______

$11\overline{)9.000}$ = 0.818 ≐ 0.82

17
Approximate $\frac{5}{7}$ with a two-place decimal. ______

$7\overline{)5.000}$ = 0.714 ≐ 0.71

18
Rounding off at two decimal places means carrying the division to three decimal places. Rounding off at three decimal places means carrying the division to ______ (how many) decimal places.

4

19
Carry the division of 0.014 by 6.2 to four decimal places and round off the quotient to three decimal places. ______

$62.\overline{)00.1400}$ = 0.0022 ≐ 0.002

20
Round off 9.132 ÷ 0.081 at one decimal place by first carrying the division to two decimal places. ______

$0081.\overline{)9132.00}$ = 112.74 ≐ 112.7

21
Round off 8.57 ÷ 3 at two decimal places. ______

$3\overline{)8.570}$ = 2.856 ≐ 2.86

22
Round off 47 ÷ 5.3 at three decimal places. ______

$53.\overline{)470.0000}$ = 8.8679 ≐ 8.868

23
Round off 4.126 ÷ 0.8 at one decimal place. ______

$08.\overline{)41.26}$ = 5.15 ≐ 5.2

24
Approximate $\frac{2}{9}$ with a three-place decimal by dividing 2 by 9. ______

$9\overline{)2.0000}$ = 0.2222 ≐ 0.222

25
Approximate $\frac{17}{14}$ with a two-place decimal. ______

$14\overline{)17.000}$ = 1.214 ≐ 1.21

FEEDBACK UNIT 7

This quiz reviews the preceding unit. Answers are at the back of the book. Begin the next unit only after all problems of this unit are understood.

In problems 1 through 4 round off the quotients to two decimal places.

1. $1.3 \div 46$
2. $4 \div 0.6$
3. $0.0035 \div 0.14$
4. $8.945 \div 7.2$

In problems 5 through 8 round off the quotients to three decimal places.

5. $0.17 \div 5.6$
6. $0.0049 \div 0.13$
7. $4.57 \div 0.27$
8. $84.34 \div 0.09$

Approximate each fraction in problems 9-12 with a two-place decimal.

9. $\frac{4}{17}$
10. $\frac{4}{7}$
11. $\frac{15}{23}$
12. $\frac{5}{32}$

UNIT 8: WORD PROBLEMS REQUIRING DIVISION

SOLVING WORD PROBLEMS USING DIVISION OF DECIMALS

Word problems using the word "each" frequently are solved by division.

When numbers are given for the total and for each item that makes up the total, the word problem is solved by dividing the "total" number by the "each" number.

("total" number) ÷ ("each" number) = answer

1
Beer is sold by the case or by the can, but the cost is the same. If each can costs \$.60 and a case (total) costs \$14.40, find the number of cans in a case. ______

$14.40 \div .60 = 24$ cans

2
Homer has a loan at the bank which costs him $.13 interest each day. When his total interest is $4.55 how many days have passed? ______

4.55 ÷ .13 = 35 days

3
A pile of coins weighs 243.6 grams (gm). If each coin weighs 1.2 gm, how many coins are in the pile? ______

243.6 ÷ 1.2 = 203 coins

4
A carnival sold tickets for $.75 each. At the end of the performance the box office had a total of $51.00. How many tickets had been sold? ______

51 ÷ .75 = 68 tickets

5
A manufacturing plant uses 4.2 cm of wire on each item. If 3,549 cm of wire are used in a day, how many items were manufactured? ______

3549 ÷ 4.2 = 845 items

6
A new carpet is purchased for $9 a square yard. The total bill was $725.40. How many square yards were purchased? ______

725.40 ÷ 9 = 80.6 sq yds

AVERAGES ARE FOUND BY USING DIVISION

To find the average of a group of numbers, first add the numbers and then divide the sum by the number of addends.

(total or sum) ÷ (number of addends) = average

If the division does not end with a zero remainder then round off the quotient at two decimal places.

7
Tom has taken four math tests and received scores of 92, 85, 67, and 82. Find his average by adding the numbers and dividing by 4. ______

326 ÷ 4 = 81.5

8
The seven linemen on the football team weigh 183 lbs., 202 lbs, 175 lbs, 210 lbs, 194 lbs, 183 lbs, and 198 lbs. Find the average weight by adding the weights and dividing by 7. ______ 1345 ÷ 7 = 192.14 lbs

9
Janice earned three pay checks over her Christmas vacation of $75.63, $81.48, and $95.14. What was her average pay check? ______ 252.25 ÷ 3 = $84.08

10
Scott took a 5-day trip and registered his miles traveled each day. His miles were 450, 292, 525, 139, and 387. What was his average miles traveled each day? ______ 1793 ÷ 5 = 358.6 miles

11
Six runners ran in a 3,000 meter race. Their times, in minutes, were 10.3, 12.2, 19.6, 15.4, 14.7, and 11.8. What was the average time of the runners? ______ 84.0 ÷ 6 = 14.0 min

12
The pollution in a stream was measured on five consecutive days. The measures, in grams per liter, were 14, 8, 25, 43, and 38. Find the average pollution each day. ______ 128 ÷ 5 = 25.6 gm per liter

13
Sara mailed 45 packages at a total cost of $37.52. What was the average mailing cost per package? ______ 37.52 ÷ 45 ≐ $.83

14
Seven passengers boarded a plane. Their total weight was 1,083 pounds. What was the average weight per passenger? ______ 1083 ÷ 7 ≐ 154.71 lbs

15
Twelve packages weighing a total of 369 lbs were loaded on a plane. What was the average weight per package? ______

$369 \div 12$
$= 30.75$ lbs

16
Sally earned \$78, \$93, \$150, \$225, and \$392 on successive days as a waitress. What was her average earnings per day? ______

\$938 $\div$ 5 = \$187.60

FEEDBACK UNIT 8

This quiz reviews the preceding unit. Answers are at the back of the book. Take the chapter mastery test only after all problems of this unit are understood.

1. A small theater sold tickets for \$.63 each, tax included, and collected a total of \$569.52. Find the number of tickets sold.

2. The heights of the members of the basketball team were, in meters, 1.84, 2.10, 1.92, 1.89, and 1.95. Find the average height of a player.

3. The druggist had sales of \$243.85 last Monday from 95 separate purchases. What was the average purchase price?

4. Smith bought 6.5 square yards of carpet for \$54.60. What was the price for each square yard of carpet?

CHAPTER 10 MASTERY TEST

This test provides an opportunity for assuring that **all** of the objectives of Chapter 10 have been learned. Each problem is accompanied by the symbol (U) which indicates the Unit in which that type of problem was studied. If any problems are missed, they need to be corrected. Answers are at the back of the book.

For problems 1-4, multiply. (3)

1. 0.57 x 4.3

2. 9.643 x 0.78

3. 105.6 x 0.094

4. 0.0789 x 100

5. Farmer Hank discovered that 0.43 of his grain had spoiled. Of his 60,000 bushels of wheat, how much had spoiled? (4)

6. A pharmacist needs 25 doses of phenobarbital. Each dose is to be 5.72 milligrams. How much of this medicine is needed to fill the order? (4)

7. The National Bank charges $0.115 per year for each dollar it loans. Burt and Lonny need a $30,000 loan. How much will the bank charge to borrow $30,000 for one year? (4)

8. The fuel tank of a car holds 12 gallons. If the manufacturer uses thinner oil in the gearbox, the car will save 0.04 of its fuel. How much fuel from a full tank will be saved by using the lighter oil? (4)

For problems 9 and 10, divide until a zero remainder is obtained.

9. 0.1813 ÷ 7 (5)

10. 4.89 ÷ 0.06 (6)

For problems 11 and 12, divide and round off at two decimal places. (7)

11. 95.8 ÷ 83

12. 38.46 ÷ 0.65

13. Round off 0.785 ÷ 4.7 at three decimal places. (7)

Approximate the fractions of problems 14-17 with a two-place decimals. (7)

14. $\frac{5}{13}$

15. $\frac{12}{7}$

16. $\frac{9}{16}$

17. $\frac{47}{85}$

18. Christmas cards are on sale for \$.18 each. How many can Pete buy with \$16.92? (8)

19. Mariah received four orders for \$45.92, \$87.50, \$19.95, and \$32.10. What was the average of these four orders? (8)

20. Find the area of the rectangle shown below. (4)

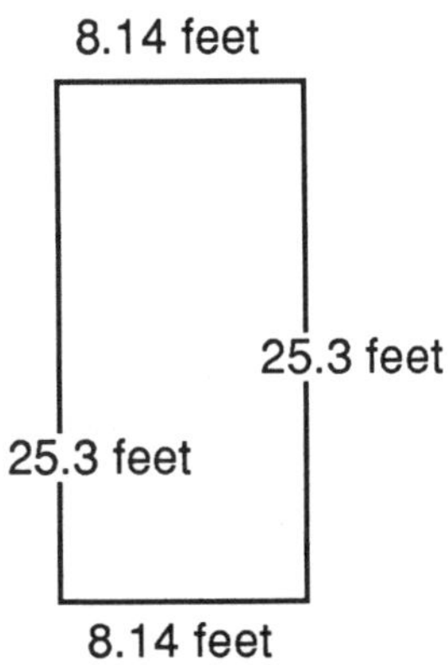

21. Find the area of the triangle shown below. (4)

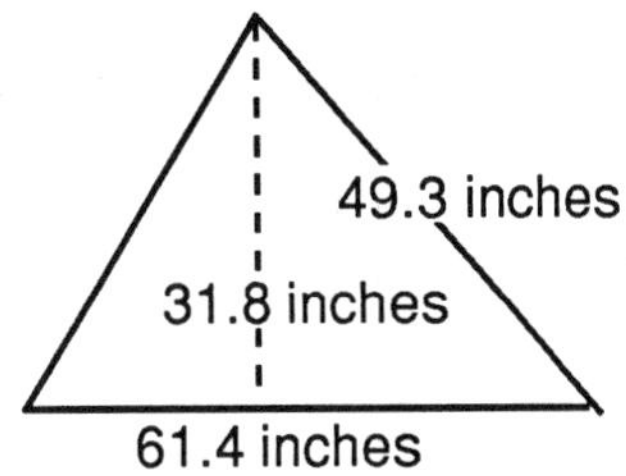

22. Find the circumference of the circle shown below. Use 3.14 as an approximation for π. (4)

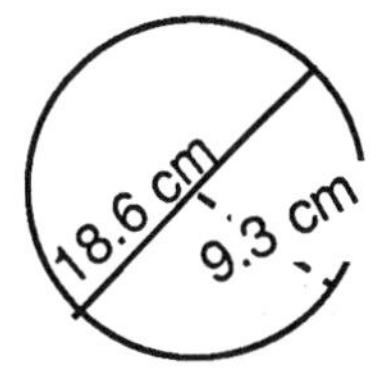

23. Find the area of the circle shown below. Use 3.14 as an approximation for π. (4)

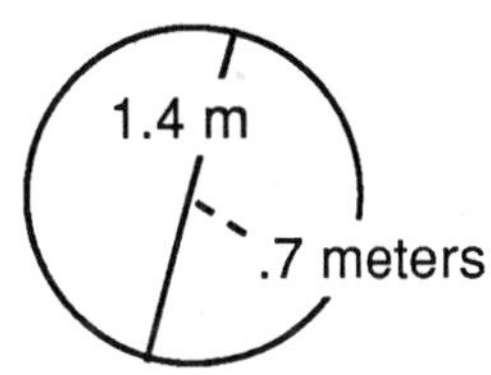

Chapter 11 Objectives

The following problems describe, by example, the skills and knowledge studied in this chapter. Each problem is accompanied by a symbol, (U), which indicates the Unit where the problem is studied. Answers are at the back of the book.

Show the following ratios in simplest terms.

1. 4 to 12 (1)

2. 26 to 13 (1)

3. 22 to 32 (1)

4. 86 kilometers to 2 liters (1)

5. Solve $\frac{n}{8} = \frac{33}{22}$ (4)

6. Solve $\frac{8}{n} = \frac{4}{5}$ (4)

7. Solve $\frac{36}{32} = \frac{45}{n}$ (4)

8. Solve $\frac{6}{5} = \frac{n}{12}$ (4)

9. A magazine kept the ratio of advertising to story copy in a 4 to 5 ratio. Find the amount of advertising that would be used with 60 pages of story copy. (5)

10. If rain occurs 8 times out of 21 windy days, how many rainy days should occur with 105 windy days? (5)

11. A nuclear plant inspector finds that a pressure guage will malfunction 3 times for every 100 days of inspection. Find the number of times it will malfunction in 267 days. (5)

12. A bank charges 11 cents for every \$2 loaned for 6 months. How much will be charged for a \$4,200 loan for 6 months? (5)

13. A 2 kgm sample of flour contained 1 mg of debris. Find the amount of debris that should be in 100 kgm of the flour. (5)

14. Three milligrams of a drug are to be given for 14 kilograms of body weight. Find the size of a person requiring 8 mg of the drug. (5)

CHAPTER 11
RATIOS AND PROPORTIONS

UNIT 1: THE MEANING OF A RATIO

RATIO

A ratio is a comparison of two numbers. Three ways to show the ratio of 2 to 3 are listed below.

2 to 3   2:3 $\frac{2}{3}$

This unit will be concerned with the first and third method of showing a ratio.

1 ✍

The ratio of 4 to 5 can be expressed as the fraction $\frac{4}{5}$. 4 is the numerator and 5 is the denominator. Show the ratio of 7 to 4 as a fraction. ______ $\frac{7}{4}$

2 ✍

The ratio of 7 to 6 is the fraction with the numerator 7 and the denominator 6. Show the ratio of 7 to 6.

______ $\frac{7}{6}$

3

The ratio of 3 to 11 is the fraction $\frac{3}{11}$. Which fraction, $\frac{8}{7}$ or $\frac{7}{8}$, is the ratio of 8 to 7? ______ $\frac{8}{7}$

4

Does $\frac{9}{6}$ show the ratio of 6 to 9? ______ No, 9 to 6

5

Show the ratio of 10 to 32 as a fraction. ______ $\frac{10}{32}$

6

The ratio of 28 to 18 is shown by the fraction ______. $\frac{28}{18}$

7

The ratio of 2.7 to 0.8 is shown by the fraction ______. $\frac{2.7}{0.8}$

8

The ratio of 8 to 7.8 is the fraction ______. $\frac{8}{7.8}$

9

When n and d represent numbers, the ratio of n to d is shown by the fraction ______. $\frac{n}{d}$

10

$\frac{11}{7}$ shows the ratio of 11 to 7.

$\frac{9}{23}$ shows the ratio of ______ to ______. 9 to 23

11

Does $\frac{0.4}{8}$ show the ratio of 0.4 to 8? ______ Yes

12
Does $\frac{1.2}{135}$ show the ratio of 135 to 1.2? ______

No, 1.2 to 135

13
Does $\frac{.008}{0.01}$ show the ratio of 0.008 to 0.01? ______

Yes

14
The ratio of 12 sophomores to 8 freshmen is $\frac{12}{8}$ which simplifies to $\frac{3}{2}$. Show the ratio of 4 books to $48 in simplest terms. ______

$\frac{1}{12} = \frac{4}{48}$

15
Show the ratio of 25 grams to 20 liters in simplest terms. ______

$\frac{5}{4} = \frac{25}{20}$

16
Show the ratio of 180 kilometers to 6 hours in simplest terms. ______

$\frac{30}{1} = \frac{180}{6}$

17
Show the ratio of 40 units to 80 kilograms. ______

$\frac{1}{2} = \frac{40}{80}$

18
Show the ratio of 210 cycles to 60 minutes. ______

$\frac{7}{2} = \frac{210}{60}$

19
Show the ratio of $20 to 5 liters of whisky. ______

$\frac{4}{1} = \frac{20}{5}$

20
Show the ratio of 4,000 particles to 8,000 particles.
______

$\frac{1}{2} = \frac{4000}{8000}$

21
Show the ratio of 4,000 millirems of radiation to 6 hours. ______

$\frac{2000}{3} = \frac{4000}{6}$

22

The ratio $\frac{0.1}{3.2}$ can be simplified to a fraction of whole numbers. To eliminate the decimal points, multiply both numerator and denominator by 10. The multiplier 10 is used because one digit is to the right of the decimal.

$$\frac{0.1}{3.2} = \frac{0.1}{3.2} \times \frac{10}{10} = \frac{1}{32}$$

Are both the numerator and denominator of $\frac{1}{32}$ whole numbers? _____

Yes

23

Simplify the ratio of 2.8 kilometers to 2 seconds by multiplying both numerator and denominator by 10. This will make the ratio a fraction of whole numbers.

$$\frac{2.8}{2} = \frac{2.8}{2} \times \frac{10}{10} = \frac{28}{20}$$

Complete the simplification of $\frac{2.8}{2}$ by simplifying $\frac{28}{20}$.

_____

$\frac{7}{5}$

24

Simplify the ratio of 3.8 grams to 1.25 liters by multiplying both numerator and denominator by 100. This will make the ratio a fraction of whole numbers.

$$\frac{3.8}{1.25} = \frac{3.8}{1.25} \times \frac{100}{100} = \frac{380}{125}$$

Complete the simplification of $\frac{3.8}{1.25}$ by simplifying $\frac{380}{125}$. _____

$\frac{76}{25}$

25

Complete the simplification of the ratio of 4 eggs to 1.25 cups of milk that is shown below.

$$\frac{4}{1.25} = \frac{4}{1.25} \times \frac{100}{100} = \_\_\_\_\_$$

$\frac{400}{125} = \frac{16}{5}$

26

Complete the simplification of the ratio of 1.362 mg to 4 liters that is shown below.

$$\frac{1.362}{4} = \frac{1.362}{4} \times \frac{1000}{1000} = \_\_\_\_\_$$

$\frac{1362}{4000} = \frac{681}{2000}$

27
Simplify the ratio of 0.3 cm to 5 kilometers.______

$\frac{0.3}{5} \times \frac{10}{10} = \frac{3}{50}$

28
Simplify the ratio of 7.5 kilometers to 0.5 centimeters. ______

$\frac{7.5}{0.5} \times \frac{10}{10} = \frac{15}{1}$

29
Simplify the ratio of 4.5 grams to 5.5 milliliters. ______

$\frac{4.5}{5.5} \times \frac{10}{10} = \frac{9}{11}$

30
Simplify the ratio of \$2.61 to \$2.

______ ______

$\frac{2.61}{2} \times \frac{100}{100} = \frac{261}{200}$

31
Simplify the ratio of 0.76 mg to 1.8 liters. ______

$\frac{.76}{1.8} \times \frac{100}{100} = \frac{19}{45}$

Feedback Unit 1

This quiz reviews the preceding unit. Answers are at the back of the book. Begin the next unit only after all problems of this unit are understood.

1. Does $\frac{7}{13}$ show the ratio of 7 to 13?
2. Show the ratio of 6 to 13.
3. Simplify the ratio of 1.3 to 7.
4. Simplify the ratio of 0.8 to 0.3.
5. Simplify the ratio of 8 to 10.
6. Simplify the ratio of 240 km to 2 hrs.
7. Simplify the ratio of 18 problems correct to 20 problems total.
8. Simplify the ratio of \$1,260 total price to 4 refrigerators.
9. Simplify the ratio of 3.4 grams to 0.5 cu. cc.
10. Simplify the ratio of 403 grams to 5.51 liters.
11. Simplify the ratio of 0.58 kilograms to 23 cu. cc.
12. Simplify the ratio of 4,756 kilometers to 2.3 hrs.

UNIT 2: THE MEANING OF PROPORTIONS

PROPORTIONS

A proportion is an equality of two ratios.

Three examples of proportions are given below.

$$\frac{2}{3} = \frac{4}{6} \qquad \frac{6}{4} = \frac{18}{12} \qquad \frac{100}{50} = \frac{34}{17}$$

1 ✍

Two ratios with an equal sign between them form a proportion.

$\frac{2}{3} = \frac{4}{6}$ is a proportion because it consists of two ratios, $\frac{2}{3}$ and $\frac{4}{6}$, with an equal sign between them.

Is $\frac{17}{12}$ a proportion? ______

No, it is one ratio

2

The statement $\frac{60}{100} = \frac{3}{5}$ is a proportion.

Is $\frac{4}{5} = \frac{8}{10}$ a proportion? ______

Yes

3 ✍

$\frac{5}{8} = \frac{15}{24}$ is a **true** proportion because $\frac{5}{8}$ is equal to $\frac{15}{24}$. Is $\frac{1}{2} = \frac{1}{3}$ a true proportion? ______

No, $\frac{1}{2}$ does not equal $\frac{1}{3}$

4

Is $\frac{600}{1000} = \frac{6}{10}$ a true proportion? ______

Yes

5 ✍

$\frac{5}{8} = \frac{12}{24}$ is a **false** proportion because $\frac{5}{8}$ is not equal to $\frac{12}{24}$. Is $\frac{1}{2} = \frac{2}{4}$ a false proportion? ______

No, $\frac{1}{2}$ equals $\frac{2}{4}$

6
Is $\frac{3}{5} = \frac{6}{10}$ a true proportion? ______ Yes

7 ✍
$\frac{n}{8} = \frac{1}{2}$ is a proportion with the letter n holding the place for a number. If n is replaced by 4, the proportion $\frac{4}{8} = \frac{1}{2}$ is obtained. Is the new proportion true? ______ Yes

8
$\frac{n}{8} = \frac{1}{2}$ is a proportion with the letter n holding the place for a number. If n is replaced by 3, the proportion $\frac{3}{8} = \frac{1}{2}$ is obtained. Is the new proportion true? ______ No

9 ✍
The proportion $\frac{n}{8} = \frac{1}{2}$ is neither true nor false. It becomes true or false when n is replaced by a number. If n is replaced by 5, the proportion $\frac{n}{8} = \frac{1}{2}$ is ______ (true, false). false

10
The proportion $\frac{n}{10} = \frac{4}{5}$ is neither true nor false. It becomes true or false when n is replaced by a number. If n is replaced by 8, the proportion $\frac{n}{10} = \frac{4}{5}$ is ______ (true, false). true

THE MEANS AND EXTREMES OF A PROPORTION

For the proportion $\frac{3}{8} = \frac{12}{32}$ the first numerator, 3, and the second denominator, 32, are called **extremes**.

The first denominator, 8, and the second numerator, 12, are called **means**.

11

For the proportion $\frac{8}{10} = \frac{4}{5}$ the **means** are 10 and 4, and the **extremes** are 8 and 5. The product of the means is 10 x 4 or 40. The product of the extremes is ______.

8 x 5 = 40

12

For the proportion $\frac{3}{4} = \frac{12}{16}$ the means are 4 and 12 and the extremes are 3 and 16. Find the product of the means and then compare it with the product of the extremes. ______

4 x 12 = 48
3 x 16 = 48
The products are equal

13

In the proportion $\frac{3}{12} = \frac{1}{4}$ the means are 12 and 1 and the extremes are ______ and ______.

3, 4

14

What are the extremes of the proportion $\frac{15}{18} = \frac{5}{6}$? ______

15, 6

15

Two ratios form a true proportion if the product of the extremes is equal to the product of the means. Is $\frac{12}{16} = \frac{4}{5}$ a true proportion? ______

No, 12 x 5 = 60
16 x 4 = 64

16

Is $\frac{5}{8} = \frac{15}{24}$ a true proportion? ______

Yes, 5 x 24 = 120
and 8 x 15 = 120

17

Decide if $\frac{7}{14} = \frac{9}{18}$ is a true proportion by comparing the products of the extremes and the means. ______

A true proportion
7 x 18 = 126
14 x 9 = 126

18
Is $\frac{3}{5} = \frac{4}{6}$ a true proportion? ______

No, $3 \times 6 = 18$
but $5 \times 4 = 20$

19
Is $\frac{13}{52} = \frac{1}{4}$ a true proportion? ______

Yes, $13 \times 4 = 52$
and $52 \times 1 = 52$

20
Is $\frac{3}{5} = \frac{39}{65}$ a true proportion? ______

Yes, $3 \times 65 = 195$
and $5 \times 39 = 195$

21
Is $\frac{4}{7} = \frac{21}{35}$ a true proportion? ______

No, $4 \times 35 = 140$
and $7 \times 21 = 147$

22
Is $\frac{6}{51} = \frac{2}{17}$ a true proportion? ______

Yes, $6 \times 17 = 51 \times 2$

23
Is $\frac{4.5}{9} = \frac{3}{6}$ a true proportion? ______

Yes, $4.5 \times 6 = 9 \times 3$

24
Is $\frac{3}{9} = \frac{6.6}{20}$ a true proportion? ______

No, $3 \times 20 \neq 9 \times 6.6$
[$\neq$ means "does not equal"]

25
$\frac{1}{2} = \frac{3}{6}$ is a true proportion because
$1 \times 6 = 2 \times 3$.
$\frac{3}{4} = \frac{6}{8}$ is a true proportion because ______________.

$3 \times 8 = 4 \times 6$

26
$\frac{2}{7} = \frac{11}{28}$ is a false proportion because
$2 \times 28 \neq 7 \times 11$.
$\frac{8}{9} = \frac{7}{10}$ is a false proportion because ______________.

$8 \times 10 \neq 9 \times 7$

27

$\frac{1}{6} = \frac{5}{30}$ is a _____ proportion because _______________.

true, $1 \times 30 = 6 \times 5$

28

$\frac{5}{10} = \frac{7}{14}$ is a _____ proportion because _______________.

true, $5 \times 14 = 10 \times 7$

29

$\frac{12}{8} = \frac{2}{3}$ is a _____ proportion because _______________.

false, $12 \times 3 \neq 8 \times 2$

30

$\frac{n}{6} = \frac{8}{12}$ will become a true proportion if $n \times 12 = 6 \times 8$.

$\frac{n}{6} = \frac{8}{12}$ will become a false proportion if _______________.

$n \times 12 \neq 6 \times 8$

31 ✍

$\frac{7}{n} = \frac{11}{14}$ will become a false proportion if $7 \times 14 \neq n \times 11$.

$\frac{8}{13} = \frac{9}{n}$ will become a true proportion if _______________.

$8 \times n = 13 \times 9$

32

$\frac{2}{9} = \frac{n}{18}$ will become a true proportion if _______________.

$2 \times 18 = 9 \times n$

33

$\frac{47}{n} = \frac{28}{33}$ will become a true proportion if _______________.

$47 \times 33 = n \times 28$

34

$\frac{7}{8} = \frac{14}{n}$ will become a true proportion if _______________.

$7 \times n = 8 \times 14$

35

$\frac{n}{19} = \frac{23}{12}$ will become a true proportion if ____________.

$n \times 12 = 19 \times 23$

36

$\frac{4}{n} = \frac{21}{47}$ will become a true proportion if ____________.

$4 \times 47 = n \times 21$

FEEDBACK UNIT 2

This quiz reviews the preceding unit. Answers are at the back of the book. Begin the next unit only after all problems of this unit are understood.

In problems 1-7 use the products of the extremes and means to determine if the ratios form a true proportion.

1. $\frac{1}{3} = \frac{13}{39}$
2. $\frac{3}{8} = \frac{15}{40}$
3. $\frac{28}{99} = \frac{2}{7}$
4. $\frac{9}{18} = \frac{13}{26}$
5. $\frac{11}{22} = \frac{15}{32}$
6. $\frac{2.5}{5} = \frac{2.4}{4.8}$
7. $\frac{4}{8} = \frac{0.2}{4}$
8. $\frac{n}{8} = \frac{2}{16}$ if ________ = ________
9. $\frac{37}{18} = \frac{n}{10}$ if ________ = ________
10. $\frac{16}{20} = \frac{5}{n}$ if ________ = ________
11. $\frac{7}{n} = \frac{1}{2}$ if ________ = ________
12. $\frac{3}{n} = \frac{7}{13}$ if ________ = ________

UNIT 3: SOLVING SIMPLE MULTIPLICATION EQUATIONS

SOLVING STATEMENTS SUCH AS $5 \times n = 75$

To **solve** $5 \times n = 75$ means to find the number that can replace n to make a true statement.

The equality $5 \times n = 75$ is called an **equation**. Every equation has two sides which are separated by the equals (=) sign.

The **left side** of the equation $5 \times n = 75$ is: $5 \times n$

The **right side** of the equation $5 \times n = 75$ is: 75

The most important principle to follow in solving equations is:

Always treat the two sides of an equation in exactly the same way. If a number is added/subtracted to the left side of an equation then it also must be added/subtracted to the right side of the equation. If a number is multiplied/divided by the left side of an equation then it also must be multiplied/divided by the right side of the equation.

1

To **solve** $2 \times n = 24$, find the number that can replace n and make a true statement. What number can be multiplied by 2 to give 24? ______

12; $2 \times 12 = 24$ is true

2

To **solve** $20 = 4 \times n$, find the number that can replace n and make a true statement. What number can be multiplied by 4 to give 20? ______

5; $20 = 4 \times 5$ is true

3

To **solve** $3 \times n = 18$, find the number that can replace n and make a true statement.
Solve $3 \times n = 18$. _____

6; $3 \times 6 = 18$ is true

4
Solve n x 5 = 15 by finding the number that can replace n and make a true statement. ___________

3; 3 x 5 = 15 is true

5
Solve 4 x n = 24. ______

6

6
Solve 38 = 2 x n. ______

19

7
Solve 3 x n = 12. ______

4

8
Solve n x 2 = 28. ______

14

9
Solve 36 = n x 9. ______

4

10
Solve 8 = 2 x n. ______

4

11
Solve n x 5 = 45. ______

9

12 ✍
To solve 5 x n = 42, it is necessary to find a number that is multiplied by 5 to give 42. Is the number a whole number? ______

No

13
The solution for 5 x n = 42 is the fraction $\frac{42}{5}$. What is the answer when 5 is multiplied by $\frac{42}{5}$?

______

42; $5 \times \frac{42}{5} = 42$

14

To solve $3 \times n = 20$ both sides of the equation are divided by 3.

a) $3 \times n$ divided by 3 is: $\frac{3 \times n}{3} = \frac{\overset{1}{\cancel{3}} \times n}{\underset{1}{\cancel{3}}} = n$

b) 20 divided by 3 is: $\frac{20}{3} = 6\frac{2}{3}$

The solution for $3 \times n = 20$ is ________. $\frac{20}{3}$ or $6\frac{2}{3}$

15

To solve $4 \times n = 27$ both sides of the equation are divided by 4.

a) $4 \times n$ divided by 4 is: $\frac{4 \times n}{4} = \frac{\overset{1}{\cancel{4}} \times n}{\underset{1}{\cancel{4}}} = n$

b) 27 divided by 4 is: $\frac{27}{4} = 6\frac{3}{4}$

The solution for $4 \times n = 27$ is ________. $\frac{27}{4} = 6\frac{3}{4}$

16

Solve $6 \times n = 31$ by dividing both sides of the equation by 6. ______ $\frac{31}{6} = 5\frac{1}{6}$

17

Solve $9 \times n = 21$ by dividing both sides of the equation by 9. ______ $\frac{21}{9} = 2\frac{1}{3}$

18

Solve $6 \times n = 5$ by dividing both sides of the equation by 6. ______ $\frac{5}{6}$

19

Solve $8 \times n = 33$. ______ $\frac{33}{8} = 4\frac{1}{8}$

20

Solve $6 \times n = 47$. ______ $\frac{47}{6} = 7\frac{5}{6}$

21
Solve $2 \times n = 19$. ______

$\frac{19}{2} = 9\frac{1}{2}$

22
Solve $7 \times n = 4$. ______

$\frac{4}{7}$

23
Solve $4 \times n = 19$. ______

$\frac{19}{4} = 4\frac{3}{4}$

24
Solve $8 \times n = 44$. ______

$\frac{44}{8} = 5\frac{1}{2}$

25
Solve $8 \times n = 6$. ______

$\frac{6}{8} = \frac{3}{4}$

26
Solve $6 \times n = 50$. ______

$\frac{50}{6} = 8\frac{1}{3}$

27 ✍
The equation $14 = 3 \times n$ can have its two sides interchanged and become $3 \times n = 14$. Solve $14 = 3 \times n$ by first interchanging its two sides. ______

$\frac{14}{3} = 4\frac{2}{3}$

28
The equation $16 = 5 \times n$ can be written as $5 \times n = 16$. Solve $5 \times n = 16$. ______

$\frac{16}{5} = 3\frac{1}{5}$

29
The equation $29 = 7 \times n$ can be written as $7 \times n = 29$. Solve $7 \times n = 29$. ______

$\frac{29}{7} = 4\frac{1}{7}$

30
The equation $10 = 15 \times n$ can be written as $15 \times n = 10$. Solve $15 \times n = 10$. ______

$\frac{10}{15} = \frac{2}{3}$

31
Solve $9 \times n = 13$. _____ $\frac{13}{9} = 1\frac{4}{9}$

32
Solve $32 = 7 \times n$. _____ $\frac{32}{7} = 4\frac{4}{7}$

33
Solve $12 \times n = 4$. _____ $\frac{4}{12} = \frac{1}{3}$

FEEDBACK UNIT 3

This quiz reviews the preceding unit. Answers are at the back of the book. Begin the next unit only after all problems of this unit are understood.

1. To solve $5 \times n = 18$:
   a. divide 5 by 18
   b. divide 18 by 5
2. To solve $8 = 11 \times n$:
   a. divide 8 by 11
   b. divide 11 by 8
3. Solve $12 = 2 \times n$.
4. Solve $17 \times n = 21$.
5. Solve $7 \times n = 63$.
6. Solve $6 \times n = 20$.
7. Solve $37 = 5 \times n$.
8. Solve $9 \times n = 4$.
9. Solve $16 = 7 \times n$.
10. Solve $8 \times n = 28$.

UNIT 4: SOLVING PROPORTIONS

THE PRODUCTS OF THE EXTREMES AND MEANS ARE EQUAL

From Unit 2, recall that a proportion is true only when the product of the extremes is equal to the product of the means.

To solve $\frac{7}{n} = \frac{9}{10}$,

1. Use the products of the extremes and means to write a multiplication equation.

$$\frac{7}{n} = \frac{9}{10}$$

$$7 \times 10 = n \times 9$$

2. Solve the multiplication equation as it was done in Unit 3.

$$70 = 9 \times n$$

$$\frac{70}{9} = \frac{9 \times n}{9}$$

$n = \frac{70}{9}$ or $7\frac{7}{9}$

1

To **solve** the proportion $\frac{n}{5} = \frac{11}{6}$ first write a new equation using the products of the extremes and means.

$\frac{n}{5} = \frac{11}{6}$ gives $n \times 6 =$ ________

5×11 or 55

2

$\frac{n}{5} = \frac{11}{6}$ gives $n \times 6 = 55$.

Solve $n \times 6 = 55$. _____

$\frac{55}{6}$ or $9\frac{1}{6}$

3 ✍

To **solve** the proportion $\frac{n}{7} = \frac{4}{9}$ first write a new equation using the products of the extremes and means. What is the new multiplication equation?

______________________.

$n \times 9 = 7 \times 4$

4

$\frac{n}{7} = \frac{4}{9}$ gives n x 9 = 28.

Solve n x 9 = 28. _____

$\frac{28}{9}$ or $3\frac{1}{9}$

5

Write the proportion $\frac{8}{n} = \frac{10}{13}$ as an equation using the products of the extremes and means.

_______________.

104 = n x 10

6

$\frac{8}{n} = \frac{10}{13}$ gives 104 = n x 10.

Solve 104 = n x 10. _____

$\frac{104}{10}$ or $\frac{52}{5}$ or $10\frac{2}{5}$

7

Solve $\frac{7}{11} = \frac{5}{n}$ by first writing an equation using the products of the extremes and means. _______________.

7 x n = 55

$\frac{55}{7}$ or $7\frac{6}{7}$

8

Solve $\frac{n}{8} = \frac{3}{5}$ by first writing an equation using the products of the extremes and means. _______________.

n x 5 = 24

$\frac{24}{5}$ or $4\frac{4}{5}$

9

Solve $\frac{n}{2} = \frac{3}{11}$ by first writing an equation using the products of the extremes and means. _______________.

n x 11 = 6

$\frac{6}{11}$

10

Solve $\frac{5}{9} = \frac{n}{3}$ by first writing an equation using the products of the extremes and means. _______________.

15 = 9 x n

$\frac{15}{9}$ or $1\frac{2}{3}$

11
Solve $\frac{6}{11} = \frac{9}{n}$ ________________.

$\frac{99}{6}$ or $16\frac{1}{2}$

12
Solve $\frac{8}{7} = \frac{1}{n}$ ________________.

$\frac{7}{8}$

13
Solve $\frac{6}{5} = \frac{1}{n}$ ________________.

$\frac{5}{6}$

14
Solve $\frac{4}{3} = \frac{n}{7}$ ________________.

$\frac{28}{3}$ or $9\frac{1}{3}$

15
Solve $\frac{12}{5} = \frac{2}{n}$ ________________.

$\frac{10}{12} = \frac{5}{6}$

FEEDBACK UNIT 4

This quiz reviews the preceding unit. Answers are at the back of the book. Begin the next unit only after all problems of this unit are understood.

1. Solve $\frac{5}{n} = \frac{8}{10}$
2. Solve $\frac{9}{14} = \frac{n}{7}$
3. Solve $\frac{6}{7} = \frac{5}{n}$
4. Solve $\frac{3}{n} = \frac{11}{2}$
5. Solve $\frac{n}{7} = \frac{3}{5}$
6. Solve $\frac{n}{8} = \frac{1}{15}$
7. Solve $\frac{5}{n} = \frac{2}{7}$
8. Solve $\frac{6}{11} = \frac{2}{n}$

UNIT 5: SOLVING WORD PROBLEMS WITH PROPORTIONS

A WORD PROBLEM PROCESS

In this section word problems are solved using proportions. Each problem involves a three step approach.

1. Write two ratios showing **similar** comparisons.
2. Write a proportion making the two ratios equal.
3. Solve the proportion.

1

The ratio of boys to girls in the sophomore class is 6 to 5. The way of showing this comparison with a fraction is ________.

$\frac{6 \text{ boys}}{5 \text{ girls}}$

2

If n represents the number of boys in a class and 45 is the number of girls, write the ratio comparing boys to girls.

$\frac{n \text{ boys}}{45 \text{ girls}}$

3 ✍

To solve word problems using proportions, always label each numerator and denominator. Both numerators must have the same label and both denominators must have the ________ label.

same

4

The ratio $\frac{1}{2}$ may be used to show "One pound of hamburger will feed 2 teenagers." Label the numerator and denominator of the ratio $\frac{1}{2}$. ________

$\frac{1 \text{ lb of hamburger}}{2 \text{ teenagers}}$

5

The ratio $\frac{n}{8}$ may be used to show "Some number of pounds of hamburger will feed 8 teenagers." Label the numerator and denominator of the ratio. _______

$\frac{\text{n lb of hamburger}}{\text{8 teenagers}}$

6

To solve the problem

> One pound of hamburger will feed 2 teenagers.
> How many pounds are needed to feed 8 teenagers?

use the proportion $\frac{1}{2} = \frac{n}{8}$. Do both ratios of the proportion have pounds of hamburger as the labels of their numerators? _______

Yes

7

To solve the problem

> One pound of hamburger will feed 2 teenagers.
> How many pounds are needed to feed 8 teenagers?

use the proportion $\frac{1}{2} = \frac{n}{8}$. Solve the proportion. ______

4 lbs of hamburger

8 ✍

If 2 inches on a map represents 15 miles, how many miles are represented by 11 inches?

a. One ratio in the problem is $\frac{2}{15}$ which compares inches to miles. Write another ratio that also compares inches to miles. _______

$\frac{\text{11 inches}}{\text{n miles}}$

b. Write the proportion in which both ratios compare inches to miles. _______

$\frac{2}{15} = \frac{11}{n}$

c. Solve the proportion and use the label for n to solve the problem. _______

$82\frac{1}{2}$ miles

9

If 4 inches of insulation cost $2, what is the cost of 10 inches of insulation?

a. One ratio in the problem is $\frac{4}{2}$ which compares inches to dollars. Write another ratio that also compares inches to dollars. _______ $\frac{\text{10 inches}}{\text{n dollars}}$

b. Write the proportion in which both ratios compare inches to dollars. _______ $\frac{4}{2} = \frac{10}{n}$

c. Solve the proportion and use the label for n to solve the problem. _______ 5 dollars

10

If 6 yards of cloth cost $9, how much does 20 yards of cloth cost?

a. One ratio in the problem is $\frac{6}{9}$ which compares yards of cloth to dollars. Write another ratio that makes the same comparison. _______ $\frac{\text{20 yards of cloth}}{\text{n dollars}}$

b. Write the proportion in which both ratios compare yards of cloth to dollars. _______ $\frac{6}{9} = \frac{20}{n}$

c. Solve the proportion and use the label for n to solve the problem. _______ 30 dollars

11

If 3 long distance runners run a total of 51 miles, how many runners are needed to run a total of 85 miles?

a. One ratio in the problem is $\frac{3}{51}$ which compares runners to miles. Write another ratio that makes the same comparison. _______ $\frac{\text{n runners}}{\text{85 miles}}$

b. Write the proportion in which both ratios compare runners to miles. _______ $\frac{3}{51} = \frac{n}{85}$

c. Solve the proportion and use the label for n to solve the problem. _______ 5 runners

12
If a mixture of nuts contains 7 pounds of pecans for every 4 pounds of cashews, how many pounds of pecans are needed for 11 pounds of cashews?

a. Write two ratios comparing pecans to cashews. _______ $\frac{7}{4}$ and $\frac{n}{11}$

b. Write a proportion for the problem. _______ $\frac{7}{4} = \frac{n}{11}$

c. Solve the proportion. _______ $19\frac{1}{4}$ lbs of pecans

13
A solution of 3 parts medicine to 20 parts alcohol is to be made. The total amount of alcohol will be 240 milliliters (ml). How much medicine is needed?

a. Write a proportion for the problem. _______ $\frac{3}{20} = \frac{n}{240}$ or $\frac{20}{3} = \frac{240}{n}$

b. Solve the problem. _______ 36 ml of medicine

14
The ratio of profit to cost for a bakery is 1 to 9. If the profit is $16,200 what is the cost?

a. Write a proportion for the problem. _______ $\frac{1}{9} = \frac{16200}{n}$ or $\frac{9}{1} = \frac{n}{16200}$

b. Solve the problem. _______ cost is $145,800

15
A mixture is made of 5 grains of drug with 160 milliliters (ml) of solution. In a sample containing 2 grains of the drug, how much solution will be used?

a. Write a proportion for the problem. _______ $\frac{5}{160} = \frac{2}{n}$ or $\frac{160}{5} = \frac{n}{2}$

b. Solve the problem. _______ 64 ml of solution

16
A recipe calls for 1 cup of sugar with 3 cups of flour. How much sugar is needed with 10 cups of flour?

a. Write a proportion for the problem. _______ $\frac{1}{3} = \frac{n}{10}$ or $\frac{3}{1} = \frac{10}{n}$

b. Solve the problem. _______ $3\frac{1}{3}$ cups of sugar

17
An intravenous solution needs 2 milliters (ml) of glucose mixed with 17 units of blood. How much glucose is needed for 35 units of blood?

a. Write a proportion for the problem. _______ $\frac{2}{17} = \frac{n}{35}$ or $\frac{17}{2} = \frac{35}{n}$

b. Solve the problem. _______ $4\frac{2}{17}$ ml of glucose

18
A doctor's order calls for 8 drams of drug to be used with 100 milliliters (ml) of water. How much water will be needed for 18 drams of drug?

a. Write a proportion for the problem. _______ $\frac{8}{100} = \frac{18}{n}$ or $\frac{100}{8} = \frac{n}{18}$

b. Solve the problem. _______ 225 ml of water

19
Two centimeters (cm) on a map represents 10 kilometers (km). How many centimeters are needed to show 37 kilometers?

a. Write a proportion for the problem. _______ $\frac{2}{10} = \frac{n}{37}$ or $\frac{10}{2} = \frac{37}{n}$

b. Solve the problem. _______ $7\frac{2}{5}$ cm

20
A car used 13 liters (l) of fuel traveling 340 kilometers (km). How many liters of fuel will be used traveling 85 kilometers?

a. Write a proportion for the problem. _______ $\frac{13}{340} = \frac{n}{85}$ or $\frac{340}{13} = \frac{85}{n}$

b. Solve the problem. _______ $3\frac{1}{4}$ liters of fuel

21
If 20 gallons of paint will cover 210 square feet of wall, how many square feet will 15 gallons of paint cover?

a. Write a proportion for the problem. _______ $\frac{20}{210} = \frac{15}{n}$ or $\frac{210}{20} = \frac{n}{15}$

b. Solve the problem. _______ $157\frac{1}{2}$ square feet

22
A store offers a discount of 12 cents for every $1 of price. A refrigerator was priced at $300. How much is the discount?

a. Write a proportion for the problem. _______ $\frac{12}{1} = \frac{n}{300}$ or $\frac{1}{12} = \frac{300}{n}$

b. Solve the problem. _______ 3600 cents or $36

23
If 1,500 20-year-olds die each year per 1,000,000 20-year olds, how many will die out of 2,000 20-year-olds?

a. Write a proportion for the problem. _______ $\frac{1500}{1000000} = \frac{n}{2000}$

b. Solve the problem. _______ 3 will die

24
If 2 nuclear accidents can occur every 360 days, how many may occur in 900 days?

a. Write a proportion for the problem. _______ $\frac{2}{360} = \frac{n}{900}$

b. Solve the problem. _______ 5 accidents

FEEDBACK UNIT 5

This quiz reviews the preceding unit. Answers are at the back of the book.
Take the Chapter Mastery Test only after all problems of this unit are understood.

1. Ace Packing Co. makes a profit of 3 cents on every $200 in sales. What is the profit on a $1000 sale?

2. A nurse is ordered to make a mixture containing 7 tablespoons of drug for each 10 liters of solvent. If 2 liters of solvent are used, how much drug is needed?

3. A student gets 5 out of every 6 problems correct on a test. If there are 180 total problems on the test, how many did the student answer correctly?

4. A business found that there were 2 cents advertising cost for every 9 cents in sales. On an item selling for $18, what was the advertising cost?

5. A photograph was 3 inches high and 5 inches wide. If it is enlarged to be 7 inches high, how wide will it be?

6. In a 250 kilogram block of alloy, 2 grams out of every 5 grams of alloy are gold. How much gold is in the block?

7. The National Bank charges interest of 13 cents for every $1 loaned for one year. How much interest will be charged for a loan of $150 for one year?

8. The ratio of the width to length of a book is to be 8 to 11. How wide should a page be if it is 16 cm long?

CHAPTER 11 MASTERY TEST

This test provides an opportunity for assuring that **all** of the objectives of Chapter 11 have been learned. Each problem is accompanied by the symbol (U) which indicates the Unit in which that type of problem was studied. If any problems are missed, they need to be corrected. Answers are at the back of the book.

1. Write and simplify the ratio of 3 to 12. (1)

2. Write and simplify the ratio of 28 to 14. (1)

3. Write and simplify the ratio of 20 to 25. (1)

4. Write and simplify the ratio of 32 kilometers to 28 kilometers. (1)

5. Solve $\frac{n}{8} = \frac{8}{16}$ (4)

6. Solve $\frac{9}{n} = \frac{3}{5}$ (4)

7. Solve $\frac{18}{20} = \frac{27}{n}$ (4)

8. Solve $\frac{12}{5} = \frac{n}{20}$ (4)

9. If the F.C.C. ruled that each TV station must have a ratio of advertising time to program time of 2 to 5, how much program time is needed to air 16 minutes of advertising? (5)

10. A weather forecaster found that 3 rainy days occur for every 8 cloudy days. In 20 cloudy days, about how many rainy days should occur? (5)

11. It is found that 2 malfunctions occur for every 45 days of operating a nuclear power plant. How many days may pass for 3 malfunctions to occur? (5)

12. A bank charges $11 interest on a $100 loan for one year. How much could be borrowed for a year for $220 interest? (5)

13. A 2 liter sample of river water contained 13 particles of pollution. How many particles would be in 50 liters of the water? (5)

14. Three drams of a drug is to be given for each 12 kilograms of body weight. Find the weight of a person that would require 14 drams of the drug. (5)

Chapter 12 Objectives

The following problems describe, by example, the skills and knowledge studied in this chapter. Each problem is accompanied by a symbol, (U), which indicates the Unit where the problem is studied. Answers are at the back of the book.

1. Write 5% as a ratio. (1)
2. Write 120% as a ratio. (1)
3. Write 5.7% as a ratio. (1)
4. The ratio $\frac{7}{100}$ is the percent ______. (1)
5. The ratio $\frac{39}{100}$ is the percent ______. (1)
6. Write 56% as a decimal. (2)
7. Write 8.5% as a decimal. (2)
8. Write .3 as a percent. (2)
9. Write .79 as a percent. (2)
10. Write $\frac{5}{9}$ to the nearest one percent. (3)
11. Write $\frac{13}{18}$ to the nearest one percent. (3)
12. Find the base (b), rate (r), and part (p) for: 55% of 80 is 44. (4)
13. Find the base (b), rate (r), and part (p) for: 85 is 17% of 500. (4)
14. Find the rate to the nearest whole number when the base is 89 and part is 60. (5)
15. Find the base when the part is 243 and rate is 30. (5)
16. Find the part when the base is 746 and rate is 7. (5)
17. ____% of 70 is 35. (6)
18. 4% of _____ is 12. (6)
19. 11% of 300 is _____. (6)
20. The interest rate on a $40,000 loan for one year is 11%. Find the amount of interest. (6)
21. A 30 liter sample of water contains 3 liters of pollution. What percent of the sample is pollution? (6)
22. It was found that 3% of a shipment of potatoes was spoiled. If 45 kilograms of potatoes was spoiled, what was the weight of the whole shipment? (6)
23. Find 132% of 52. (6)
24. 6 is 120% of what number? (6)

CHAPTER 12
PERCENT

UNIT 1: THE MEANING OF PERCENT

1
The large square at the right has been divided into 100 smaller squares. Each of the smaller squares represents ________ (what fraction) of the large square.

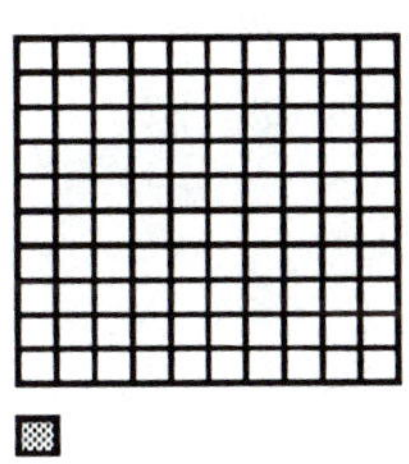

$\frac{1}{100}$

2
Each small square represents $\frac{1}{100}$ of the large square. What fraction of the large square shown at the right is shaded? ________

$\frac{7}{100}$

3
The large square shown at the right has been divided into 100 smaller squares. Each small square represents $\frac{1}{100}$ of the large square. What fraction of the large square shown at the right is shaded? _______

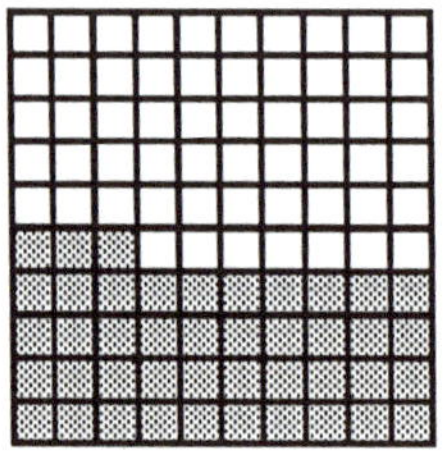

$\frac{43}{100}$

4
If the large square at the right represents the total land area of the United States, then the shaded area of the figure represents ________ of the land area of the United States.

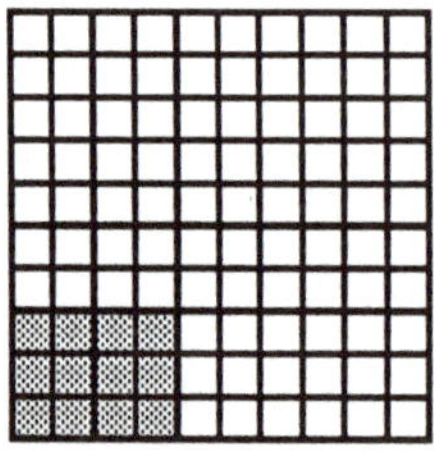

$\frac{12}{100}$

5
If the large square at the right represents the income of a doctor in 1985, then the shaded area of the figure represents ________ of the doctor's 1985 income.

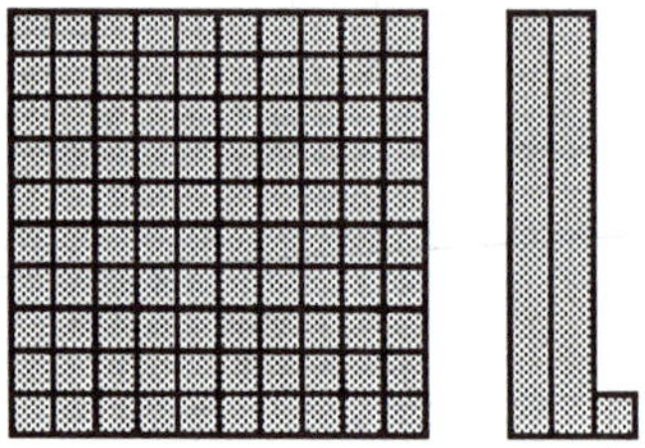

$\frac{121}{100}$

6
If the large square at the right represents all of the engineers in Texas, then the shaded area of the figure represents ______ of the Texas engineers.

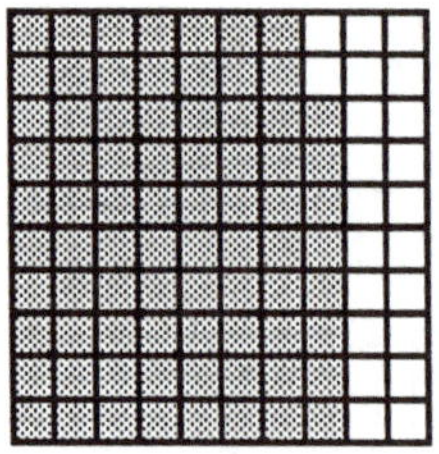

$\frac{78}{100}$

PERCENT

In Chapter 11, ratios were studied. In this chapter, a special ratio called percent is studied.

Percent is a ratio ($\frac{r}{100}$) with a denominator of 100.

Every percent represents a situation similar to a large square that has been divided into 100 smaller squares.

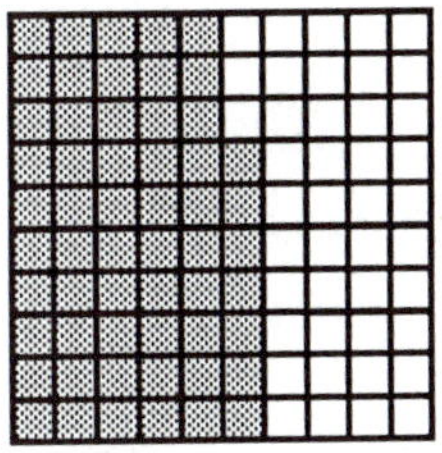

The large square represents 100%.
Each small square represents 1%.

The ratio $\frac{57}{100}$ is a percent because it has a denominator of 100. The ratio $\frac{57}{100}$ indicates that something has been divided into 100 equal parts and there is special interest in 57 of those parts.

7

A special ratio, called a percent, has a denominator of 100.
The ratio $\frac{81}{100}$ is a percent and can be read as "81 percent."
Is $\frac{42}{100}$ a percent? ______

Yes, 42 percent

8

The ratio $\frac{13}{100}$ is 13 one-hundredths or 13 percent.
The ratio $\frac{63}{100}$ is 63 one-hundredths or ___ _________.

63 percent

9

The ratio $\frac{7}{100}$ is 7 one-hundredths or 7 percent.
The ratio $\frac{11}{100}$ is 11 one-hundredths or ___ _________.

11 percent

10

The ratio $\frac{23}{100}$ is 23 percent.
The ratio $\frac{47}{100}$ is ___ _________.

47 percent

11 ✍

Percent is a ratio with 100 as its denominator. 5 percent means $\frac{5}{100}$ and 8 percent means ______. $\frac{8}{100}$

12 ✍

The symbol for percent is "%." 67% = $\frac{67}{100}$ and 19% = ____. $\frac{19}{100}$

13

The symbol for percent is ______. %

14

Use the symbol for percent to write $\frac{96}{100}$. 96%

15

Write the ratio $\frac{9}{100}$ as a percent. ______ 9%

16

Write 74% as a ratio. ______ $\frac{74}{100}$

17

To write 6.7% as a ratio, 6.7 is the numerator and _____ is the denominator. 100

18

6.7% is the ratio $\frac{6.7}{100}$.
5.8% is the ratio _____. $\frac{5.8}{100}$

19

To write 133% as a ratio, 133 is the numerator and _____ is the denominator. 100

20

133% is the ratio $\frac{133}{100}$.
245% is the ratio ______. $\frac{245}{100}$

21
The ratio $\frac{6}{100}$ is 6%.
The ratio $\frac{38}{100}$ is ______. 38%

22
The ratio $\frac{537}{100}$ is ______. 537%

23
The ratio $\frac{9.3}{100}$ is ______. 9.3%

24
66% is the ratio ______. $\frac{66}{100}$

25
403% is the ratio ______. $\frac{403}{100}$

26
4.2% is the ratio ______. $\frac{4.2}{100}$

FEEDBACK UNIT 1

This quiz reviews the preceding unit. Answers are at the back of the book. Begin the next unit only after all problems of this unit are understood.

1. Write 17% as a ratio.
2. Write 49% as a ratio.
3. Write 57% as a ratio.
4. Write 143% as a ratio.
5. Write 7% as a ratio.
6. Write 8.4% as a ratio.
7. Percent is a ratio with denominator ______.
8. The ratio $\frac{19}{100}$ is ______ percent.
9. The ratio $\frac{81}{100}$ is ______ percent.
10. The ratio $\frac{3}{100}$ is ______ percent.
11. The ratio $\frac{181}{100}$ is ______ percent.
12. The ratio $\frac{6.7}{100}$ is ______ percent.

UNIT 2: PERCENT AND DECIMAL EQUIVALENTS

DIVISION BY 100

Division by 100 can be accomplished by simply moving the decimal point two places in the dividend.

$47 \div 100 = .47$ $865 \div 100 = 8.65$ $6 \div 100 = .06$

1
The decimal point in 83 is to the right of the 3. 83 = 83.0
Write 47 with a decimal point. ______

47.0 or 47.

2
Write 836 with a decimal point. ______

836.0 or 836.

3
Write 5 with a decimal point. ______

5.0 or 5.

4

17% is the ratio $\frac{17}{100}$ or $\frac{17.}{100}$
214% is the ratio $\frac{214}{100}$ or ______. (Write the numerator with a decimal point.

$\frac{214.}{100}$

5

23% is the ratio $\frac{23}{100}$ or $\frac{23.}{100}$.
7% is the ratio $\frac{7}{100}$ or ______. (Write the numerator with a decimal point.

$\frac{7.}{100}$

6
Division by 100 is accomplished by moving the decimal point in the dividend two places to the left.

$57 \div 100 = .57$ and $923 \div 100 =$ ______

9.23

7
To divide a number by 100, move its decimal point ____ (how many) places. 2

8

$23\% = \frac{23}{100} = \frac{23.}{100}$.

Since $\frac{23}{100}$ means $23 \div 100$, $\frac{23}{100} =$ _____ .23

9

$17\% = \frac{17}{100} = \frac{17.}{100}$

Since $\frac{17}{100}$ means $17 \div 100$, $\frac{17}{100} =$ _____ .17

10

$214\% = \frac{214}{100} = \frac{214.}{100}$

Since $\frac{214}{100}$ means $214 \div 100$, $\frac{214}{100} =$ _____ 2.14

11

$7\% = \frac{7}{100} = \frac{7.}{100}$

Since $\frac{7}{100}$ means $7 \div 100$, $\frac{7}{100} =$ _____ .07

12 ✍
Because percent requires division by 100 and division by 100 is accomplished by moving the decimal point 2 places, any percent may be changed to a decimal just by moving its decimal point two places to the left.

52% = .52 and 43% = _____ .43

13
To change 9% to a decimal, move its decimal point 2 places to the left. 9% = .09 and 7% = _____ .07

14
To change 137% to a decimal, move its decimal point 2 places to the left. 137% = 1.37 and 322% = _____ 3.22

15
To change 6.9% to a decimal, move its decimal point 2 places to the left.
6.9% = .069 and 3.8% = _____ .038

16
Write 95% as a decimal. ______ .95

17
Write 8% as a decimal. ______ .08

18
Write 645% as a decimal. ______ 6.45

19
Write 57.9% as a decimal. ______ .579

20
Write .6% as a decimal. ______ .006

CHANGING DECIMALS TO PERCENTS

To change a percent to a decimal, the decimal point is moved two places to the left.

To change a decimal to a percent, just reverse the procedure. Move the decimal point two places to the right.

21 ✍
To change a decimal to a percent, move its decimal point 2 places to the right.
.63 = 63% and .45 = ______ 45%

22
To change a decimal to a percent, move its decimal point 2 places to the right.
5.34 = 534% and 1.23 = ______ 123%

23
To change a decimal to a percent, move its decimal point 2 places to the right.
.03 = 3% and .05 = ______

5%

24
To change a decimal to a percent, move its decimal point 2 places to the right.
.098 = 9.8% and .307 = ______

30.7%

25
Write .71 as a percent. ______

71%

26
Write 1.56 as a percent. ______

156%

27
Write .2 as a percent. ______

20%

28
Write 6.2 as a percent. ______

620%

29
Write .413 as a percent. ______

41.3%

30
Write .0075 as a percent. ______

.75%

31
Write 1.25 as a percent. ______

125%

32
Write .0241 as a percent. ______

2.41%

33
Write .12 as a percent. ______

12%

FEEDBACK UNIT 2

This quiz reviews the preceding unit. Answers are at the back of the book. Begin the next unit only after all problems of this unit are understood.

1. Write 27% as a decimal.
2. Write 85% as a decimal.
3. Write 56% as a decimal.
4. Write 4% as a decimal.
5. Write 37% as a decimal.
6. Write 163% as a decimal.
7. Write 92% as a decimal.
8. Write 5.8% as a decimal.
9. Write 16% as a decimal.
10. Write 47.3% as a decimal.
11. Write .26 as a percent.
12. Write .89 as a percent.
13. Write .08 as a percent.
14. Write .63 as a percent.
15. Write 2.4 as a percent.
16. Write .52 as a percent.
17. Write 1.45 as a percent.
18. Write .22 as a percent.
19. Write .043 as a percent.
20. Write 1.41 as a percent.
21. Write .01 as a percent.
22. Write .025 as a percent.

UNIT 3: PERCENT PROBLEMS ARE PROPORTIONS

PROPORTIONS

In Chapter 11, proportions such as $\frac{3}{4} = \frac{n}{7}$ were solved.

$\frac{3}{4} = \frac{n}{7}$ became $3 \times 7 = 4 \times n$ or $21 = 4 \times n$

After dividing both sides by 4, $n = \frac{21}{4}$.

In this unit, percent problems are solved as proportions.

1
Percent is a ratio which has a denominator of 100. To change $\frac{2}{7}$ to a percent, use the proportion $\frac{2}{7} = \frac{r}{100}$. The proportion is used to write an equation,
$2 \times 100 =$ ______.

$7 \times r$

2
To write $\frac{2}{7}$ as a percent, use the proportion $\frac{2}{7} = \frac{r}{100}$. Then solve the equation, $2 \times 100 = 7 \times r$. What is the solution of the equation? ______

$200 = 7 \times r$
$r = \frac{200}{7} = 28\frac{4}{7}$

3
To write $\frac{2}{7}$ as a percent, the proportion $\frac{2}{7} = \frac{r}{100}$ is solved to give $r = 28\frac{4}{7}$. The fraction $\frac{4}{7}$ is ______ (more, less) than $\frac{1}{2}$.

more

4
$28\frac{4}{7}\%$ is closer to 29% than it is to 28% because $\frac{4}{7}$ is more than $\frac{1}{2}$. The fraction $\frac{2}{7}$ is approximately equal to ___%.

29

5
To find the percent approximation of $\frac{3}{11}$, solve the proportion $\frac{3}{11} = \frac{r}{100}$. What is the solution for this proportion? ______

$300 = 11 \times r$
$r = \frac{300}{11} = 27\frac{3}{11}$

6
$\frac{3}{11} = \frac{r}{100}$ is solved to give $r = \frac{300}{11} = 27\frac{3}{11}$.
Is $27\frac{3}{11}\%$ closer to 27% or 28%? ______

27%

7

$\frac{3}{11} = \frac{r}{100}$ is solved to give $r = \frac{300}{11} = 27\frac{3}{11}$.

To the nearest one percent $\frac{3}{11}$ = ______ 27%

8

Begin the process for changing $\frac{4}{7}$ to a percent by writing the proportion and solving it. ______

$\frac{4}{7} = \frac{r}{100}$

$r = \frac{400}{7} = 57\frac{1}{7}$

9

To write $\frac{4}{7}$ as a percent:

a. Write the proportion $\frac{4}{7} = \frac{r}{100}$

b. Solve the proportion $r = 57\frac{1}{7}$

c. Is $57\frac{1}{7}$% closer to 57% or 58%? ______ 57%

10

Write $\frac{4}{7}$ to the nearest one percent. ______ 57%

11

To write $\frac{5}{11}$ to the nearest one percent :

a. Write the problem as a proportion. ______ $\frac{5}{11} = \frac{r}{100}$

b. Solve the proportion. ______ $r = 45\frac{5}{11}$

c. Round off the answer to the nearest one percent. ______ 45%

12

To write $\frac{2}{3}$ to the nearest one percent :

a. Write the problem as a proportion. ______ $\frac{2}{3} = \frac{r}{100}$

b. Solve the proportion. ______ $r = 66\frac{2}{3}$

c. Round off the answer to the nearest one percent. ______ 67%

13

To write $\frac{5}{8}$ to the nearest one percent :

a. Write the problem as a proportion. ______ $\frac{5}{8} = \frac{r}{100}$

b. Solve the proportion. _____ $r = 62\frac{1}{2}$

c. Round off the answer to the nearest one percent. _____ 63%

14

To write $\frac{5}{6}$ to the nearest one percent :

a. Write the problem as a proportion. ______ $\frac{5}{6} = \frac{r}{100}$

b. Solve the proportion. _____ $r = 83\frac{1}{3}$

c. Round off the answer to the nearest one percent. _____ 83%

15

To write $\frac{1}{2}$ to the nearest one percent :

a. Write the problem as a proportion. ______ $\frac{1}{2} = \frac{r}{100}$

b. Solve the proportion. _____ $r = 50$

c. Does this answer require rounding off? _____ No, 50%

16

To write $\frac{7}{10}$ to the nearest one percent :

a. Write the problem as a proportion. ______ $\frac{7}{10} = \frac{r}{100}$

b. Solve the proportion. _____ $r = 70$

c. Does this answer require rounding off? _____ No, 70%

COMMON FRACTION — PERCENT EQUIVALENTS

Any fraction can be changed to a percent approximation. However, some fractions occur frequently enough that their percent equivalents are memorized.

$\frac{1}{2} = 50\%$ $\frac{1}{3} = 33\frac{1}{3}\%$ $\frac{1}{4} = 25\%$ $\frac{1}{5} = 20\%$

$\frac{1}{6} = 16\frac{2}{3}\%$ $\frac{1}{7} = 14\frac{2}{7}\%$ $\frac{1}{8} = 12\frac{1}{2}\%$ $\frac{1}{10} = 10\%$

17
Write $\frac{9}{10}$ to the nearest one percent. ______ 90%

18
Write $\frac{3}{8}$ to the nearest one percent. ______ 38%

19
Write $\frac{3}{4}$ to the nearest one percent. ______ 75%

20
Write $\frac{57}{80}$ to the nearest one percent. ______ 71%

21
Write $\frac{15}{47}$ to the nearest one percent. ______ 32%

22
The whole number 1 is equal to 100%. Any improper fraction will have a percent greater than _____. 100%

23
Write $\frac{9}{5}$ to the nearest one percent. ______ 180%

24
Write $\frac{15}{8}$ to the nearest one percent. ______ 188%

25
Write $\frac{23}{9}$ to the nearest one percent. ______ 256%

25
Write $\frac{63}{58}$ to the nearest one percent. ______ 109%

FEEDBACK UNIT 3

This quiz reviews the preceding unit. Answers are at the back of the book.
Begin the next unit only after all problems of this unit are understood.

1. Write $\frac{3}{8}$ to the nearest one percent.
2. Write $\frac{5}{9}$ to the nearest one percent.
3. Write $\frac{4}{5}$ to the nearest one percent.
4. Write $\frac{1}{3}$ to the nearest one percent.
5. Write $\frac{3}{10}$ to the nearest one percent.
6. Write $\frac{15}{29}$ to the nearest one percent.
7. Write $\frac{41}{85}$ to the nearest one percent.
8. Write $\frac{19}{43}$ to the nearest one percent.
9. Write $\frac{57}{48}$ to the nearest one percent.
10. Write $\frac{6}{145}$ to the nearest one percent.

UNIT 4: THREE NUMBERS IN A PERCENT PROBLEM

BASE, RATE, AND PART

Every percent problem is a proportion like the one shown at the right in which the percent ratio has denominator 100.

$$\frac{r}{100} = \frac{p}{b}$$

The other three numbers in the proportion are the rate (r), base (b), and part (p). In this unit, methods are explained for identifying these three numbers.

1

r% of b is p
There are three numbers in: 15% of 200 is 30.
The numbers are: 15%, 200, and _____.

30

2

r% of b is p
How many numbers are in the statement below? ______
12% of 400 is 48.

3
12%, 400, 48

3

p is r% of b

How many numbers are in the statement below? ______ 3

60 is 5% of 1200. 60, 5%, 1200

4 ✍

The percent proportion $\frac{r}{100} = \frac{p}{b}$ involves three letters, r, b, and p.

The letter r represents the **rate** and needs the % symbol.

The letter p represents the **part**.

The letter b represents the **base**.

The letter ____ represents the rate. r

5

The percent proportion $\frac{r}{100} = \frac{p}{b}$ involves three letters, r, b, and p.

The letter r represents the **rate** and needs the % symbol.

The letter p represents the **part**.

The letter b represents the **base**.

The letter ____ represents the part. p

6

The percent proportion $\frac{r}{100} = \frac{p}{b}$ involves three letters, r, b, and p.

The letter r represents the **rate** and needs the % symbol.

The letter p represents the **part**.

The letter b represents the **base**.

The letter ____ represents the base. b

7 ✍

To find the base number (b) in a percent problem look for the number that follows the word "of." In the statement below, what number appears directly after "of." _____ 500

90% of 500 is 450.

8
The base number in
"90% of 500 is 450"
is 500 and it appears directly after the word "of." Find the base number in the statement below. ______ 60
24 is 40% of 60.

9
The base number in a percent problem usually follows the word "of." To find the base (b) in a percent problem, look for the word _____. of

10
The word "of" in a percent problem is a good clue for identifying the base (b). To find the base of a percent problem look for the number following the word ____. of

11
Find the base (b) of the percent statement below by choosing the number directly following "of." _____ b = 900
8% of 900 is 72.

12
Find the base (b) of the percent statement below by choosing the number directly following "of." _____ b = 44
25% of 44 is 11.

13
Find the base (b) of the percent statement below. _____ b = 40
30 is 75% of 40.

14
Find the base (b) of the percent statement below. _____ b = 40
125% of 40 is 50.

15
Find the base (b) of the percent statement below. _____ b = 50
3 is 6% of 50.

16
Find the base (b) of the percent statement below. _____ b = 200
19% of 200 is 38.

17 ✍
In the percent statement below the base is 35 and the number with the percent symbol (%) is _____. 60
21 is 60% of 35.

18
In the percent statement below the base is ____ and the number with the percent symbol (%) is _____. b = 60 25
25% of 60 is 15.

19
The rate (r) in a percent problem is the number with the symbol %. To find the rate number, look for the symbol ____. %

20
In the percent statement below the base is ____ and the rate number with the percent symbol (%) is _____. b = 70 30
30% of 70 is 21.

21
Find the base (b) and rate (r) for:
16 is 50% of 32.

b = 32
r = 50

22
Find the base (b) and rate (r) for:
18% of 300 is 54.

b = 300
r = 18

23
Find the base (b) and rate (r) for:
45 is 150% of 30.

b = 30
r = 150

24
In a percent problem the number that comes after "of" is the _____ and the number with the percent symbol (%) is the _______.

base
rate

25
Find the base (b) and rate (r) for:
60% of 120 is 72.

b = 120
r = 60

26
The third number in a percent problem is its part (p). Find the base and rate. Then the third number is the _____.

part

27
Find the part for: 60% of 120 is 72.

72

28
Find the base (b), rate (r), and part (p) for:
45 is 150% of 30.

b = 30
r = 150
p = 45

29
Find the base (b), rate (r), and part (p) for:
55% of 80 is 44.

b = 80
r = 55
p = 44

30
Find the base (b), rate (r), and part (p) for:
85 is 17% of 500.

b = 500
r = 17
p = 85

31
Find the base (b), rate (r), and part (p) for:
150% of 90 is 135.

b = 90
r = 150
p = 135

32
Find the base (b), rate (r), and part (p) for:
67% of 200 is 134.

b = 200
r = 67
p = 134

33
Find the base (b), rate (r), and part (p) for:
4 is 200% of 2.

b = 2
r = 200
p = 4

34
Find the base (b), rate (r), and part (p) for:
3% of 8 is 0.24

b = 8
r = 3
p = 0.24

35
Find the base (b), rate (r), and part (p) for:
14 is 20% of 70.

b = 70
r = 20
p = 14

FEEDBACK UNIT 4

This quiz reviews the preceding unit. Answers are at the back of the book. Begin the next unit only after all problems of this unit are understood.

1. The three letters in the percent proportion $\frac{r}{100} = \frac{p}{b}$ represent the ______ (b), ______ (r), and ______ (p).
2. To find the base (b) of percent problem look for ______.
3. To find the rate (r) of a percent problem look for ______.
4. Find the base (b), rate (r), and part (p) for: 53% of 500 is 265.
5. Find the base (b), rate (r), and part (p) for: 17 is 85% of 20.
6. Find the base (b), rate (r), and part (p) for: 125% of 60 is 75.
7. Find the base (b), rate (r), and part (p) for: 144 is 80% of 180.

UNIT 5: SOLVING PERCENT PROBLEMS

THE PRODUCTS OF THE EXTREMES AND MEANS ARE EQUAL

From Chapter 11, recall that a proportion is solved using the fact that the product of the extremes is equal to the product of the means.

To solve any proportion $\frac{r}{100} = \frac{p}{b}$

1. Use the products of the extremes and means to write a multiplication equation.

   $r \times b = 100 \times p$

2. Solve the multiplication equation.

1

To **solve** $\frac{r}{100} = \frac{p}{b}$ when the base is 40 and the rate is 15, replace b by 40 and r by 15. What proportion is obtained from these replacements? ______

$\frac{15}{100} = \frac{p}{40}$

2

To **solve** the proportion $\frac{15}{100} = \frac{p}{40}$ first write a new equation using the products of the extremes and means. What is the new multiplication equation?

____

$15 \times 40 = 100 \times p$

3

To **solve** $\frac{r}{100} = \frac{p}{b}$ when b = 40 and r = 15, the process shown at the right is used.

$\frac{r}{100} = \frac{p}{b}$

$\frac{15}{100} = \frac{p}{40}$

$15 \times 40 = 100 \times p$

Complete the solution by dividing both sides of $600 = 100 \times p$ by 100.

$600 = 100 \times p$

_____ $= p$

$\frac{600}{100} = \frac{100 \times p}{100}$

6

4

Use the percent proportion $\frac{r}{100} = \frac{p}{b}$ to write the proportion when the base (b) is 84 and the rate (r) is 25. ______

$\frac{25}{100} = \frac{p}{84}$

5

Simplify the fraction $\frac{25}{100}$ and write another proportion in the place of $\frac{25}{100} = \frac{p}{84}$. ________

$\frac{1}{4} = \frac{p}{84}$

6

Solve $\frac{1}{4} = \frac{p}{84}$ using the products of the extremes and means. Find the part (p) when the base is 84 and rate is 25. _____

$1 \times 84 = 4 \times p$

$p = 21$

7

Use the percent proportion $\frac{r}{100} = \frac{p}{b}$ to write the proportion when the base (b) is 810 and the rate (r) is 30. ______

$\frac{30}{100} = \frac{p}{810}$ or $\frac{3}{10} = \frac{p}{810}$

8

Solve $\frac{3}{10} = \frac{p}{810}$ using the products of the extremes and means. Find the part (p) when the base is 810 and rate is 30. _____

$3 \times 810 = 10 \times p$

$p = 243$

9

Use the percent proportion $\frac{r}{100} = \frac{p}{b}$ to find the part (p) when the base (b) is 60 and the rate (r) rate (r) is 45. ______

$\frac{45}{100} = \frac{p}{60}$

$p = 27$

10

Use the percent proportion $\frac{r}{100} = \frac{p}{b}$ to find the part (p) when the base (b) is 42 and the rate (r) is 19. ______

$\frac{19}{100} = \frac{p}{42}$

$p = 7.98$

11

To **solve** $\frac{r}{100} = \frac{p}{b}$ when the base is 437 and the part is 85, replace b by 437 and p by 85. What proportion is obtained from these replacements? ______

$\frac{r}{100} = \frac{85}{437}$

12

To **solve** the proportion $\frac{r}{100} = \frac{85}{437}$ first write a new equation using the products of the extremes and means. What is the new multiplication equation?

______________.

$r \times 437 = 100 \times 85$

13 ✍

To **solve** $\frac{r}{100} = \frac{p}{b}$ when $b = 437$ and $p = 85$, the process shown at the right is used.

$$\frac{r}{100} = \frac{p}{b}$$

$$\frac{r}{100} = \frac{85}{437}$$

$$r \times 437 = 100 \times 85$$

$$r \times 437 = 8500$$

$$r = \_\_\_\_\_$$

Complete the solution by dividing both sides of $r \times 437 = 8500$ by 437 and rounding off the answer to the nearest whole number. _______

$\frac{r \times 437}{437} = \frac{8500}{437}$

r is approximately equal to 19.4 which rounds off to 19%

14

Use the percent proportion $\frac{r}{100} = \frac{p}{b}$ to write the proportion when the base (b) is 812 and the part (p) is 56. ______

$\frac{r}{100} = \frac{56}{812}$

15

Simplify the fraction $\frac{56}{812}$ and write another proportion in the place of $\frac{r}{100} = \frac{56}{812}$. _________

$\frac{r}{100} = \frac{2}{29}$

16

Solve $\frac{r}{100} = \frac{2}{29}$ using the products of the extremes and means. Find the rate (r) to the nearest one percent when the base is 812 and part is 56. ______

$r \times 29 = 100 \times 2$

r is approximately 6.8 which rounds off to 7%

17

Use the percent proportion $\frac{r}{100} = \frac{p}{b}$ to write the proportion when the base (b) is 258 and the part (p) is 46. ______

$\frac{r}{100} = \frac{46}{258}$ or $\frac{r}{100} = \frac{23}{129}$

18

Solve $\frac{r}{100} = \frac{23}{129}$ using the products of the extremes and means. Find the rate to the nearest whole number when the base is 258 and the part is 46. ______

$r \times 129 = 100 \times 23$

r is approximately 17.8 which rounds off to 18%

19

Use the percent proportion $\frac{r}{100} = \frac{p}{b}$ to find the rate to the nearest whole number when the base (b) is 75 and the part (p) is 112. ______

$\frac{r}{100} = \frac{112}{75}$

$r \times 75 = 100 \times 112$

r is approximately 149.3 which rounds off to 149%

20

Use the percent proportion $\frac{r}{100} = \frac{p}{b}$ to find the rate to the nearest whole number when the base (b) is 50 and the part (p) is 46. ______

$\frac{r}{100} = \frac{46}{50}$ or $\frac{r}{100} = \frac{23}{25}$

$r = 92\%$

21

To **solve** $\frac{r}{100} = \frac{p}{b}$ when the part is 115.44 and the rate is 37, replace p by 115.44 and r by 37. What proportions is obtained from these replacements?

______

$\frac{37}{100} = \frac{115.44}{b}$

22
To **solve** the proportion $\frac{37}{100} = \frac{115.44}{b}$ first write a new equation using the products of the extremes and means. What is the new multiplication equation? ______________.

$37 \times b = 100 \times 115.44$

23
To **solve** $\frac{r}{100} = \frac{p}{b}$ when p = 115. 44 and r = 37, the process shown at the right is used.

$$\frac{r}{100} = \frac{p}{b}$$

$$\frac{37}{100} = \frac{115.44}{b}$$

$$37 \times b = 100 \times 115.44$$

Complete the solution by dividing both sides of 37 x b = 11,544 by 37.

$$37 \times b = 11{,}544$$

$$b = \_\_\_\_$$

$$\frac{37 \times b}{37} = \frac{11{,}544}{37}$$

312

24
Use the percent proportion $\frac{r}{100} = \frac{p}{b}$ to write the proportion when the part (p) is 60 and the rate (r) is 15. ______

$\frac{15}{100} = \frac{60}{b}$

25
Simplify the fraction $\frac{15}{100}$ and write another proportion in the place of $\frac{15}{100} = \frac{60}{b}$. ________

$\frac{3}{20} = \frac{60}{b}$

26
Solve $\frac{3}{20} = \frac{60}{b}$ using the products of the extremes and means. Find the base (b) when the part is 60 and the rate is 15. _____

$3 \times b = 20 \times 60$

$b = 400$

27
Use the percent proportion $\frac{r}{100} = \frac{p}{b}$ to write the proportion when the part (p) is 48 and the rate (r) is 75. ______

$\frac{75}{100} = \frac{48}{b}$ or $\frac{3}{4} = \frac{48}{b}$

28

Solve $\frac{3}{4} = \frac{48}{b}$ using the products of the extremes and means. Find the base (b) when the part is 48 and rate is 25. _____

$3 \times b = 4 \times 48$

$b = 64$

29

Use the percent proportion $\frac{r}{100} = \frac{p}{b}$ to find the base (b) when the part (p) is 7.8 and the rate (r) is 12. ______

$\frac{12}{100} = \frac{7.8}{b}$

$b = 65$

30

Use the percent proportion $\frac{r}{100} = \frac{p}{b}$ to find the base (b) when the part (p) is 252.2 and the rate (r) is 52. ______

$\frac{52}{100} = \frac{252.2}{b}$

$b = 485$

FEEDBACK UNIT 5

This quiz reviews the preceding unit. Answers are at the back of the book. Begin the next unit only after all problems of this unit are understood.

Use the percent proportion $\frac{r}{100} = \frac{p}{b}$.

1. Find the part when the base is 542 and rate is 6.
2. Find the rate to the nearest whole number when the part is 97 and base is 438.
3. Find the base when the part is 21 and rate is 7.
4. Find the part when the rate is 84 and base is 894.
5. Find the rate to the nearest whole number when the base is 64 and part is 46.
6. Find the base when the part is 97.2 and rate is 18.
7. Find the part when the base is 20 and rate is 130.
8. Find the rate to the nearest whole number when the part is 8.64 and base is 96.
9. Find the base when the part is 275.6 and rate is 130.
10. Find the part when the rate is 19 and base is 413.

UNIT 6: SOLVING PERCENT WORD PROBLEMS

THE FORM OF A PERCENT WORD PROBLEM

Simple word problems involving percent can always be stated in the form

____% of _____ is _____

or

r % of b is p

where r is the rate, b is the base, and p is the part.

---

1 ✍

To solve a percent word problem state it in the form:
r % of b is p where r is rate, b is base, p is part. To solve
56% of 334 is _____
it is necessary to find the _____ (rate, base, or part).

part
r = 56, b = 334

2

Use the proportion $\frac{r}{100} = \frac{p}{b}$ to solve
56% of 334 is _____.

$\frac{56}{100} = \frac{p}{334}$
p = 187.04

3

To solve a percent word problem state it in the form
r % of b is p where r is rate, b is base, p is part. To solve
_____% of 248 is 58
it is necessary to find the _____ (rate, base, or part).

rate
b = 248, p = 58

4

Use the proportion $\frac{r}{100} = \frac{p}{b}$ to solve
_____% of 248 is 58

$\frac{r}{100} = \frac{58}{248}$
r is rounded off to the nearest whole number. r = 23%

5

To solve a percent word problem state it in the form

r % of b is p where r is rate, b is base, p is part. To solve

5 % of _____ is 43.6

it is necessary to find the _____ (rate, base, or part).

base

r = 5, p = 43.6

6

Use the proportion $\frac{r}{100} = \frac{p}{b}$ to solve

5 % of _____ is 43.6

$\frac{5}{100} = \frac{43.6}{b}$

b = 872

7

Solve. 40% of 240 is _____.

$\frac{40}{100} = \frac{p}{240}$

p = 96

8

Solve. _____% of 46 is 84.

$\frac{r}{100} = \frac{84}{46}$

r is approximately 183%

9

Solve. 52 % of _____ is 19.24

$\frac{52}{100} = \frac{19.24}{b}$

b = 37

10

Fill in the blanks below with letters or numbers to fit the following problem.

A class of 46 students has 19 boys. What percent of the class is boys?

_____% of _____ is _____

r% of 46 is 19

11

Solve the problem.

A class of 46 students has 19 boys. What percent of the class is boys? _____

$\frac{r}{100} = \frac{19}{46}$

r is approximately 41.

The answer is 41%

12
Notice that each number in a percent problem has a label.

> A class of 46 students has 19 boys. What percent of the class is boys?

The label for 46 is "students," the label for 19 is "boys," and the label for the answer is ______.

percent or %

13
Fill in the blanks below with letters or numbers to fit the following problem.

> 75% of a test with 60 questions are true-false questions. How many true-false questions are on the test?
>
> _____% of ______ is _______

75% of 60 is p

14
Solve the problem.

> 75% of a test with 60 questions are true-false questions. How many true-false questions are on the test? _____

$\frac{75}{100} = \frac{p}{60}$

$p = 45$

15
Each number in the percent problem has a label.

> 75% of a test with 60 questions are true-false questions. How many true-false questions are on the test? _____

The label for 75 is "percent," the label for 60 is "questions" and the label for the answer is ______.

true-false questions

16
Fill in the blanks below with letters or numbers to fit the following problem.

> An \$80 bicycle was reduced \$8 for a sale. What was the percent of reduction?
>
> _____% of ______ is _______

r % of 80 is 8

17
Solve the problem.

> An \$80 bicycle was reduced \$8 for a sale. What was the percent of reduction?
>
> _____

$\frac{r}{100} = \frac{8}{80}$

$r = 10\%$

18

Each number in the percent problem has a label.

An $80 bicycle was reduced $8 for a sale.
What was the percent of reduction?

The label for 80 is "price in dollars," the label for 8 is "reduction in dollars" and the label for the answer is ______.

percent or %

19

Fill in the blanks below with letters or numbers to fit the following problem.

22% of a solution of rubbing alcohol is water. If there were 3.96 liters of water in the solution, what is the amount of liters in the solution itself?

_____% of ______ is _______

22% of b is 3.96

20

Solve the problem.

22% of a solution of rubbing alcohol is water. If there were 3.96 liters of water in the solution, what is the amount of liters in the solution itself? _____

$$\frac{22}{100} = \frac{3.96}{b}$$

b = 18

21

Each number in the percent problem has a label.

22% of a solution of rubbing alcohol is water. If there were 3.96 liters of water in the solution, what is the amount of liters in the solution itself?

The label for 22 is "percent," the label for 3.96 is "liters of water" and the label for the answer is ______.

liters of solution

22

Fill in the blanks below with letters or numbers to fit the following problem.

Last year, Ms. Gottza paid $27,000 for her income taxes. This was 38% of her income. What was her income for the year?

_____% of ______ is _______

38 % of b is 27,000

23
Solve the problem.

Last year, Ms. Gottza paid $27,000 for her income taxes. This was 38% of her income. What was her income for the year? ______

To the nearest penny, $71,052.63 income

24
Fill in the blanks below with letters or numbers to fit the following problem.

2% of the students in a school have hearing impairments. If there are 1500 students, how many have hearing impairments?

______% of ______ is ______

2 % of 1500 is p

25
Solve the problem.

2% of the students in a school have hearing impairments. If there are 1500 students, how many have hearing impairments? ______

30 students have hearing impairments

26
Fill in the blanks below with letters or numbers to fit the following problem.

What percent of a ring weighing 20 grams is silver if there are 8 grams of silver in the ring?

______% of ______ is ______

r % of 20 is 8

27
What percent of a ring weighing 20 grams is silver if there are 8 grams of silver in the ring? ______

40%

28
Fill in the blanks below with letters or numbers to fit the following problem.

What is the weight of a rock if 3% of its weight is 36 kilograms?

______% of ______ is ______

3 % of b is 36

29
What is the weight of a rock if 3% of its weight is 36 kilograms? _____

1200 kgm

30
Fill in the blanks below with letters or numbers to fit the following problem.

How many girls are in a class of 450 students if 52% of the class is girls?

_____% of _____ is _____ 52 % of 450 is p

31
How many girls are in a class of 450 students if 52% of the class is girls? _____ 234 girls

32
Fill in the blanks below with letters or numbers to fit the following problem.

The interest on a loan of $2000 is $360 for one year. Find the annual rate of interest.

_____% of _____ is _____ r % of 2000 is 360

33
The interest on a loan of $2000 is $360 for one year. Find the annual rate of interest. _____ 18%

34
Fill in the blanks below with letters or numbers to fit the following problem.

30% of the patients in a hospital had heart problems. If 210 patients had heart problems, how many patients are in the hospital?

_____% of _____ is _____ 30 % of b is 210

35
30% of the patients in a hospital had heart problems. If 210 patients had heart problems, how many patients are in the hospital? _____ 700 hospital patients

36
Fill in the blanks below with letters or numbers to fit the following problem.

A store makes a 6% profit on its total sales of $2,000,000. Find its profit.

_____% of ______ is _______

6 % of 2,000,000 is p

37
A store makes a 6% profit on its total sales of $2,000,000. Find its profit. ___________

$120,000 profit

38
A container of milk has 2.61 liters of water. If 87% of the milk is water, how much milk is in the container?
_____

87% of b is 2.61
3 liters of milk

39
A 12 liter bottle of medicine contains 2.6% active ingredients. How much of the bottle is active ingredients? _____

2.6% of 12 is p
.312 liters

40
The body weight of a man is 92% water. How much weight of a 160 pound man is water? ______

92% of 160 is p
147.2 pounds

41
At a sale, slacks were sold for $15 which was 80% of the original price. What was the original price? _____

80% of b is 15
$18.75

42
A three-year-old house is worth 125% of its original price of $36,000. How much is the house worth? _____

125% of 36,000 is p
$45,000

43
About 22,000,000 American workers are union members. If 25% of the labor force is union workers, what is the size of the labor force? ________

25% of b is 22,000,000
88,000,000 workers

44
A $600 refrigerator was sold for $360 during a sale. This was a reduction of $240. The reduction in price is what percent of the original price? ______

r % of 600 is 240
40%

45
The pollution count in a bay increased from 46 parts per million to 54 parts per million. The increase is what percent of the former rate? ______

r % of 46 is 8
17% to the nearest one percent

46
A microwave oven uses 35% less power than a conventional oven. A conventional oven requires 800 watts to cook a roast. How much power would the microwave oven require for the same roast? ______

65% of 800 is p
520 watts

FEEDBACK UNIT 6

This quiz reviews the preceding unit. Answers are at the back of the book.
Take the Chapter Mastery Test only after all problems of this unit are understood.

1. 58% of 938 is _____.
2. 16 is _____% of 298.
3. 50.66 is 34% of _____.
4. _____% of 38 is 10.
5. _____ is 6% of 47.
6. 155% of _____ is 93.
7. A 6% commission is paid on sales of $40,000. What is the commission?
8. A team won 30 games which was 40% of the games it played. How many games were played?
9. 18 girls are in a class of 42 students. What percent of the class is girls?
10. 80% of a 40 member biology class passed the course. How many students passed the course?
11. A sample of uranium ore weighs 90 kilograms. How much of it is uranium if 2.5% of the ore is uranium?
12. A lawyer's fee for a lawsuit was 35% of the award for the case. The fee was $10,500. Find the size of the award.

CHAPTER 12 MASTERY TEST

This test provides an opportunity for assuring that all of the objectives of Chapter 12 have been learned. Each problem is accompanied by the symbol (U) which indicates the Unit in which that type of problem was studied. If any problems are missed, they need to be corrected. Answers are at the back of the book.

1. Write 2% as a ratio. (1)

2. Write 110% as a ratio. (1)

3. Write 7.9% as a ratio. (1)

4. Percent is a ratio with denominator _____. (1)

5. The ratio $\frac{31}{100}$ is the percent ______. (1)

6. The ratio $\frac{5.2}{100}$ is the percent ______. (1)

7. Write 55% as a decimal. (2)

8. Write 13.7% as a decimal. (2)

9. Write .2 as a percent. (2)

10. Write .58 as a percent. (2)

11. Write $\frac{5}{8}$ to the nearest one percent. (3)

12. Write $\frac{17}{28}$ to the nearest one percent. (3)

13. Find the base (b), rate (r), and part (p) for: 110% of 70 is 77. (4)

14. Find the base (b), rate (r), and part (p) for: 51 is 75% of 68. (4)

15. Find the rate to the nearest whole number when the part is 33 and base is 68. (5)

16. Find the base when the part is 180.2 and rate is 85%. (5)

17. Find the part when the rate is 41% and base is 348. (5)

18. _____% of 80 is 40. (6)

19. 3% of _____ is 9. (6)

20. 10% of 200 is _____. (6)

21. The interest rate on a $50,000 loan for one year is 12%. Find the interest. (6)

22. A 20 liter sample of water contains 1 liter of pollution. What percent of the sample is pollution? (6)

23. An inspector found that 2% of a shipment of meat was spoiled. If 46 kilograms of the meat were spoiled, how much meat was in the shipment? (6)

24. Find 152% of 46. (6)

25. 10 is 125% of what number? (6)

Chapter 13 Objectives

The following problems describe, by example, the skills and knowledge studied in this chapter. Each problem is accompanied by a symbol, (U), which indicates the Unit where the problem is studied. Answers are at the back of the book.

1. (-7) + (+4) = _____ (1)

2. (-2) + (-6) = _____ (1)

3. (-7) – (+3) = _____ (2)

4. (+3) – (-2) = _____ (2)

5. (-5) x (+4) = _____ (2)

6. (-8) x (-3) = _____ (2)

7. (-10) ÷ (+5) = _____ (2)

8. (-15) ÷ (-3) = _____ (2)

Simplify the numerical expressions in problems 9–14 following the correct order of operations.

9. 7 + 5 x 8 – 10 ÷ 5 (3)

10. 6 – 30 ÷ 5 ÷ 2 – 8 (3)

11. 9 x 4 ÷ 6 x 2 + 5 (3)

12. 24 – 17 + 12 ÷ 3 – 7 (3)

13. 3 x 7 + (9 – 2 x 4) (3)

14. [12 – 7] x (14 – 22 ÷ 2) (3)

Convert the temperatures given in problems 15-16.

15. 50° C = ___° F (4)

16. 212° F = ___° C (4)

Chapter 13
Signed Numbers

Unit 1: Adding Signed Numbers

The Number Line

The figure shown below is a number line. It is marked with locations for 2, 4, and 6 which are positive numbers and -2, -4, and -6 which are negative numbers.

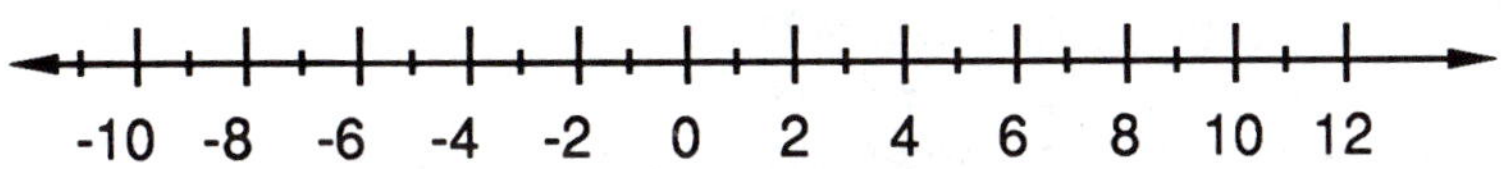

All of the numbers indicated on the number line are called **signed numbers, directed numbers**, or **integers**. In this unit they will be referred to as integers.

Every integer, except zero, consists of two parts:

1. A sign (+ or -) which indicates the integer's direction from zero.

    Positive integers are usually to the right of zero on the number line.
    Positive integers may or may not have a plus sign (+) written preceding them.

    Negative integers are to the left of zero on the number line.
    A negative integer always has a minus sign (–) preceding it.

2. A magnitude (size) which indicates the integer's distance from zero.

1

The integer +6 (which may be written as 6) is positive and has a magnitude (distance from zero) of 6. What is the magnitude of +11. ______ 11

2

The integer -4 is four units left of zero on the number line. The integer +7 is _____ units _____ of zero. seven, right

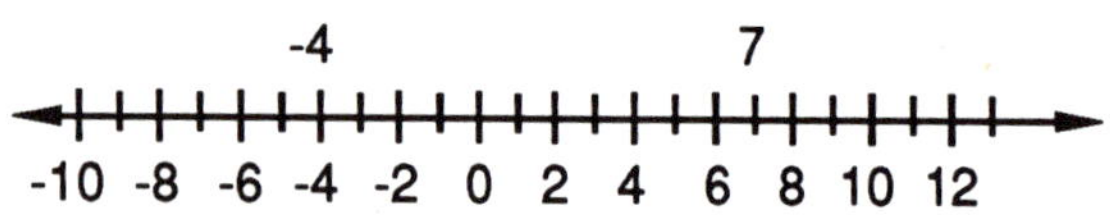

3

The integer 5 is five units right of zero on the number line. The integer -8 is _____ units _____ of zero. eight, left

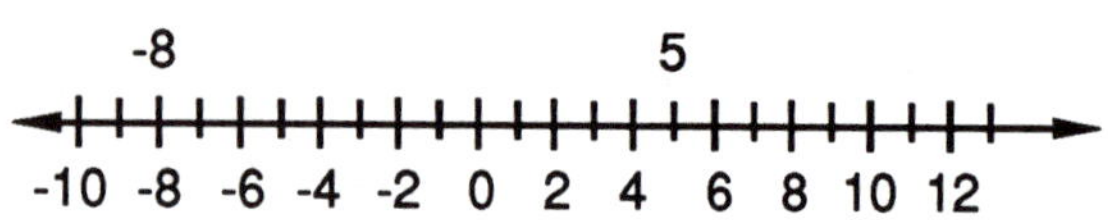

4

The **addition** of the integers -5 and -2 can be seen on the number line. Both -5 and -2 are negative and are represented by arrows that go to the left. The -5 arrow means "left 5" and the -2 arrow means "left 2." The end of the second arrow is at ______. -7 or "left 7"

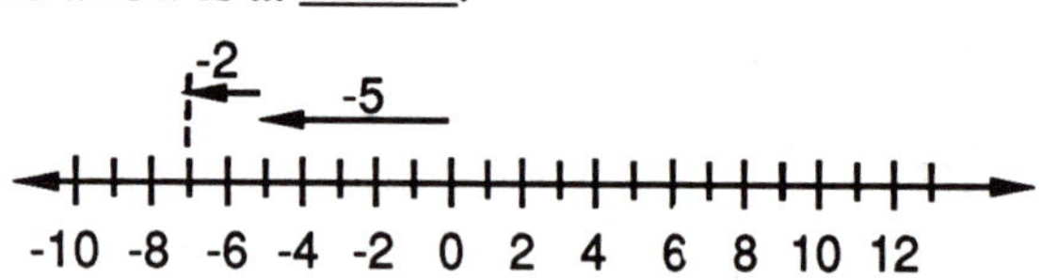

5

The **addition** of the integers -6 and +4 can be seen on the number line. The -6 arrow means "left 6" and the +4 arrow means "right 4." The end of the second arrow is at ______. -2 or "left 2"

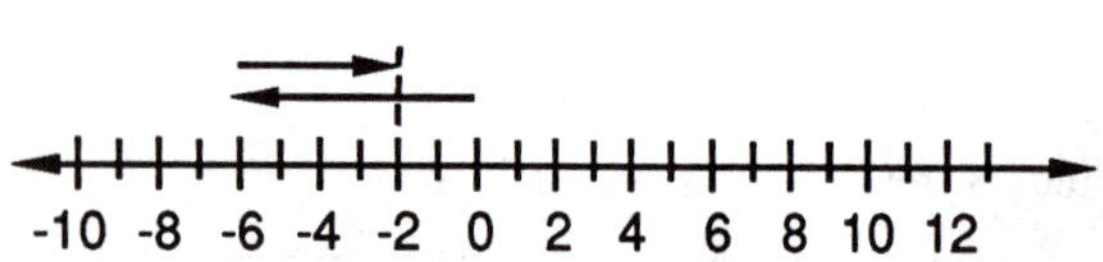

6 ✍

The addition of -5 and 8 involves a "left 5" arrow and a "right 8" arrow. The end of the second arrow is at ______.

3 or "right 3"

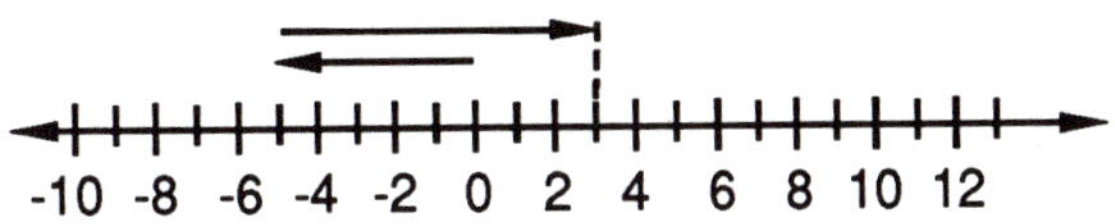

7 ✍

The addition of 7 and -5 can be seen as the connection of two arrows as shown below. The sum of 7 and -5 is ______.

2

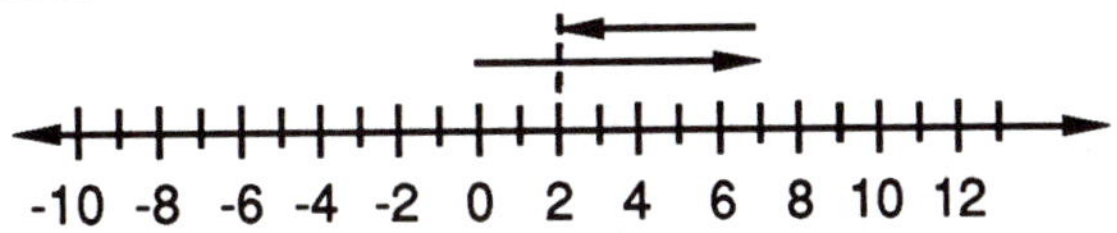

ADDING INTEGERS

To add two integers means to:

1. Begin at 0 on the number line and go the direction and distance of the first integer.

2. Begin at the end of the first integer arrow and go the direction and distance of the second integer.

3. The answer (sum) of the addition is the point on the number line where the second integer arrow ends.

Notice that the addition of integers does not necessarily make the answer greater than either of the addends. In fact, addition of integers will frequently seem like a form of subtraction.

8

Find the sum of -6 and 2. If needed, draw arrows on a number line. -6 + 2 = ______

-4

9
Find the sum of -5 and -3. If needed, draw arrows on a number line. -5 + (-3) = ______ -8

10
Find the sum of 7 and 4. If needed, draw arrows on a number line. (+7) + (+4) = ______ 11

11
Find the sum of -4 and 4. If needed, draw arrows on a number line. (-4) + (+4) = ______ 0

12
(-5) + (+6) = ______ 1

13
(+6) + (-8) = ______ -2

14
(-1) + (+3) = ______ 2

15
(-2) + (-3) = ______ -5

16
(14) + (-9) = ______ 5

17
(-9) + (-5) = ______ -14

18
(-11) + (4) = ______ -7

FEEDBACK UNIT 1

This quiz reviews the preceding unit. Answers are at the back of the book. Begin the next unit only after all problems of this unit are understood.

If needed, draw number line arrows to add.

1. (-7) + (-3) = ______
2. (6) + (-1) = ______
3. (4) + (5) = ______
4. (-5) + (9) = ______
5. (-3) + (1) = ______
6. (3) + (-8) = ______
7. (-4) + (-2) = ______
8. (-11) + (11) = ______

UNIT 2: SUBTRACTING, MULTIPLYING, AND DIVIDING SIGNED NUMBERS

SUBTRACTING INTEGERS

To subtract two integers such as (-4) – (-7):

1. Change the subtraction sign (–) to addition (+) by changing the sign of the subtrahend. (+ to –) or (– to +)   (-4) – (-7)   (-4) + (+7)
2. Follow the new signs and add the integers.   (-4) + (+7) = 3

1 ✍

To **subtract** (-2) – (-7), make the following two changes:

a. Change the subtraction to an addition, and
b. Change the sign of the subtrahend.

Make the two required changes for (-2) – (-7).

______

(-2) + (+7)

2
Complete the following subtraction.
(-2) – (-7) = (-2) + (+7) = ______

5

3 ✍
To **subtract** (+5) – (+3), make the following two changes:
a. Change the subtraction to an addition, and
b. Change the sign of the subtrahend.
Make the two required changes for (+5) – (+3).
______

(+5) + (-3)

4
Complete the following subtraction.
(+5) – (+3) = (+5) + (-3) = ______

2

5 ✍
To **subtract** (+6) – (-3), make the following two changes:
a. Change the subtraction to an addition, and
b. Change the sign of the subtrahend.
Make the two required changes for (+6) – (-3).
______

(+6) + (+3)

6
Complete the following subtraction.
(+6) – (-3) = (+6) + (+3) = ______

9

7
Subtract. (-4) – (+5). ______

(-4) + (-5) = -9

8
Subtract. (+5) – (+8). ______

(+5) + (-8) = -3

9
(-6) – (+6) = ______

(-6) + (-6) = -12

10
(9) – (-1) = ______

(9) + (+1) = 10

MULTIPLYING/DIVIDING INTEGERS

To multiply/divide two integers,

1. The sign of the product/quotient is:

   Plus (+) if both integers have the same sign.

   Minus (-) if the integers have opposite signs.

2. The magnitude of the answer is the product/quotient of the magnitudes.

   (-6) x (-3) = +18 (-12) ÷ (+4) = -3

11 ✍
The product of two integers with the same sign is positive.
(-2) x (-7) = +14 and (-3) x (-8) = ____ +24

12 ✍
The quotient of two integers with opposite signs is negative.
(-21) ÷ (+7) = -3 and (+32) ÷ (-8) = ____ -4

13 ✍
If two integers in a division problem have the same sign, the quotient is _________ (positive, negative). positive

14 ✍
If two integers in a division problem have different signs, the quotient is _________ (positive, negative). negative

15
The rules for signs for multiplication and division of integers are _____________ (the same, different). the same

16
(-6) x (-2) = +12 (-6) x (+2) = -12
If the signs of the integers in a multiplication problem are the same, the product is ____________ (positive, negative). positive

17

(-8) ÷ (-4) = +2 (-8) ÷ (+4) = -2

If the signs of the integers in a division problem are different, the quotient is ____________ (positive, negative). negative

18

(-15) x (-2) = +30 and (-15) x (+2) = _____ -30

19

(-10) ÷ (+5) = -2 and (-10) ÷ (-5) = _____ +2

20

(30) x (-5) = _____ -150

21

(+7) x (+3) = _____ 21

22

(-8) x (-7) = _____ 56

23

(9) x (-7) = _____ -63

24

(-20) ÷ (5) = _____ -4

25

(18) ÷ (9) = _____ 2

26

(24) ÷ (-6) = _____ -4

27

(-40) ÷ (-5) = _____ 8

28

Any number multiplied by zero gives zero.

(-9) x 0 = 0 and (+13) x 0 = ______ 0

29

Zero divided by any non–zero number gives zero.

0 ÷ (-13) = 0 and 0 ÷ (+8) = ______ 0

30

0 ÷ (-2) = ______ 0

31

(-12) x 0 = ______ 0

32

0 x (+25) = ______ 0

FEEDBACK UNIT 2

This quiz reviews the preceding unit. Answers are at the back of the book. Begin the next unit only after all problems of this unit are understood.

1. (-4) – (-3) = ______
2. (-4) – (+7) = ______
3. (+6) – (8) = ______
4. (2) – (-1) = ______
5. (-6) x (5) = ______
6. (4) x (9) = ______
7. (-3) x (-1) = ______
8. (4) x (-4) = ______
9. (-18) ÷ (9) = ______
10. (26) ÷ (13) = ______
11. (-16) ÷ (-8) = ______
12. (35) ÷ (-7) = ______

UNIT 3: THE ORDER OF OPERATIONS

1
Addition, subtraction, multiplication, and division are operations that are performed on **two** numbers. Is it possible to add three numbers in a **single** step? ______ — No

2
To add three numbers, first find the sum of two of them. Then add the third number to that sum. How many numbers are added in each step? ______ — Two

3 ✍
Addition, subtraction, multiplication, and division may be performed with only _____ numbers at a time. — two

4
To add 5 + 3 + 7,
a. Add 5 and 3.
b. Add 7 to the sum.
The final answer is _____.

$$5 + 3 + 7$$
$$(5 + 3) + 7$$
$$8 + 7$$

15

5
Add 9 + 4 + 11 by first adding 9 and 4. ______ — $13 + 11 = 24$

6
To subtract 15 − 8 − 3,
a. Subtract 8 from 15.
b. Subtract 3 from the difference.
The final answer is _____.

$$15 - 8 - 3$$
$$(15 - 8) - 3$$
$$7 - 3$$

4

7
Subtract 19 − 4 − 8 by first subtracting 4 from 19. ______ — $15 - 8 = 7$

8

To multiply $5 \times 8 \times 3$,
a. Multiply 5 and 8.
b. Multiply 3 and the product.
The final answer is _____.

$5 \times 8 \times 3$
$(5 \times 8) \times 3$
40×3

120

9

Multiply $2 \times 7 \times 5$ by first multiplying 2 and 7. ______

$14 \times 5 = 70$

10

To divide $50 \div 5 \div 2$,
a. Divide 50 by 5.
b. Divide the quotient by 2
The final answer is _____.

$50 \div 5 \div 2$
$(50 \div 5) \div 2$
$10 \div 2$

5

11

Divide $80 \div 2 \div 5$ by first dividing 80 by 2.

______

$40 \div 5 = 8$

12

When a problem involves only one operation it is done from left to right. For example, the problem $19 + 6 + 7$ involves only one operation, addition, and the work is completed from left to right.
Add $19 + 6 + 7$. ______

$(19 + 6) + 7$
$25 + 7$
32

13

When a problem involves both multiplication and addition, the multiplication is completed before the addition.
To calculate $12 + 5 \times 2$,
a. Multiply 5 and 2.
b. Add.

$12 + 5 \times 2$
$12 + 10$
22

The first step in a problem involving both addition and multiplication is to first complete the ____________.

multiplication

14

Calculate the answer for the problem shown below by first doing any multiplication and then the addition.

$$6 + 5 \times 3$$

$6 + 15$

21

15

Calculate the answer for the problem shown below by first doing any multiplication and then the addition.

$$4 \times 7 + 9 \times 2$$

$28 + 18$

46

16

Multiplication or division are completed first in calculating any numerical expression. Do just the first step in calculating $8 + 9 - 4 \div 2$. ______

$8 + 9 - 2$

17

Calculate $8 + 9 - 4 \div 2$. _____

$8 + 9 - 2$

$17 - 2$

15

18

In calculating a numerical expression,

a. First do any multiplication/division from left to right

b. Then do any addition/subtraction from left to right.

Calculate $6 - 30 \div 10$. _____

$6 - 3 = 3$

19

In calculating a numerical expression,

a. First do any multiplication/division from left to right

b. Then do any addition/subtraction from left to right.

Calculate $2 \times 10 \div 5 + 6 \div 3$. _____

$20 \div 5 + 6 \div 3$

$4 + 6 \div 3$

$4 + 2$

6

20

In calculating a numerical expression,

a. First do any multiplication/division from left to right

b. Then do any addition/subtraction from left to right.

Calculate $15 - 14 \div 2 + 4 \times 3$. _____

$15 - 7 + 4 \times 3$
$15 - 7 + 12$
$8 + 12$
20

21

Calculate $9 + 16 \div 2 \times 7 - 17$. _____

$9 + 8 \times 7 - 17$
$9 + 56 - 17$
$65 - 17$
48

22

Calculate $22 - 13 + 12 \div 12 \times 5$. _____

$22 - 13 + 1 \times 5$
$22 - 13 + 5$
$9 + 5$
14

23 ✍

If parentheses ( ) or square brackets [ ] appear in a problem, then the numerical expression contained in them is calculated first. Complete the calculation of $2 \times (9 + 20 \div 5) - 11$ which is shown below.

$$2 \times (9 + 20 \div 5) - 11$$
$$2 \times (9 + 4) - 11$$
$$2 \times 13 - 11$$

____

$26 - 11 = 15$

24

If parentheses ( ) or square brackets [ ] appear in a problem, then the numerical expression contained in them is calculated first. Complete the calculation of $[8 - 5] \times (3 + 15 \div 3)$ which is shown below.

$$[8 - 5] \times (3 + 15 \div 3)$$
$$3 \times (3 + 15 \div 3)$$
$$3 \times (3 + 5)$$

____

$3 \times 8 = 24$

25
Calculate [8 + 2 x 5] − (7 − 18 ÷ 9). ______

[8 + 10] − (7 − 2)
18 − 5
13

26
Calculate [3 x 10 ÷ 6] x (13 − 12 ÷ 4). ______

[30 ÷ 6] x (13 − 3)
5 x 10
50

27
Calculate 108 − 9 x (42 ÷ 14 x 3). ______

108 − 9 x (3 x 3)
108 − 9 x 9
108 − 81
27

FEEDBACK UNIT 3

This quiz reviews the preceding unit. Answers are at the back of the book. Begin the next unit only after all problems of this unit are understood.

1. 17 + 5 x 6 = ______
2. 19 − 3 x 14 ÷ 6 + 4 = ______
3. 20 ÷ 5 + 6 − 12 ÷ 2 = ______
4. 6 + 8 x 3 − 30 ÷ 10 = ______
5. (17 − 2 x 3) x 4 − 6 = ______
6. 14 − (35 − 4 x 7) + 5 = ______
7. 6 x (3 + 8) − [45 ÷ 9] = ______
8. [13 + 2 x 6] ÷ (19 − 7 x 2) = ______

UNIT 4: TEMPERATURE CONVERSIONS

Two methods of measuring temperature are commonly used. The Fahrenheit (F) system is the most common method in the United States, but in Europe and in most science applications the Celsius (C) method is more common.

Water freezes at 32° on the Fahrenheit scale, but at 0° on the Celsius scale. Similarly, water boils at 212° Fahrenheit, but at 100° Celsius.

The formulas for converting temperatures between the two systems are:

$$F = \frac{9}{5}C + 32 \qquad C = \frac{5}{9}(F - 32)$$

1

The formula $F = \frac{9}{5}C + 32$ is used to convert a Celsius temperature to a Fahrenheit temperature. The formula $C = \frac{5}{9}(F - 32)$ is used to convert a ________ temperature to a ______________ temperature.

Fahrenheit
Celsius

2

To change 16° C to Fahrenheit, replace C with 16 in the formula.

$$\begin{aligned} F &= \frac{9}{5}C + 32 \\ &= \frac{9}{5} \times 16 + 32 \\ &= \frac{144}{5} + 32 \\ &= 28.8 + 32 \\ &= 60.8 \end{aligned}$$

Therefore, 16° C = ____F

60.8° F

3
Complete the steps below to change 30° C to a Fahrenheit temperature. Notice the cancellation used in simplifying the computation.

$$F = \frac{9}{5}C + 32$$
$$= \frac{9}{5} \times \frac{30}{1} + 32$$
$$= \frac{9}{1} \times \frac{6}{1} + 32$$
$$= \_\_\_\_\_\_$$

54 + 32 = 86° F

4
Change 55° C to a Fahrenheit temperature. Use cancellation to simplify the computation.

$$F = \frac{9}{5}C + 32$$
$$= \frac{9}{5} \times 55 + 32$$
$$= \_\_\_\_\_\_$$

99 + 32 = 131° F

5
Change 10° C to a Fahrenheit temperature. ______

18 + 32 = 50° F

6
Change 60° C to a Fahrenheit temperature. ______

108 + 32 = 140° F

7 ✍
To change 41° F to Celsius replace F with 41 in the formula.

$$C = \frac{5}{9}(F - 32)$$
$$= \frac{5}{9}(41 - 32)$$
$$= \frac{5}{9}(9)$$
$$= \frac{5}{9} \times \frac{9}{1}$$

Therefore, 41° F = ____C

5° C

8
To change 86° F to Celsius replace F with 86 in the formula.

$$C = \frac{5}{9}(F - 32)$$
$$= \frac{5}{9}(86 - 32)$$
$$= \frac{5}{9}(54)$$
$$= \frac{5}{9} \times \frac{54}{1}$$

Therefore, 86° F = ____C

30° C

9
Complete the process shown below of changing -13° F to a Celsius temperature.

$$C = \frac{5}{9}(F - 32)$$
$$= \frac{5}{9}([-13] - 32)$$
$$= \frac{5}{9}(-45)$$

Therefore, -13° F = ____C

$\frac{5}{1} \times -5 = -25°$ C

10
Complete the process shown below of changing 14° F to a Celsius temperature.

$$C = \frac{5}{9}(F - 32)$$
$$= \frac{5}{9}(14 - 32)$$
$$= \frac{5}{9}(-18)$$

Therefore, 14° F = ____C

$\frac{5}{1} \times -2 = -10°$ C

11
Change 68° F to a Celsius temperature. ______

$\frac{5}{9} \times 36 = 20°$ C

12
Change 40° F to a Celsius temperature. ______

$\frac{5}{9} \times 8 \doteq 4.4°$ C

13
Change -40° F to a Celsius temperature. ______

$\frac{5}{9} \times -72 = -40°$ C

14
Change 90° F to a Celsius temperature. ______ $\frac{5}{9}$ x 58 ≐ 32.2° C

15
Change 212° F to a Celsius temperature. ______ $\frac{5}{9}$ x 180 = 100° C

16
Change 95° F to a Celsius temperature. ______ $\frac{5}{9}$ x 63 = 35° C

FEEDBACK UNIT 4

This quiz reviews the preceding unit. Answers are at the back of the book. Begin the next unit only after all problems of this unit are understood.

Change the following Celsius temperatures to Fahrenheit readings.

1. 40° C = ____F
2. 65° C = ____F
3. 100° C = ____F
4. 0° C = ____F

Change the following Fahrenheit temperatures to Celsius readings.

5. 86° F = ____C
6. 60° F = ____C
7. 32° F = ____C
8. 23° F = ____C

CHAPTER 13 MASTERY TEST

This test provides an opportunity for assuring that **all** of the objectives of Chapter 13 have been learned. Each problem is accompanied by the symbol (U) which indicates the Unit in which that type of problem was studied. If any problems are missed, they need to be corrected. Answers are at the back of the book.

1. (-6) + (-3) = _____ (1)

2. (-5) + (+8) = _____ (1)

3. (-4) – (-6) = _____ (2)

4. (-9) – (3) = _____ (2)

5. (-5) x (-2) = _____ (2)

6. (6) x (-3) = _____ (2)

7. (-12) ÷ (+6) = _____ (2)

8. (-24) ÷ (-3) = _____ (2)

Simplify the numerical expressions in problems 9–14 following the correct order of operations.

9. 6 + 4 x 7 – 12 ÷ 3 (3)

10. 9 – 40 ÷ 2 ÷ 5 – 8 (3)

11. 7 x 6 ÷ 14 x 3 + 6 (3)

12. 31 – 14 + 15 ÷ 5 – 3 (3)

13. 9 x 8 + (13 – 3 x 2) (3)

14. [17 – 12] x (19 – 30 ÷ 2) (3)

Convert the temperatures given in problems 15-16.

15. 35° C = ___° F (4)

16. 77° F = ___° C (4)

CHAPTER 14 OBJECTIVES

The following problems describe, by example, the skills and knowledge studied in this chapter. Each problem is accompanied by a symbol, (U), which indicates the Unit where the problem is studied. Answers are at the back of the book.

1. 200 millimeters = _____ meters (1)
2. 34.5 kilometers = _____ meters (1)
3. 256 milliliters = _____ liters (1)
4. 1,250 grams = _____ kilograms (1)
5. Identify each of the angles at the right as acute, right, or obtuse. (2)

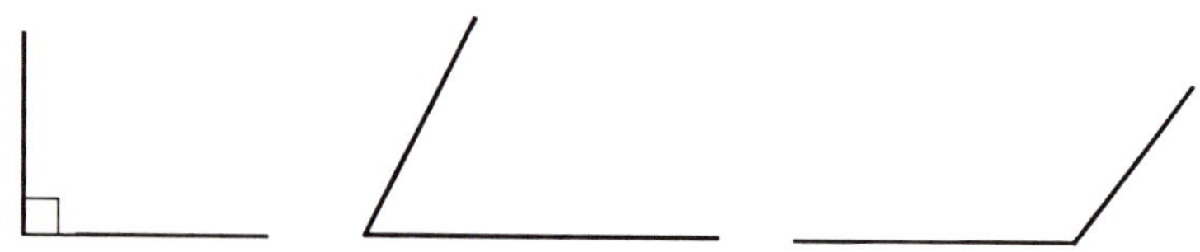

a. b. c.

6. Identify each of the quadrilaterals at right as trapezoid, parallelogram, rectangle, or square. (2)

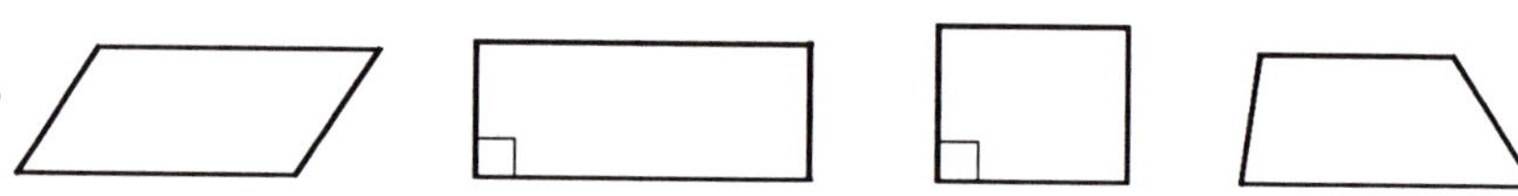

a. b. c. d.

7. Find the perimeter of the triangle. (2)

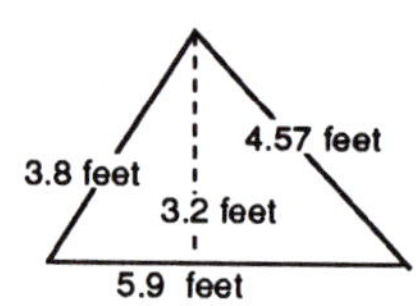

8. Find the area of the triangle. (2)

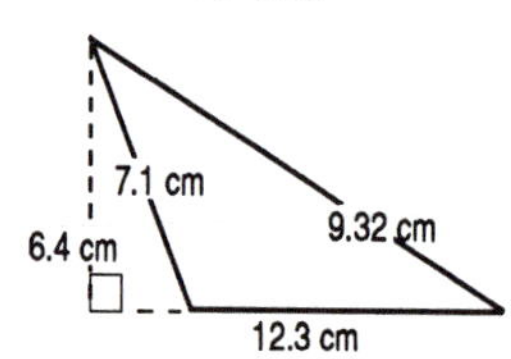

9. Find the perimeter of the quadrilateral. (2)

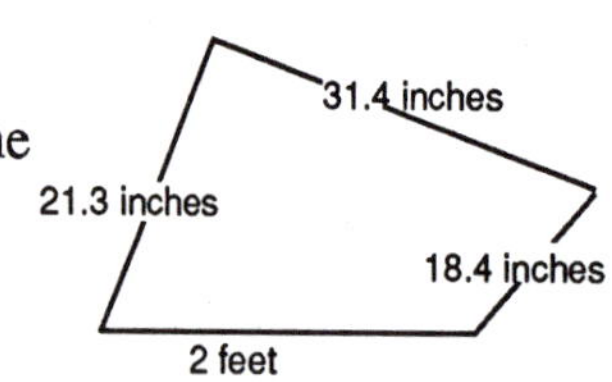

10. Find the area of the rectangle. (2)

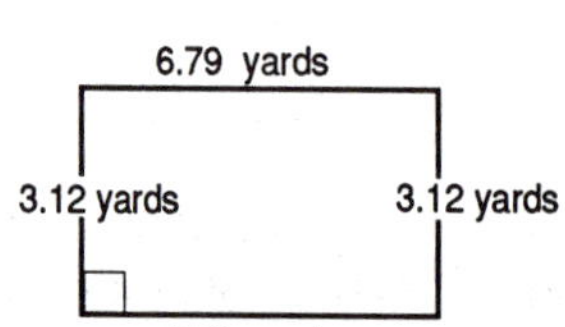

11. Find the circumference of the circle. Use 3.14 as the approximation for π. (2)

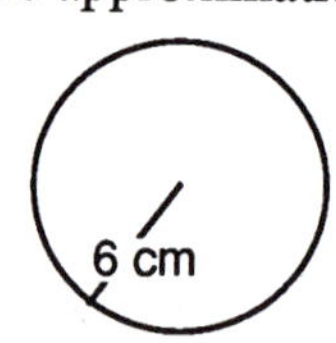

12. Find the area of the circle. Use $3\frac{1}{7}$ as the approximation for π. (2)

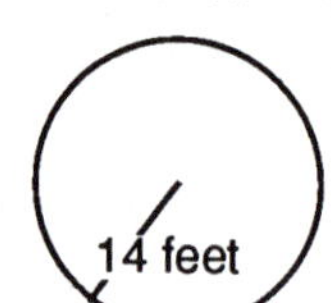

13. Find the missing side of the similar triangle. (2)

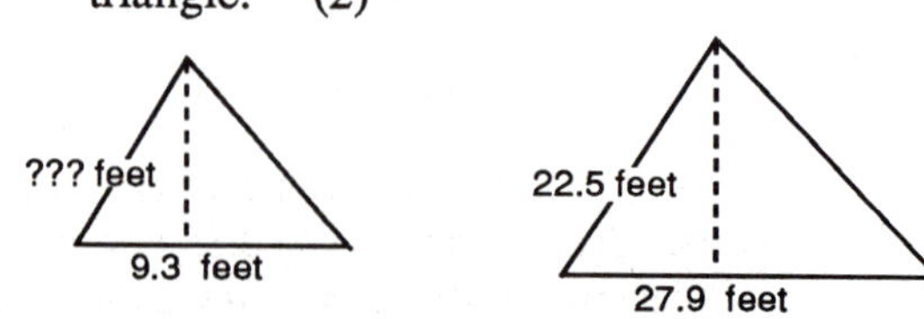

CHAPTER 14
THE METRIC SYSTEM AND GEOMETRY REVIEW

UNIT 1: METRIC MEASURES

Two different systems for measuring length or distance are often used in the United States. Most distance measurements are stated in the English system which uses units such as the inch, foot, yard, and mile. Changing from one unit of measure to another involves no constant multiple and often requires a fraction.

1 foot = 12 inches	or	$\frac{1}{12}$ foot = 1 inch
1 yard = 36 inches	or	$\frac{1}{36}$ yard = 1 inch
1 yard = 3 feet	or	$\frac{1}{3}$ yard = 1 foot
1 mile = 5,280 feet	or	$\frac{1}{5280}$ mile = 1 foot
1 mile = 1,760 yards	or	$\frac{1}{1760}$ mile = 1 yard

The metric system is also used frequently in highway signs and science classes. The metric system differs from the English system because it has only one unit of measure — the meter. All other measurements in the metric system are given in terms of the meter and are written using prefixes such as milli, centi, deci, deca, hecto, and kilo. A meter is approximately 39.37 inches or about three inches more than a yard.

1 millimeter	= .001	meters	or	1000 millimeters	=	1 meter
1 centimeter	= .01	meters	or	100 centimeters	=	1 meter
1 decimeter	= .1	meters	or	10 decimeters	=	1 meter
1 meter	= 1	meter	or	1 meter	=	1 meter
1 decameter	= 10	meters	or	.1 decameter	=	1 meter
1 hectometer	= 100	meters	or	.01 hectometer	=	1 meter
1 kilometer	= 1000	meters	or	.001 kilometer	=	1 meter

One of the most valuable aspects of the metric system is the ease of changing from one prefix to another. For example, to convert 573.6 centimeters to meters, the decimal point is moved two places to the left. This move of the decimal point gives 5.736 meters which is equal to 573.6 centimeters.

1

The conversion of 5 feet to inches depends on the fact that each one foot contains 12 inches. How many inches are in 5 feet? ______

5 x 12
60 inches

2

The conversion of 13 yards to feet depends on the fact that each one yard contains 3 feet. How many feet are in 13 yards? ______

13 x 3
39 feet

3

The conversion of 2 miles to feet depends on the fact that each one mile contains 5,280 feet. How many feet are in 2 miles? ______

2 x 5280
10,560 feet

4

The conversion of 50 inches to yards depends on the fact that each one inch contains $\frac{1}{36}$ of a yard. How many yards are in 50 inches? ______

$50 \times \frac{1}{36}$
$1\frac{7}{18}$ yards

5

Conversions in the English system often involve the types of computations shown in the problem of frame 4. Conversions in the metric system ________ (do, do not) require that type of computation.

do not

6
Conversions in the metric system require moving the decimal point, but there are no other fractions involved. Is there any conversion in the metric system that will require multiplying by $\frac{1}{3}$? _____

No

7
The conversion of 2 meters to centimeters depends on the fact that one meter contains 100 centimeters. How many centimeters are in 2 meters? ______

2 x 100
200 centimeters

8
The prefix "centi" means $\frac{1}{100}$ or .01 and refers to the fact that a centimeter is one hundredth of a meter. A centimeter is slightly more than one-third of an inch. To change 563 centimeters to meters the decimal point is moved two places to the left. Change 563 centimeters to meters. _____

563 x .01
5.63 meters

9
Change 4.2 meters to centimeters. _____

4.2 x 100
420 centimeters

10
The prefix "milli" means $\frac{1}{1000}$ or .001 and refers to the fact that a millimeter is one thousandth of a meter. A millimeter is one tenth of a centimeter. To change 376 millimeters to meters the decimal point is moved three places to the left. Change 376 millimeters to meters. ____

376 x .001
.376 meters

11
Change 3.49 meters to millimeters. _____

3.49 x 1000
3,490 millimeters

12
The prefix "deci" means $\frac{1}{10}$ or .1 and refers to the fact that a decimeter is one tenth of a meter. To change 57.8 decimeters to meters the decimal point is moved one place to the left. Change 57.8 decimeters to meters. ____

57.8 x .1
5.78 meters

13
Change .79 meters to decimeters. _____

.79 x 10
7.9 decimeters

14
The prefix "deca" means ten and refers to the fact that a decameter is 10 meters. To change 4.82 decameters to meters the decimal point is moved one place to the right. Change 4.82 decameters to meters. ____

4.82 x 10
48.2 meters

15
Change 57.3 meters to decameters. _____

57.3 x .1
5.73 decameters

16
The prefix "hecto" means one hundred and refers to the fact that a hectometer is 100 meters. To change .78 hectometers to meters the decimal point is moved two places to the right. Change .78 hectometers to meters. ____

.78x 100
78 meters

17
Change 4,387 meters to hectometers. _____

4387 x .01
43.87 hectometers

18
The prefix "kilo" means one thousand and refers to the fact that a kilometer is 1000 meters. A kilometer is about three-fifths of a mile and driving 70 kilometers per hour is close to the speed of 43 miles per hour. To change 5.6 kilometers to meters the decimal point is moved three places to the right. Change 5.6 kilometers to meters. ____

5.6 x 1000
5,600 meters

19
Change 569 meters to kilometers. _____

569 x .001
.569 kilometers

20
Change 876 meters to kilometers. _____

876 x .001
.876 kilometers

21 ✍

Metric measures for weight use the gram as the basic unit, but the prefixes are the same as for the distance measures. For example, a kilogram is 1000 grams. One kilogram is approximately 2.2 pounds. What is the approximate weight in pounds of 4 kilograms? _____

8.8 pounds

22

A kilogram is 1000 grams. A decigram is .1 gram. What is a centigram? ______

.01 gram

23

Change 573 grams to kilograms. _____

573 x .001
.573 kilograms

24 ✍

Metric measures for volume use the liter as the basic unit. A liter is approximately 1.05 quarts. What is the approximate measure in quarts of 4 liters? _____

4.2 quarts

25

As with the other metric measures, the prefixes for the volume measures are the same as for the distance measures. For example, a milliliter is .001 liter. What is a hectoliter? ______

100 liters

FEEDBACK UNIT 1

This quiz reviews the preceding unit. Answers are at the back of the book. Begin the next unit only after all problems of this unit are understood.

Complete the following conversions.

1. 1 meter to centimeters
2. 1 centimeter to meters
3. 1,000 kilometers to meters
4. 1,000 millimeters to meters
5. 10 centimeters to meters
6. 1,000 milliliters to liters
7. 1,000 meters to kilometers
8. 1,000 grams to kilograms
9. 1.65 decagrams to grams
10. 550 meters to kilometers

UNIT 2: GEOMETRY

1
The angle shown at the right is a **right angle**. Every right angle measures 90 degrees. What is the measure for an angle that is 31 degrees (31°) less than a right angle? ______

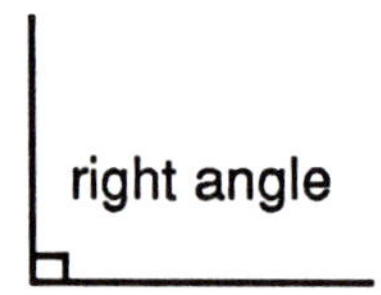

90 – 31 = 59
59°

2
An angle that is smaller than a right angle is an **acute angle**. Angles with measures of 45°, 18°, or 73° are ________ angles.

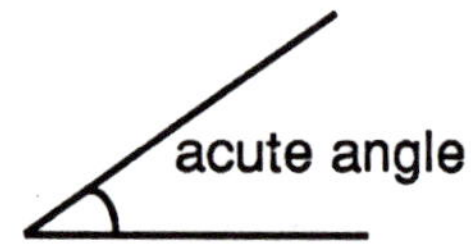

acute

3
Two acute angles are **complementary** when the sum of their measures is 90°. For example, a 60° angle and a 30° angle are complementary. 30 + 60 = 90. Find the complementary angle to a 40° angle. ______

90 – 40 = 50
50°

4
The angle shown at the right is a **straight angle**. Every straight angle measures 180°. What is the measure for an angle that is 57° less than a straight angle? ______

180 – 57 = 123
123°

5
An angle that is smaller than a straight angle and larger than a right angle is an **obtuse angle**. Angles with measures of 145°, 118°, or 93° are ________ angles.

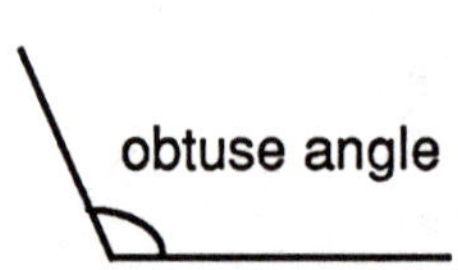

obtuse

6

Two angles are **supplementary** when the sum of their measures is 180°. For example, a 60° angle and a 120° angle are supplementary. 60 + 120 = 180. Find the supplementary angle to a 35° angle. ______

180 – 35 = 145
145°

7

The sum of the measures of any triangle's angles is 180°. For the triangle at the right, find the size of its third angle using the fact that the sum of the angles is 180°.

______

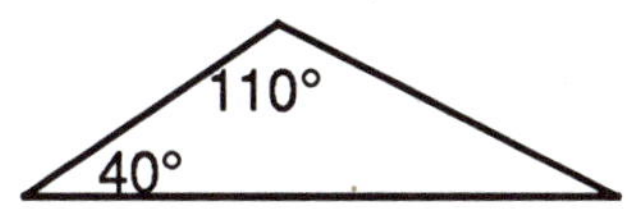

40 + 110 = 150
180 – 150 = 30
30°

8

To add two numbers they must have the same label. Is it possible to find the total of 19 cows and 483 cows? _____

Yes

9

To add 15 inches to 1 foot, both measurements need to be expressed as inches. 1 foot equals 12 inches. What is the sum of 15 inches and 1 foot? ______

27 inches

10

The **perimeter** of a geometric figure is the sum of the lengths of its sides. To find the perimeter of a triangle, _____ the lengths of its three sides.

add

11

The **perimeter** of the quadrilateral shown at the right is the sum of the lengths of the four sides. Find the perimeter. ______

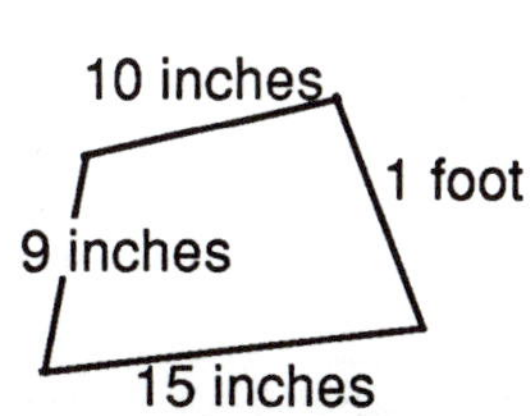

[Note: A **quadrilateral** is a four-sided figure.]

1 foot = 12 inches

12 + 15 + 9 + 10
46 inches is the perimeter

12
The perimeter of the **trapezoid** shown at the right is the sum of the lengths of the four sides. Find the perimeter. ______

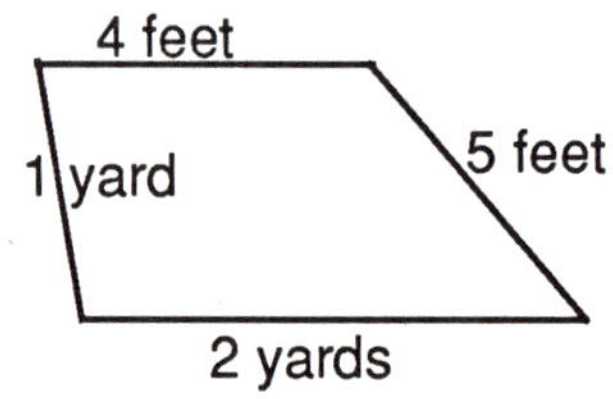

[Note: A **trapezoid** is a quadrilateral with one pair of parallel sides.]

1 yard = 3 feet

3 + 6 + 5 + 4
18 feet is the perimeter

13
The perimeter of the **parallelogram** shown at the right is the sum of the lengths of the four sides. Find the perimeter. ______

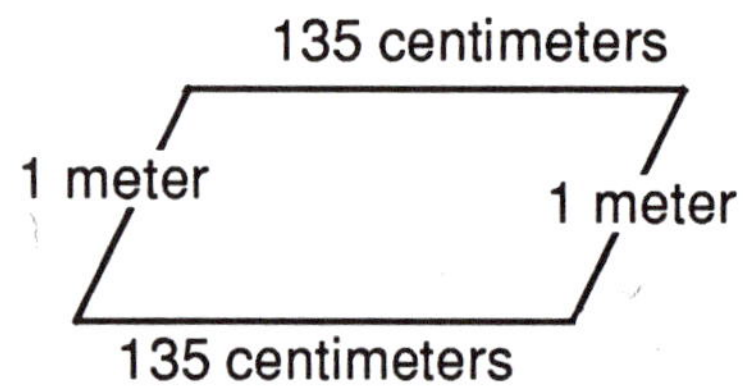

[Note: A **parallelogram** is a quadrilateral with two pairs of parallel and equal sides.]

100 cm = 1 meter

100 + 135 + 100 + 135
470 cm is the perimeter

14
Find the perimeter of the **rectangle** at the right.

______

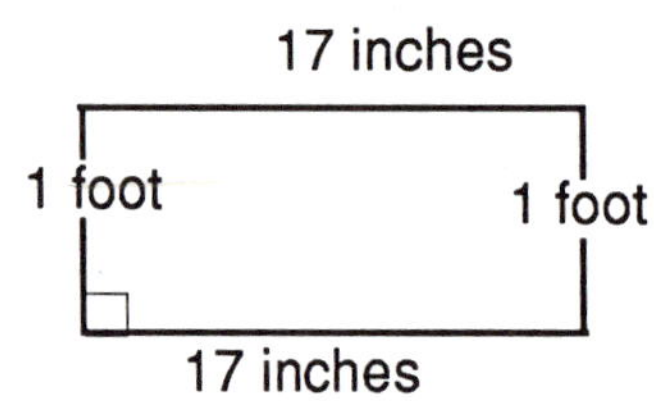

[Note: A **rectangle** is a parallelogram with four right angles. The small square in the corner indicates a right angle.]

12 + 17 + 12 + 17
58 inches is the perimeter

15
The **area** of the **rectangle** shown at the right is the product of its length and width. Find the area. ______

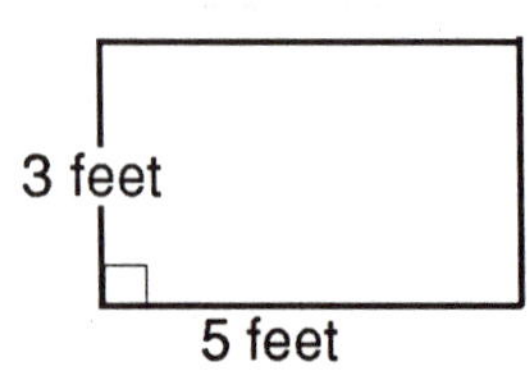

5 x 3 = 15
15 square feet is the area

16
The **area** of a rectangle is measured in terms of the number of small squares that it can contain. How many 1-inch squares are contained in the rectangle shown at the right? _____

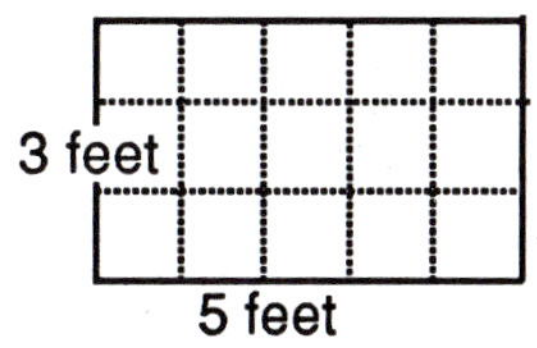

15

17
The formula for finding the area of a **rectangle** is A = LW where L is the length of the rectangle and W is the width. Use the formula to find the area of the rectangle shown at the right. ______

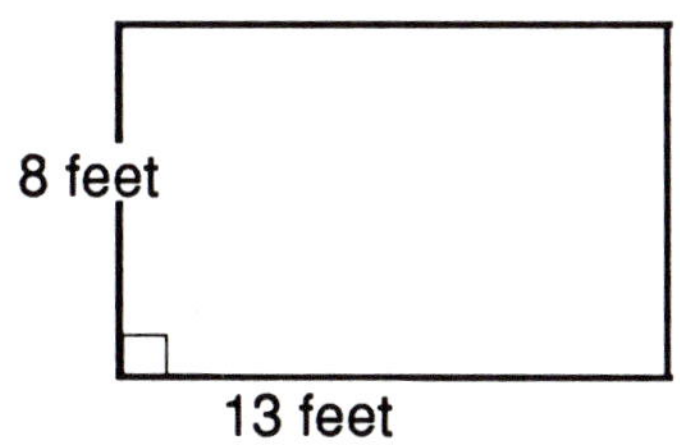

A = L x W
= 13 x 8
104 square feet is the area

18
Area is always measured in square units such as square inches, square feet, or square centimeters. Find the area of a rectangle with length 20 yards and width 9 yards.

______

A = L x W
= 20 x 9
180 square yards is the area

19
The formula for finding the area of a **parallelogram** is A = BH where B is the base of the parallelogram and H is the height. For the parallelogram shown at the right, the base is 13 meters and the height is _______.

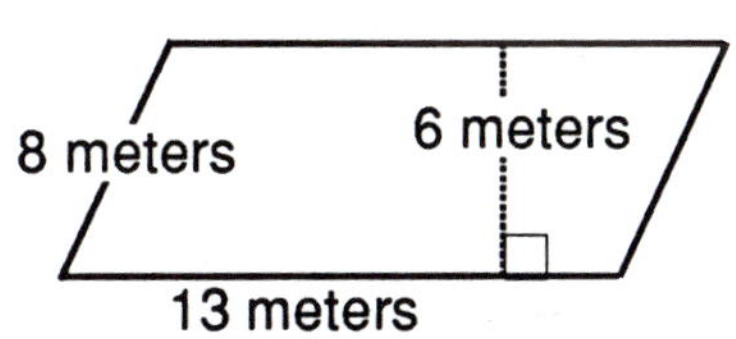

(Note: Height is perpendicular to the base and makes a right angle with it.)

6 meters is the height
8 meters is the length of a side
13 meters is the base

20
Use the formula A = BH to find the area of the parallelogram shown at the right. ______

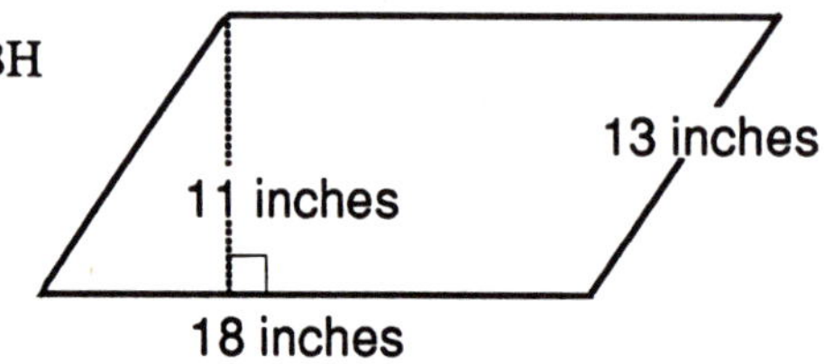

A = B x H
= 18 x 11
198 square inches is the area

21
The distance around a circle is slightly more than 3 times the length of its diameter. For the circle shown at the right, the diameter is _______.

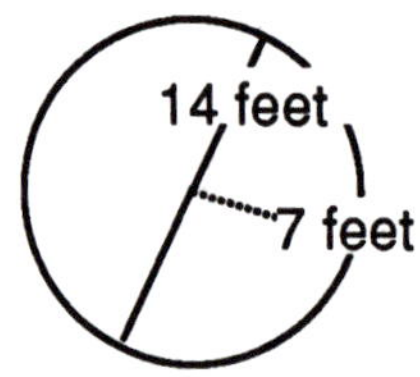

14 feet
7 feet is the radius

22
The distance around a circle is its **circumference**. The circumference can be roughly approximated by multiplying 3 times its diameter. Find a rough approximation for the circumference of the circle shown at the right.

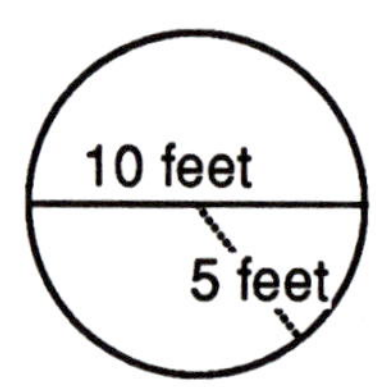

C is approximately
3 x 10 or 30 feet

23
The circumference of a circle is slightly more than 3 times its diameter. What is a rough approximation for the **circumference** of the circle shown at the right? _______

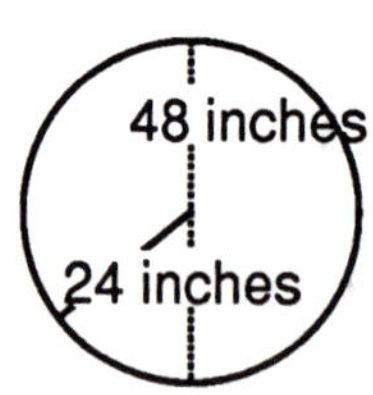

C is approximately
3 x 48 or 144 inches

24
The **perimeter** of a geometric figure is the sum of the lengths of its sides. To find the perimeter of a triangle, _____ the lengths of its three sides.

add

25
The **perimeter** of the scalene triangle shown at the right is the sum of the lengths of the three sides. Find the perimeter. ______

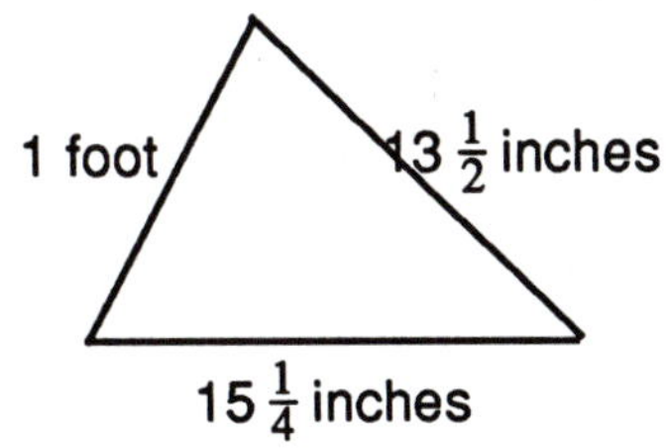

[Note: A **scalene triangle** has three sides of different lengths.]

1 foot = 12 inches

$12 + 13\frac{1}{2} + 15\frac{1}{4}$
$40\frac{3}{4}$ inches is the perimeter

26
The **perimeter** of the **isosceles triangle** shown at the right is the sum of the lengths of the three sides. Find the perimeter. ______

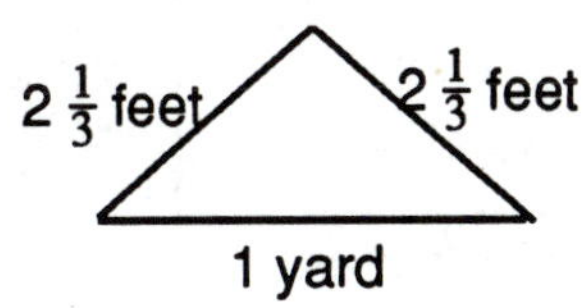

[Note: An **isosceles triangle** has two sides of the same length.]

1 yard = 3 feet

$3 + 2\frac{1}{3} + 2\frac{1}{3}$
$7\frac{2}{3}$ feet is the perimeter

27
Find the perimeter of the **equilateral triangle** shown at the right.

______

[Note: An **equilateral triangle** has three sides of the same length.]

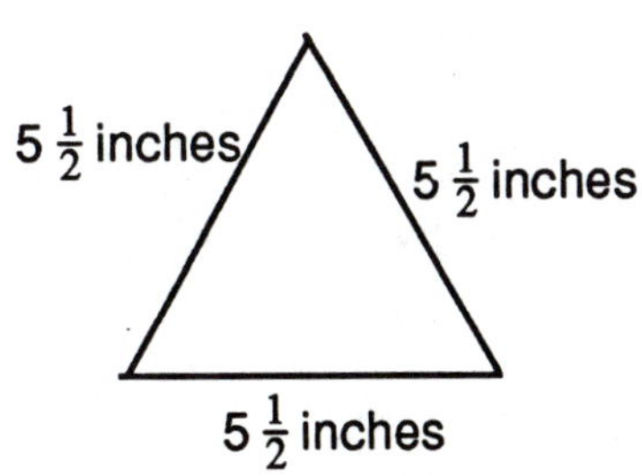

$16\frac{1}{2}$ inches

28
If a triangle has sides of $4\frac{1}{2}$ feet and $5\frac{2}{3}$ feet with a perimeter of $14\frac{1}{4}$ feet, how long is the third side? ________

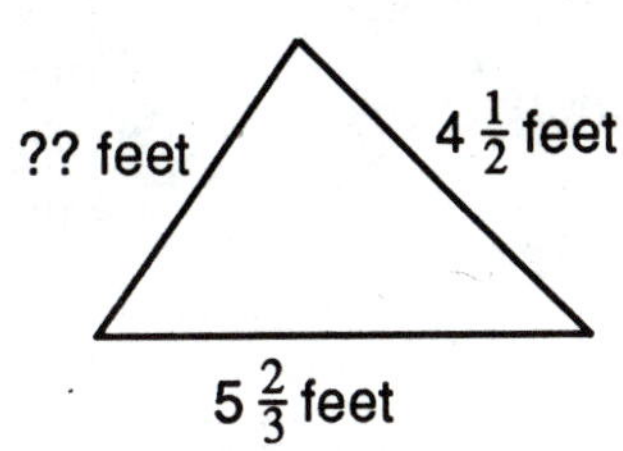

$4\frac{1}{2} + 5\frac{2}{3} = 10\frac{1}{6}$

$14\frac{1}{4} - 10\frac{1}{6} = 4\frac{1}{12}$ feet

29

The **area** of the **rectangle** shown at the right is the product of its length and width. Find the area. ______

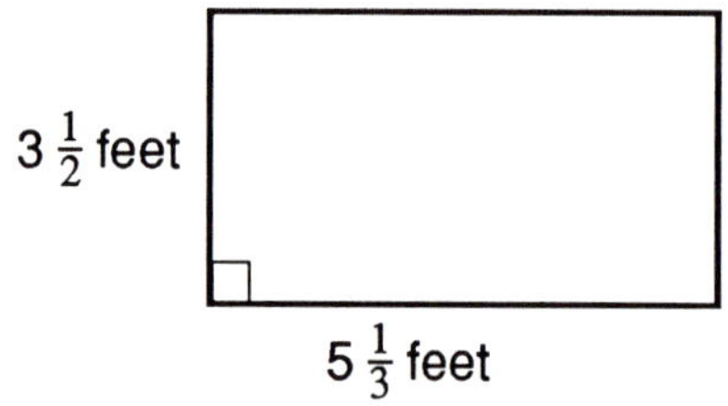

$5\frac{1}{3} \times 3\frac{1}{2} = \frac{56}{3}$

$18\frac{2}{3}$ square feet is the area

30

The formula for finding the area of a **rectangle** is $A = LW$ where L is the length of the rectangle and W is the width. Use the formula to find the area of the rectangle shown at the right. ______

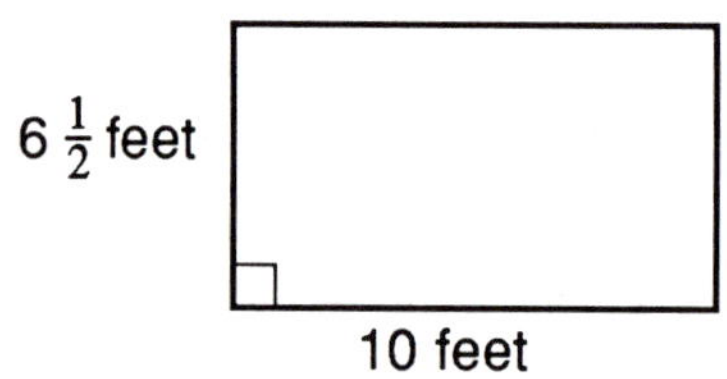

$A = L \times W$

$= 10 \times 6\frac{1}{2}$

65 square feet is the area

31

Area is always measured in square units such as square inches, square feet, or square centimeters. Find the area of a rectangle with length 12 yards and width $4\frac{1}{3}$ yards.

______

$A = L \times W$

$= 12 \times 4\frac{1}{3}$

52 square yards is the area

32

The formula for finding the area of a **triangle** is $A = \frac{1}{2}BH$ where B is the base of the triangle and H is the height. For the triangle shown at the right, the base is 14 meters and the height is ______.

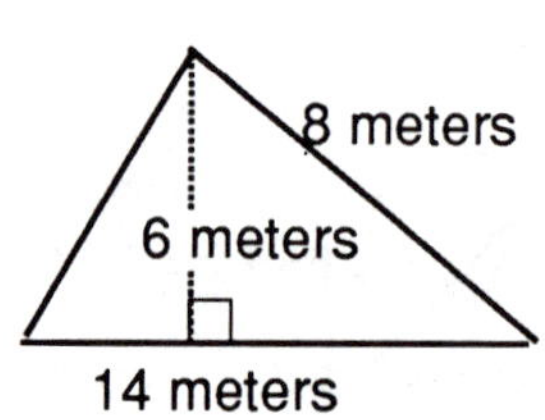

6 meters is the height
8 meters is the length of the side
14 meters is the base

33

Use the formula $A = \frac{1}{2}BH$ to find the area of the triangle shown at the right. ______

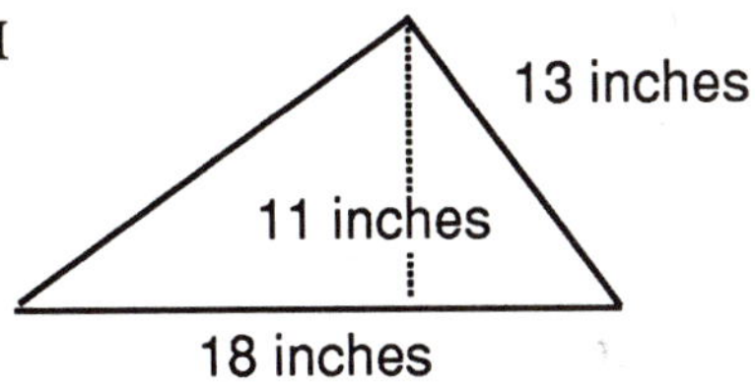

$A = \frac{1}{2} \times B \times H$

$= \frac{1}{2} \times 18 \times 11$

99 square inches is the area

34

The formula for finding the area of a **circle** is $A = \pi r^2$ where π is the number Pi (approximately $3\frac{1}{7}$), r is the radius, and r^2 means r x r. Find the area of the circle shown at the right. ______

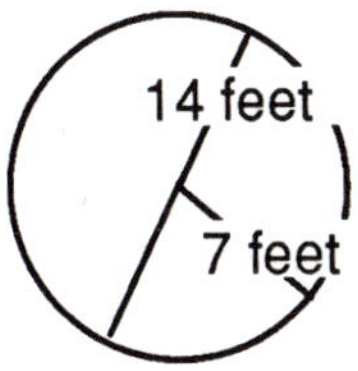

$A = \pi r^2$

$= 3\frac{1}{7} \times 7^2$

$= \frac{22}{7} \times 7 \times 7$

$= 22 \times 1 \times 7$

$= 154$ square feet

35

Use the formula $A = \pi r^2$ to find the area of the circle shown at the right. ______

[Use $3\frac{1}{7}$ as an approximation for π.]

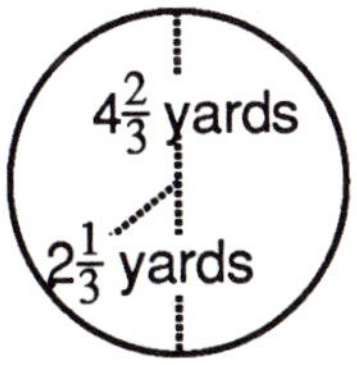

$A = \pi r^2$

$= 3\frac{1}{7} \times (2\frac{1}{3})^2$

$= \frac{22}{7} \times \frac{7}{3} \times \frac{7}{3}$

$= 22 \times \frac{1}{3} \times \frac{7}{3}$

$= \frac{154}{9}$

$= 17\frac{1}{9}$ square yards

36

The **perimeter** of a geometric figure is the sum of the lengths of its sides. To find the perimeter of a rectangle, ______ the lengths of its four sides.

add

37

The **perimeter** of the quadrilateral shown at the right is the sum of the lengths of the four sides. Find the perimeter. ______

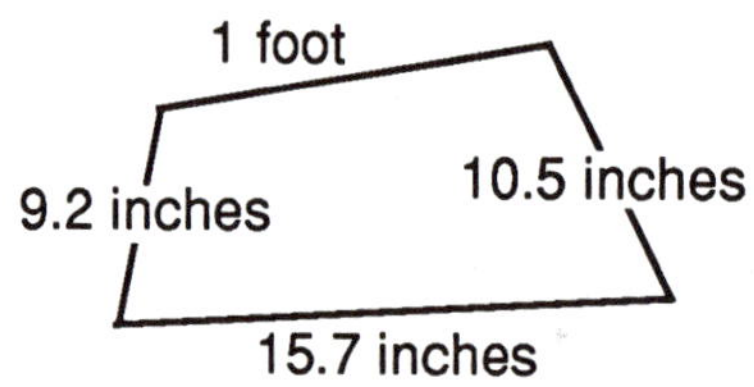

[Note: A **quadrilateral** is a four-sided figure.]

1 foot = 12 inches

12 + 9.2 + 15.7 + 10.5
47.4 inches is the perimeter

38

The perimeter of the **trapezoid** shown at the right is the sum of the lengths of the four sides. Find the perimeter. ______

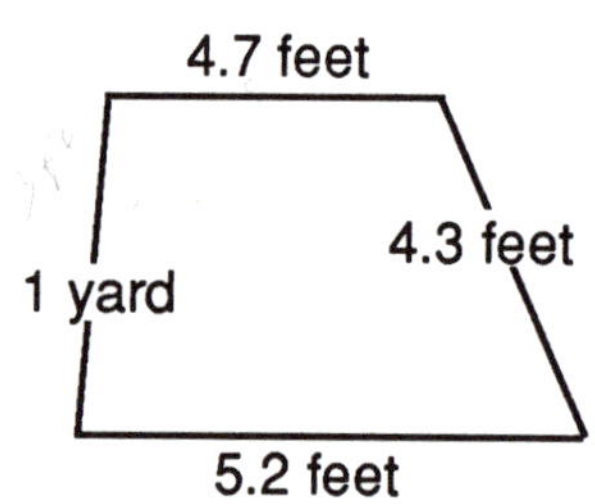

[Note: A **trapezoid** is a quadrilateral with one pair of parallel sides.]

1 yard = 3 feet

3 + 4.7 + 4.3 + 5.2
17.2 feet is the perimeter

39

The perimeter of the **parallelogram** shown at the right is the sum of the lengths of the four sides. Find the perimeter. ______

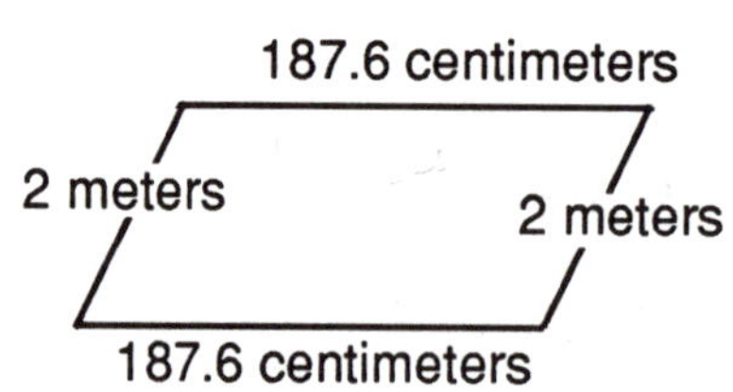

[Note: A **parallelogram** is a quadrilateral with two pairs of parallel and equal sides.]

100 cm = 1 meter

200 + 187.6 + 200 + 187.6
775.2 cm is the perimeter

40
Find the perimeter of the **rectangle** shown at the right.

______

[Note: A **rectangle** is a parallelogram with four right angles.]

1.61 + 1.24 + 1.61 + 1.24
5.7 inches is the perimeter

41
If a triangle has sides of 4.57 feet and 5.9 feet with a perimeter of 14.25 feet, how long is the third side? __________

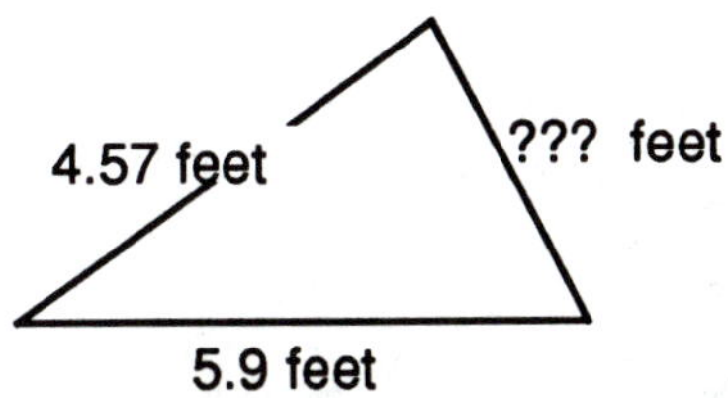

4.57 + 5.9 = 10.47

14.25 – 10.47 = 3.78 feet

42
The **area** of the **rectangle** shown at the right is the product of its length and width. Find the area. ______

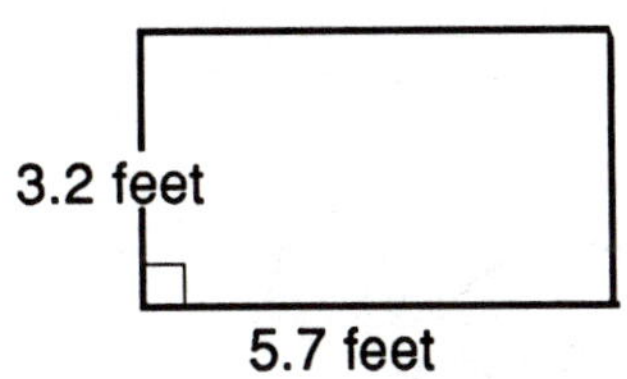

A = L x W
= 5.7 x 3.2
18.24 square feet is the area

43
The formula for finding the area of a **rectangle** is A = LW where L is the length of the rectangle and W is the width. Use the formula to find the area of the rectangle shown at the right. ______

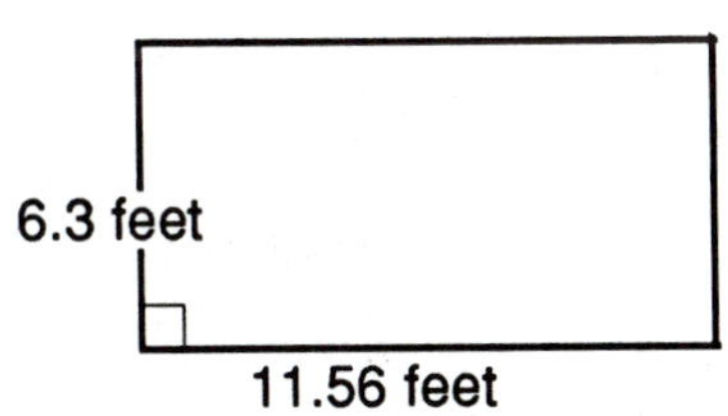

A = L x W
= 11.56 x 6.3
72.828 square feet is the area

44
Area is always measured in square units such as square inches, square feet, or square centimeters. Find the area of a rectangle with length 12.7 yards and width 4.5 yards.

______

A = L x W
= 12.7 x 4.5
A = 57.15 square yards

45

The formula for finding the area of a **triangle** is $A = \frac{1}{2}BH$ where B is the base of the triangle and H is the height. For the triangle shown at the right, the base is 1.7 meters and the height is _______.

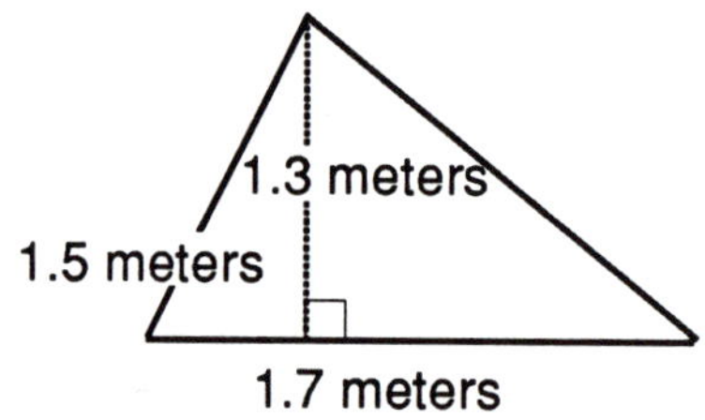

1.3 meters is the height
1.5 meters is the length of a side

46

Use the formula $A = \frac{1}{2}BH$ and the fact that $\frac{1}{2} = .5$ to find the area of the triangle shown at the right. ______

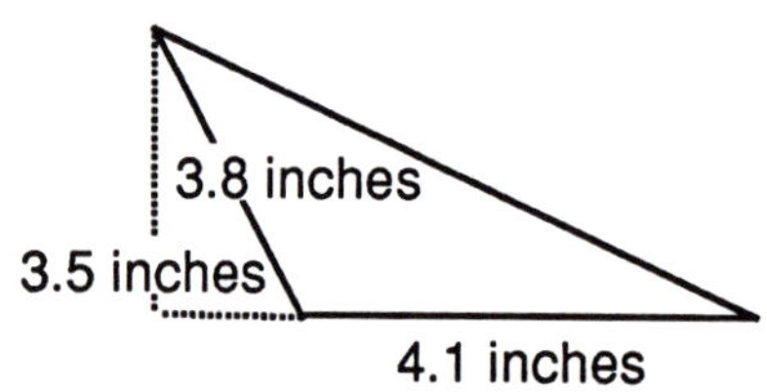

$A = \frac{1}{2} \times 3.5 \times 4.1$
$= .5 \times 4.1 \times 3.5$
7.175 square inches is the area

47

The formula for finding the **circumference** of a circle is $C = \pi D$ where π is the number Pi (approximated as 3.14) and D is the diameter. For the circle shown at the right, the circumference is _______.

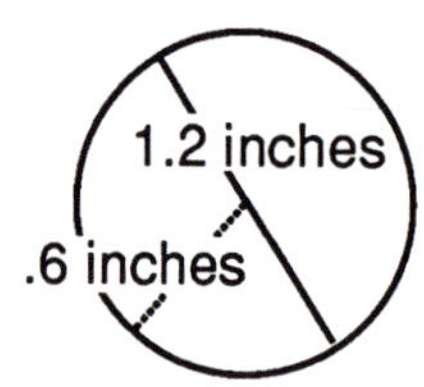

$C = \pi D$
$= 3.14 \times 1.2$
$= 3.768$
3.768 inches is the approximate circumference

48

The formula for finding the **area** of a circle is $A = \pi r^2$ where π is the number Pi (approximated as 3.14) and r is the radius. For the circle shown at the right, the area is _______.

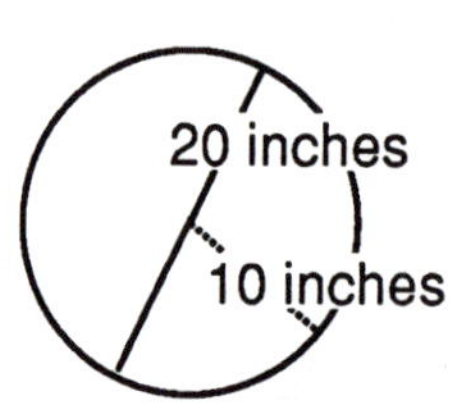

$A = \pi r^2$
$= 3.14 \times 10 \times 10$
$= 3.14 \times 100$
314 square inches is the approximate area

49

The two triangles shown at the right are **similar** triangles because they have exactly the same shape. Two similar triangles always have the same _______ (shape, size).

shape

50

The corresponding sides of two similar triangles are proportional. The ratio of the two marked sides of the small triangle is $\frac{5}{2}$. The ratio of the corresponding two sides of the larger triangle is _______.

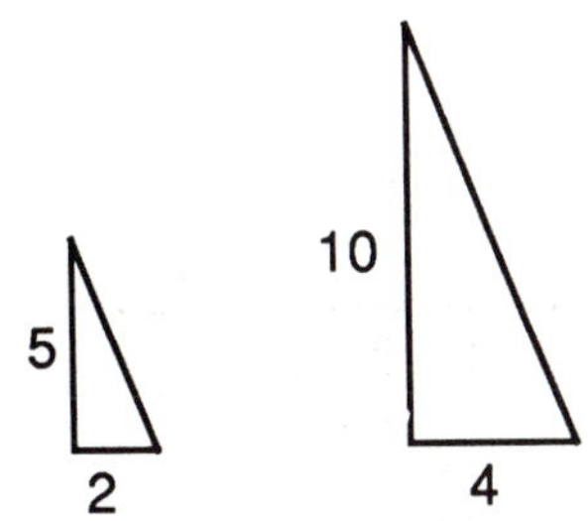

$\frac{10}{4}$ which equals $\frac{5}{2}$

51

Proportions can be used to find missing lengths of similar triangles. For the pair of similar triangles at the right, the ratio of one pair of sides is $\frac{3}{7}$. What is the matching ratio for the other triangle?

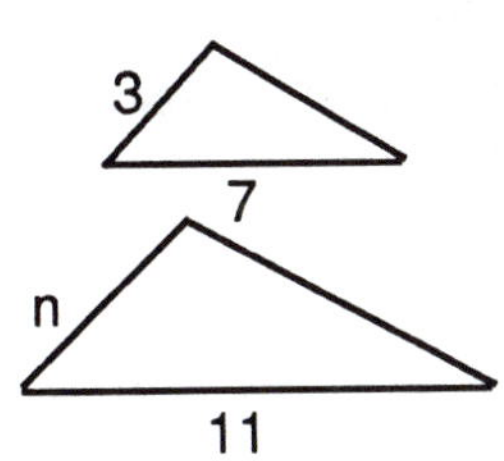

____

$\frac{n}{11}$

52

The proportion $\frac{3}{7} = \frac{n}{11}$ may be used to find the length of the missing side in frame 51. Solve the proportion and find n, the length of the missing side. ______

$4\frac{5}{7}$

53
Write a proportion that may be used to find the length of the missing side of the similar triangle shown at the right.

______

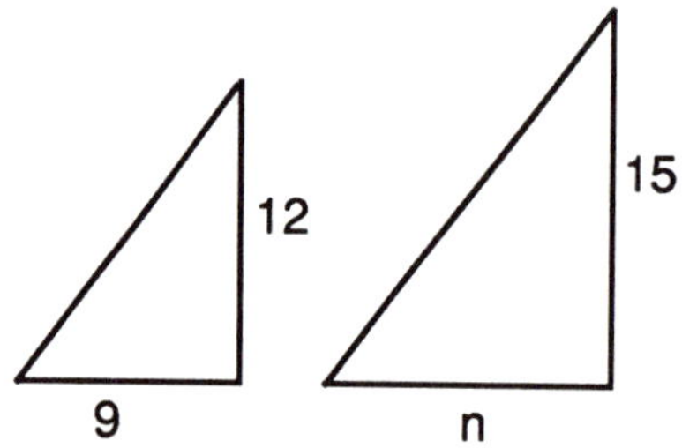

$\frac{9}{12} = \frac{n}{15}$ or $\frac{12}{9} = \frac{15}{n}$

54
Find the length of the missing side, n, for the triangle shown in frame 53. ______

$11\frac{1}{4}$

FEEDBACK UNIT 2

This quiz reviews the preceeding unit. Answers are at the back of the book.

1. Identify each of the following angles as acute, right, obtuse, or straight.

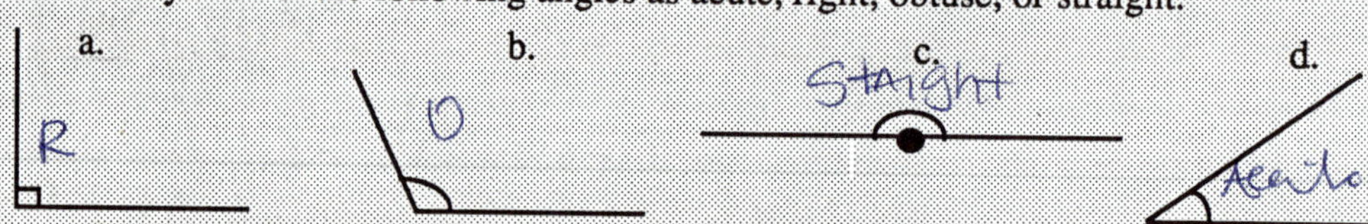

2. Find the complementary angle of a 23°angle.

3. Find the supplementary angle of a 52° angle.

4. If two angles of a triangle are 48° and 61°, what is the degree measure of the third angle?

5. Identify each of the following triangles as acute, obtuse, or equilateral.

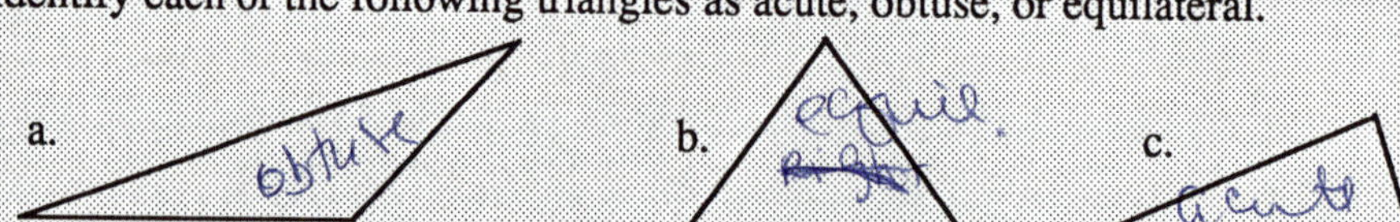

6. Find the perimeter of the isosceles triangle shown below.

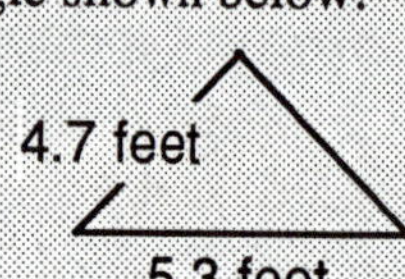

7. A triangle has a perimeter of $20\frac{3}{8}$ inches. If two of its sides are $8\frac{1}{4}$ inches and $5\frac{2}{3}$ inches long, what is the length of the third side?

(Continued on following page)

Feedback Unit 2 Cont.

8. Find the area of the triangle shown at the right.

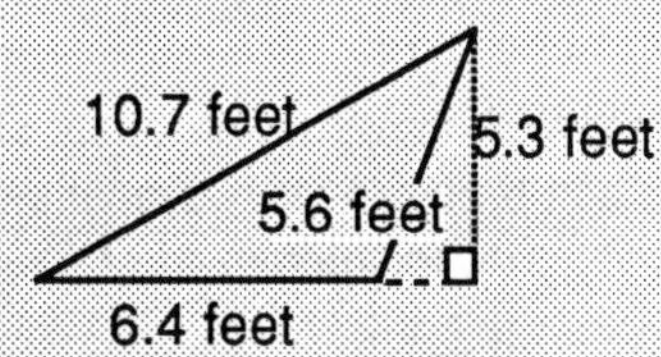

9. Identify each of the quadrilaterals below as trapezoid, parallelogram, rectangle or square.

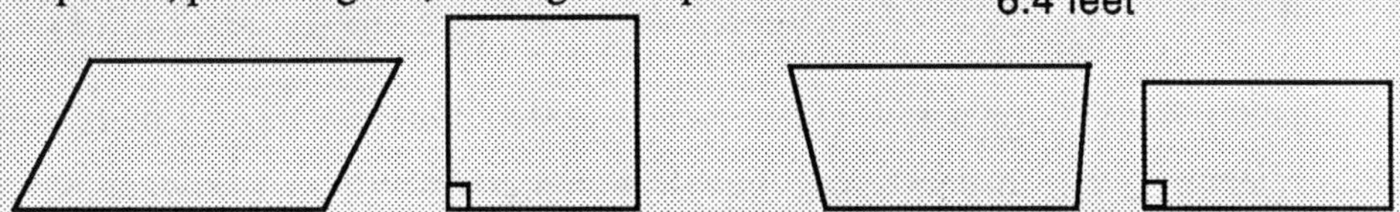

a. b. c. d.

10. Find the perimeter of the quadrilateral shown below.

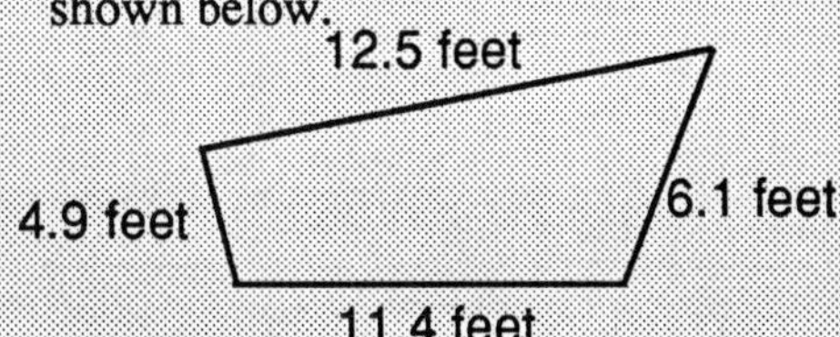

11. Find the perimeter of the rectangle shown below.

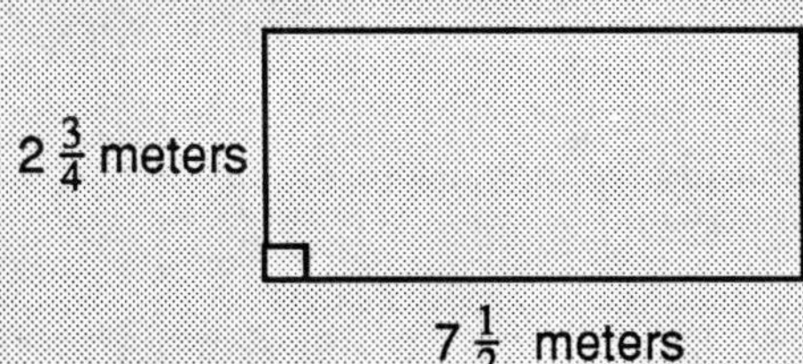

12. Find the perimeter of the parallelogram shown below.

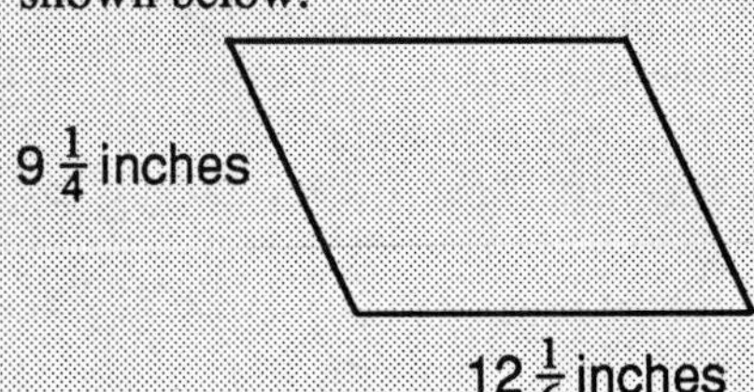

13. Find the perimeter of the square shown below.

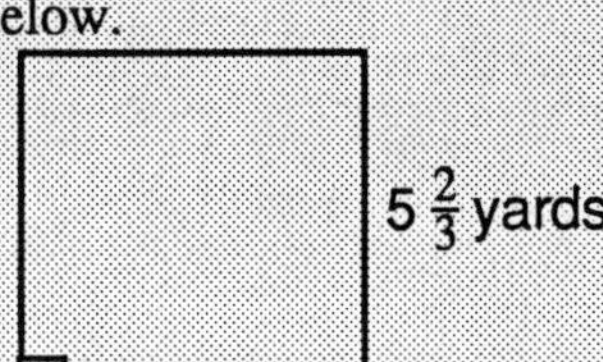

14. Find the area of the rectangle shown below.

15. Find the area of the parellelogram shown below.

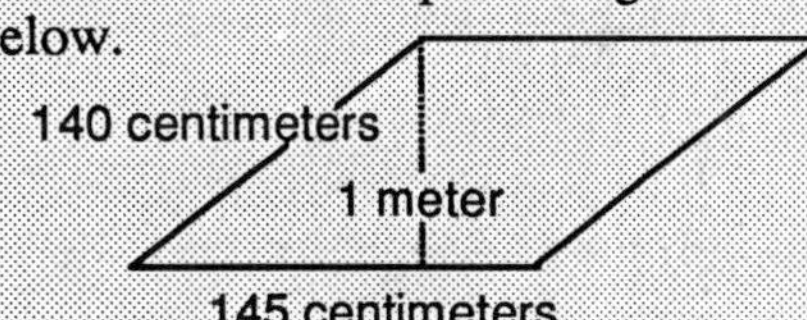

16. Find the circumference of the circle shown at the right. Use $3\frac{1}{7}$ as an approximation for π.

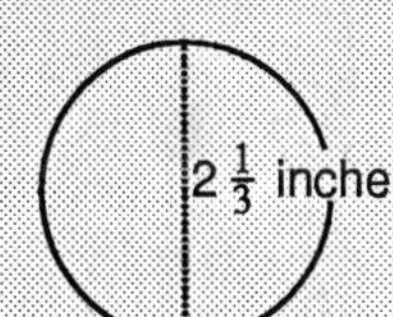

17. Find the area of the circle shown at the right. Use 3.14 as an approximation for π.

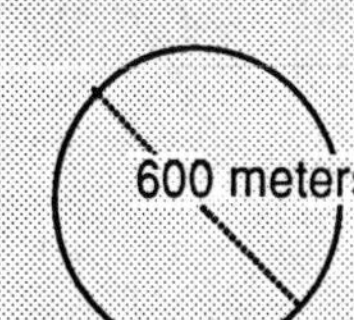

18. Find the missing side, n, of the triangle shown at the right. The triangles are similar.

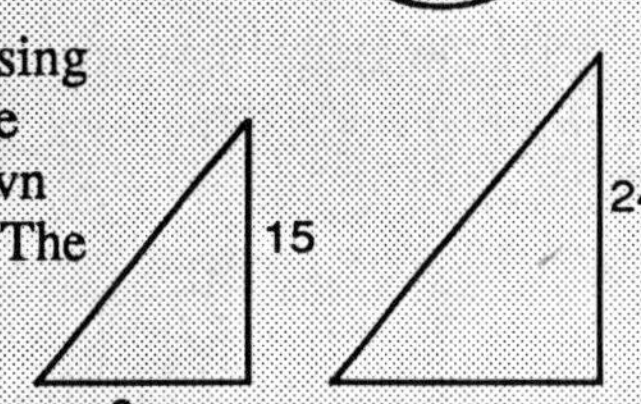

CHAPTER 14 MASTERY TEST

This test provides an opportunity for assuring that **all** of the objectives of Chapter 14 have been learned. Each problem is accompanied by the symbol (U) which indicates the Unit in which that type of problem was studied. If any problems are missed, they need to be corrected. Answers are at the back of the book.

1. 250 centimeters = _____ meters (1)
2. 16.2 kilometers = _____ meters (1)
3. 260 milliliters = _____ liters (1)
4. 584 grams = _____ kilograms (1)
5. Identify each of the angles at right as acute, right, or obtuse. (2)

a. b. c.

6. Identify each of the quadrilaterals at right as trapezoid, parallelogram, rectangle, or square. (2)

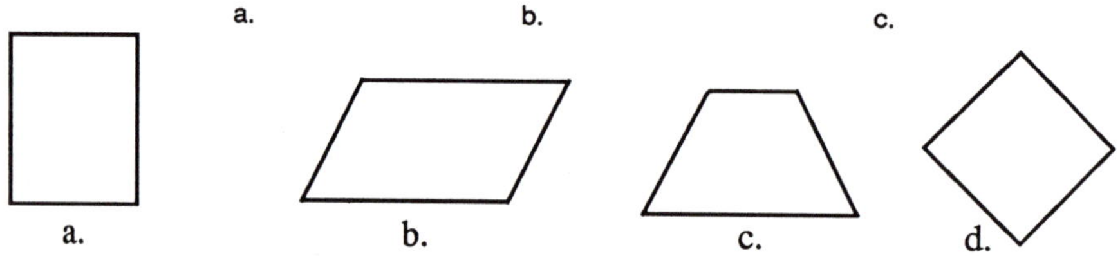

a. b. c. d.

7. Find the perimeter of the triangle. (2)

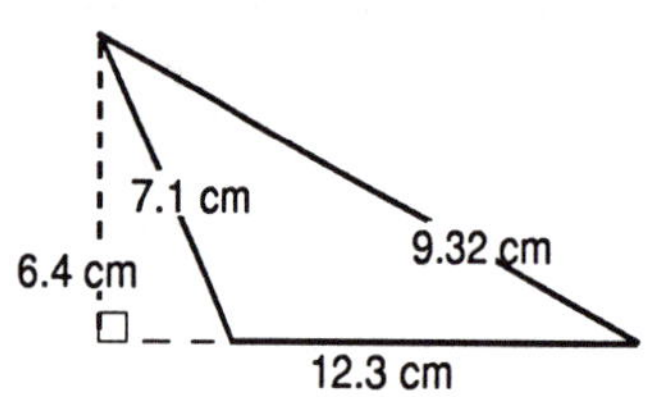

8. Find the area of the triangle. (2)

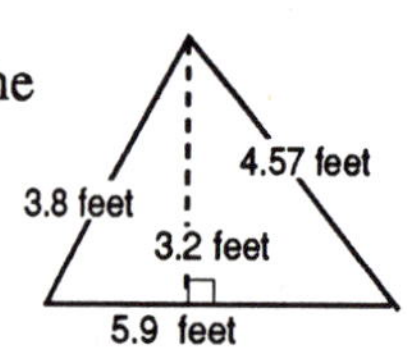

9. Find the perimeter of the quadrilateral. (2)

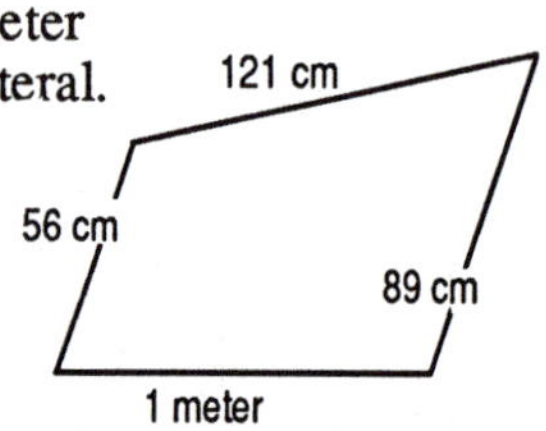

10. Find the area of the parellelogram. (2)

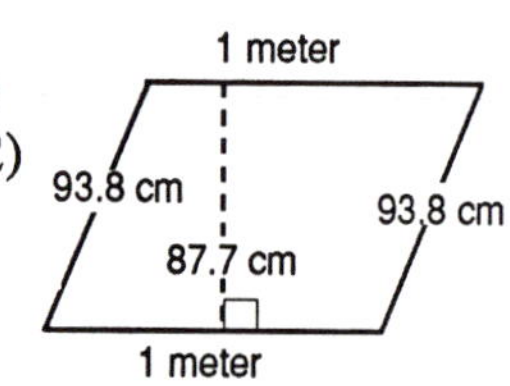

11. Find the circumference of the circle. Use 3.14 as the approximation for π. (2)

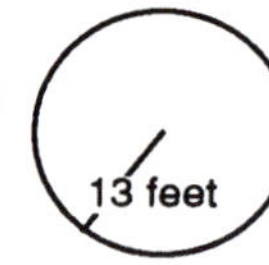

12. Find the area of the circle. Use $3\frac{1}{7}$ as the approximation for π. (2)

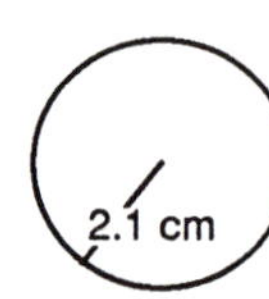

13. Find the missing side of the triangle. (2) The triangles are similar.

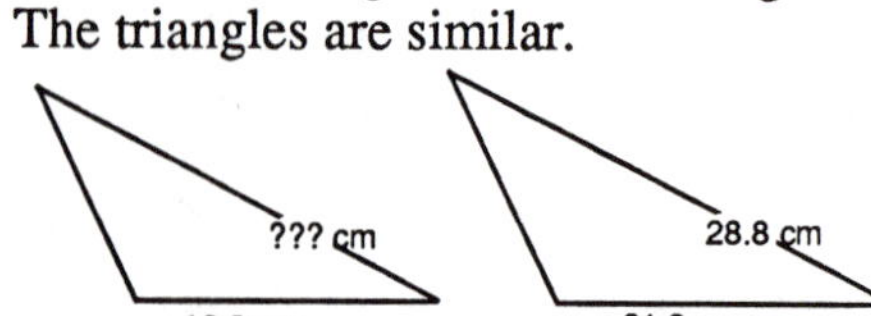

Interactive Arithmetic Answers

Chapter 1 Objectives Test

1. 9
2. 12
3. 3
4. 6
5. 643
6. 1,072
7. 424
8. 1,012
9. 1,288 prescriptions
10. 1,389 beds
11. 77,900
12. 34,933

Chapter 1, Feedback Unit 1

1. 14
2. 68
3. 21
4. 98
5. 21
6. 58
7. 18
8. 966
9. sum, total
10. addends

Chapter 1, Feedback Unit 2

1. thousands
2. 6
3. 17
4. 13
5. 11
6. 16
7. 9
8. 10

Chapter 1, Feedback Unit 3

1. 2,752
2. 10,644
3. 3,357
4. 15,679
5. 11,085
6. 76,958

Chapter 1, Feedback Unit 4

1. different
2. 75 days
3. 268 lbs
4. $56,488
5. 385 feet
6. 149 feet
7. 446 inches
8. 2,154 meters

Chapter 1 Mastery Test

1. 17
2. 19
3. 6
4. 1
5. sum, total
6. addends
7. 1,012
8. 1,432
9. 71,083
10. 90,264
11. 1,093 miles
12. $386
13. 394 feet
14. 206 inches

Chapter 2 Objectives Test

1. 67, 87, 439, 512, 913, 3062
2. 3
3. 5
4. X
5. 6
6. 3,926
7. 1,224
8. 3,864
9. 2,037
10. 41,471
11. 252 sophomores
12. 456 orders

Chapter 2, Feedback Unit 1

1. a. 2
   b. 9
   c. 5
   d. 0
2. 5, 8, 17, 39, 49, 94, 212, 763, 934, 8012
3. 2
4. 6
5. 4
6. 8
7. 4
8. 8

Chapter 2, Feedback Unit 2

1. 4
2. 8
3. X
4. X
5. 7
6. 5
7. 2
8. X
9. 2
10. 3

CHAPTER 2, FEEDBACK UNIT 3

1. 277
2. 505
3. 301
4. 218
5. X
6. 2,077
7. 5,307
8. 2,209
9. 33,178
10. 46,613

CHAPTER 2, FEEDBACK UNIT 4

1. 17
2. 12
3. 11
4. a. 3,451;
   2163 + 3451 = 5614
   b. 25,449;
   15487 + 25449 = 40936
   c. 3,302;
   2847 + 3302 = 6149
   d. 30,694;
   19306 + 30694 = 50000
5. $58
6. 133 pages

CHAPTER 2 MASTERY TEST

1. 76, 78, 139, 215, 394, 6032
2. 4
3. 5
4. X
5. 5
6. 1,217
7. 252
8. 7,596
9. 3,308
10. 29,091
11. 8,914
12. X
13. 291 miles
14. 563 acres

CHAPTER 3 OBJECTIVES TEST

1. factors
2. product
3. 376
4. 318
5. 36,211
6. 54,423
7. 6,156
8. 18,574
9. 1,706,670
10. 2,233,980
11. 612 cans
12. $2,220

CHAPTER 3, FEEDBACK UNIT 1

1. a. 2, 4, 6, 8, 10, 12, 14, 16, 18, 20
   b. 5, 10, 15, 20, 25, 30, 35, 40, 45, 50
   c. 7, 14, 21, 28, 35, 42, 49, 56, 63, 70
   d. 8, 16, 24, 32, 40, 48, 56, 64, 72, 80
2. 6
3. 9
4. 3
5. 35
6. 36

CHAPTER 3, FEEDBACK UNIT 2

1. 28
2. 54
3. 63
4. 48
5. 24
6. 45
7. 36
8. 36

CHAPTER 3, FEEDBACK UNIT 3

1. 10,346
2. 9,072
3. 43,078
4. 18,942
5. 24,680
6. 73,431
7. 27,951
8. 19,756
9. 18,249
10. 43,947

CHAPTER 3, FEEDBACK UNIT 4

1. 243,456
2. 29,952
3. 28,768
4. 25,578
5. 235,600
6. 183,124
7. 36,381
8. 32,976

9. 136,652
10. 213,306

Chapter 3, Feedback Unit 5

1. factor
2. a. 560
   b. 56,780
   c. 6,100
   d. 4,800
3. 8,208 grams
4. 442 lbs
5. 153 sq meters
6. 36 sq inches

Chapter 3 Mastery Test

1. factors
2. product
3. 520
4. 210
5. 22,225
6. 36,684
7. 4,212
8. 14,965
9. 2,318,904
10. 1,537,650
11. 4,580
12. 50,800
13. 613,900
14. 432,000
15. 672 jars
16. $3,060
17. 646 sq. inches
18. 1,260 sq. yards

Chapter 4 Objectives Test

1. 47, 2726, 58
2. 781 R2
3. 751 R6
4. 632 R7
5. 1,829 R1
6. 205 R5
7. 222 R22
8. 987 R22
9. 725 R40
10. 782 R84
11. 65
12. 32 fl. oz.

Chapter 4, Feedback Unit 1

1. 12, 42, 48
2. 12, 28, 32
3. 0, 27, 45
4. 7, 21, 42, 56
5. 9
6. 0
7. 6
8. 1
9. 4
10. 6

Chapter 4, Feedback Unit 2

1. 48
2. 40
3. 36
4. 6
5. 0
6. 1 R1
7. 0 R4
8. 0 R1
9. 2 R0
10. 1 R1

Chapter 4, Feedback Unit 3

1. 1,293 R1
2. 619 R1
3. 4,073 R0
4. 2,943 R2
5. 307 R1
6. 360 R3
7. 647 R0
8. 204 R0
9. 673 R0
10. 625 R2

Chapter 4, Feedback Unit 4

1. 125 R18
2. 497 R44
3. 214 R12
4. 522 R16
5. 337 R7
6. 111 R2
7. 198 R58
8. 264 R94
9. 609 R10
10. 88 R60
11. 172 R123
12. 113 R69

Chapter 4, Feedback Unit 5

1. 46, 912, 19, 38
2. (507 x 18) + 17 = 9,143
3. 51
4. $357
5. 1,563 cards

Chapter 4 Mastery Test

1. 46, 2,990, 65
2. 903 R3
3. 887 R4
4. 243 R2
5. 1,210 R3
6. 186 R5
7. 987 R22
8. 1,036 R46
9. 1,574 R34
10. 60
11. 40 customers

Chapter 5 Objectives Test

1. $\frac{2}{3}$
2. 4
3. 11
4. $\frac{9}{12}$
5. no
6. yes
7. no
8. yes
9. yes
10. $\frac{7}{9}$

11. $\frac{8}{15}$
12. $\frac{31}{34}$
13. $\frac{2}{3}$
14. 1
15. $1\frac{1}{9}$
16. $8\frac{4}{5}$
17. 4
18. $3\frac{5}{13}$
19. $8\frac{6}{7}$
20. $\frac{5}{2}$
21. $\frac{50}{9}$
22. $\frac{27}{7}$
23. $\frac{33}{8}$
24. $\frac{59}{10}$

CHAPTER 5, FEEDBACK UNIT 1

1. $\frac{5}{7}$
2. 5
3. 10
4. $\frac{5}{8}$
5. $\frac{5}{6}$
6. $\frac{8}{45}$

CHAPTER 5, FEEDBACK UNIT 2

1. yes
2. no
3. no
4. no
5. no
6. no
7. yes
8. no
9. no
10. no

CHAPTER 5, FEEDBACK UNIT 3

1. $\frac{3}{8}$
2. $\frac{3}{5}$
3. $\frac{1}{5}$
4. $\frac{2}{5}$
5. $\frac{13}{15}$
6. $\frac{4}{7}$
7. $\frac{6}{11}$
8. $\frac{5}{9}$
9. $\frac{3}{4}$
10. $\frac{3}{4}$

CHAPTER 5, FEEDBACK UNIT 4

1. 20
2. 8
3. 32
4. 11
5. 63
6. 18
7. 12
8. 30
9. $\frac{9}{11} = \frac{72}{88}$
10. $\frac{8}{13} = \frac{32}{52}$
11. $\frac{4}{15} = \frac{8}{30}$
12. $\frac{7}{9} = \frac{14}{18}$

CHAPTER 5, FEEDBACK UNIT 5

1. proper
2. improper
3. proper
4. improper
5. $3\frac{1}{9}$
6. $5\frac{5}{6}$
7. $5\frac{1}{8}$
8. $5\frac{5}{12}$
9. 6
10. $9\frac{3}{4}$
11. $9\frac{2}{5}$
12. 1

CHAPTER 5, FEEDBACK UNIT 6

1. $\frac{21}{8}$
2. $\frac{64}{11}$
3. $\frac{19}{4}$
4. $\frac{41}{5}$
5. $\frac{29}{9}$
6. $\frac{45}{7}$
7. $\frac{29}{5}$
8. $\frac{11}{8}$
9. $\frac{18}{7}$
10. $\frac{4}{3}$

CHAPTER 5 MASTERY TEST

1. $\frac{2}{3}$
2. 5
3. 10
4. $\frac{3}{8}$
5. no
6. yes
7. no
8. yes
9. $\frac{4}{5}$
10. $\frac{1}{3}$
11. $\frac{5}{6}$
12. $\frac{3}{4}$
13. $1\frac{2}{7}$
14. $2\frac{2}{5}$
15. $9\frac{3}{8}$
16. $5\frac{5}{7}$
17. $\frac{47}{8}$
18. $\frac{49}{13}$
19. $\frac{44}{5}$
20. $\frac{64}{9}$

Chapter 6 Objectives Test

1. 20
2. 60
3. 42
4. 12
5. $\frac{20}{32}$
6. $\frac{27}{63}$
7. $1\frac{2}{3}$
8. $1\frac{6}{13}$
9. $1\frac{1}{6}$
10. $1\frac{3}{8}$
11. $1\frac{1}{30}$
12. $3\frac{3}{8}$
13. $\frac{3}{8}$
14. $6\frac{2}{5}$
15. $\frac{4}{21}$
16. $\frac{5}{12}$
17. $5\frac{2}{5}$
18. $1\frac{11}{12}$
19. $6\frac{11}{20}$ kgm
20. $4\frac{7}{10}$ cm
21. $4\frac{1}{2}$ cups

Chapter 6, Feedback Unit 1

1. $\frac{7}{10}$
2. $\frac{7}{13}$
3. $1\frac{2}{5}$
4. $1\frac{4}{9}$
5. $1\frac{1}{2}$
6. $1\frac{1}{4}$
7. 1
8. $1\frac{2}{3}$
9. $1\frac{4}{15}$
10. $1\frac{1}{10}$
11. $\frac{2}{3}$
12. $1\frac{7}{9}$

Chapter 6, Feedback Unit 2

1. 15
2. 10
3. 14
4. 28
5. 4
6. 4
7. $\frac{24}{40}, \frac{25}{40}$
8. $\frac{21}{24}, \frac{20}{24}$
9. $\frac{4}{18}, \frac{3}{18}$
10. $\frac{5}{10}, \frac{7}{10}$

Chapter 6, Feedback Unit 3

1. 24
2. 10
3. 12
4. 18
5. 20
6. $\frac{24}{30}, \frac{5}{30}$
7. $\frac{10}{16}, \frac{3}{16}$
8. $\frac{4}{9}, \frac{6}{9}$
9. $\frac{7}{14}, \frac{10}{14}$
10. $\frac{25}{30}, \frac{21}{30}$

Chapter 6, Feedback Unit 4

1. $\frac{7}{8}$
2. $\frac{29}{35}$
3. $1\frac{5}{18}$
4. $1\frac{21}{44}$
5. $\frac{8}{9}$
6. $1\frac{1}{8}$
7. $1\frac{23}{90}$
8. $\frac{59}{56} = 1\frac{3}{56}$
9. $\frac{31}{60}$
10. $1\frac{17}{24}$
11. $\frac{29}{35}$
12. $1\frac{71}{80}$

Chapter 6, Feedback Unit 5

1. $7\frac{2}{3}$
2. $13\frac{5}{8}$
3. $3\frac{6}{7}$
4. $11\frac{4}{11}$
5. $7\frac{1}{2}$
6. $11\frac{1}{8}$
7. $11\frac{7}{10}$
8. $13\frac{5}{14}$
9. $3\frac{31}{40}$
10. $12\frac{7}{12}$
11. $92\frac{13}{15}$
12. $84\frac{3}{40}$

Chapter 6, Feedback Unit 6

1. $\frac{7}{13}$
2. X
3. $\frac{1}{2}$
4. $\frac{9}{56}$
5. $\frac{11}{24}$
6. $\frac{3}{8}$
7. X
8. $\frac{19}{56}$
9. $\frac{1}{12}$
10. X

11. $\frac{2}{15}$
12. $\frac{29}{70}$

CHAPTER 6, FEEDBACK UNIT 7

1. $2\frac{3}{8}$
2. $7\frac{1}{16}$
3. $2\frac{3}{56}$
4. $3\frac{2}{5}$
5. $4\frac{11}{15}$
6. $3\frac{2}{63}$
7. $\frac{1}{2}$
8. $2\frac{3}{4}$
9. $4\frac{13}{18}$
10. $7\frac{11}{24}$
11. $4\frac{5}{8}$
12. $3\frac{31}{35}$

CHAPTER 6, FEEDBACK UNIT 8

1. difference
2. sum
3. $4\frac{3}{8}$ inches
4. $3\frac{5}{12}$ yds
5. $\$20\frac{5}{8}$
6. $9\frac{4}{5}$ kgm
7. $1\frac{1}{2}$ gm
8. $3\frac{87}{100}$ m
9. $17\frac{5}{6}$ inches
10. $6\frac{11}{24}$ inches

CHAPTER 6 MASTERY TEST

1. 15
2. 24
3. 60
4. 16
5. $\frac{21}{24}$
6. $\frac{27}{45}$
7. 2
8. $1\frac{3}{4}$
9. 1
10. $\frac{3}{5}$
11. $\frac{7}{10}$
12. $\frac{17}{21}$
13. $3\frac{9}{28}$
14. X
15. $\frac{1}{2}$
16. $1\frac{1}{4}$
17. $\frac{3}{8}$
18. $1\frac{1}{12}$
19. $2\frac{11}{15}$
20. $5\frac{7}{9}$
21. $4\frac{1}{4}$ kgm
22. $4\frac{3}{4}$ m
23. $1\frac{3}{4}$ cups
24. $20\frac{11}{12}$ m
25. $7\frac{1}{6}$ inches

CHAPTER 7 OBJECTIVES TEST

1. $1\frac{1}{20}$
2. $1\frac{3}{11}$
3. $\frac{2}{7}$
4. $3\frac{6}{7}$
5. $6\frac{3}{4}$
6. $1\frac{1}{2}$
7. $\frac{14}{65}$
8. $10\frac{2}{5}$
9. $\frac{1}{4}$
10. 9
11. $4\frac{1}{6}$
12. $3\frac{1}{30}$
13. $8\frac{1}{2}$
14. $1\frac{13}{15}$
15. $2\frac{1}{7}$
16. $1\frac{5}{6}$
17. $\frac{7}{10}$
18. 2
19. 28
20. $2\frac{1}{2}$
21. $\frac{9}{8}, \frac{2}{3}, \frac{3}{4}$
22. $2\frac{5}{8}$
23. $\$1\frac{4}{5}$ million
24. 66 spoons
25. $2,800

CHAPTER 7, FEEDBACK UNIT 1

1. $\frac{8}{45}$
2. $\frac{15}{77}$
3. $\frac{4}{15}$
4. $\frac{18}{35}$
5. $\frac{4}{5}$
6. $\frac{8}{45}$
7. $1\frac{7}{11}$
8. $5\frac{2}{5}$
9. $\frac{16}{33}$
10. $1\frac{7}{9}$
11. $3\frac{1}{9}$
12. $6\frac{3}{4}$

CHAPTER 7, FEEDBACK UNIT 2

1. $\frac{3}{10}$
2. $\frac{8}{21}$

3. $\frac{15}{19}$
4. $\frac{5}{27}$
5. $\frac{5}{24}$
6. $\frac{4}{5}$
7. $\frac{5}{6}$
8. $\frac{2}{3}$
9. $\frac{3}{10}$
10. $\frac{2}{5}$
11. $\frac{15}{28}$
12. $\frac{3}{20}$

Chapter 7, Feedback Unit 3

1. $\frac{6}{7}$
2. $\frac{4}{5}$
3. $\frac{24}{25}$
4. 1
5. $2\frac{1}{4}$
6. $\frac{2}{3}$
7. $\frac{15}{16}$
8. 3
9. $1\frac{1}{2}$
10. $\frac{5}{7}$
11. $2\frac{1}{4}$
12. $\frac{1}{3}$

Chapter 7, Feedback Unit 4

1. $1\frac{1}{8}$
2. $2\frac{4}{7}$
3. $\frac{2}{5}$
4. $13\frac{1}{3}$
5. $\frac{3}{7}$
6. $3\frac{3}{7}$
7. 4
8. $3\frac{9}{10}$
9. 2
10. 3
11. $\frac{9}{10}$
12. $2\frac{1}{4}$

Chapter 7, Feedback Unit 5

1. $1\frac{3}{7}$
2. $\frac{1}{2}$
3. 12
4. 18 ($\frac{1}{3}$ years)
5. 16 bottles
6. 12 battalions
7. 100 coins
8. 105 blocks
9. $4\frac{1}{5}$ tons
10. 16 shares
11. 33 sq ft
12. $6\frac{3}{10}$ sq yd
13. $4\frac{5}{18}$ sq in

Chapter 7 Mastery Test

1. $\frac{8}{45}$
2. $\frac{10}{63}$
3. $\frac{6}{55}$
4. $3\frac{1}{5}$
5. $2\frac{5}{8}$
6. $\frac{3}{10}$
7. $\frac{7}{36}$
8. $\frac{1}{4}$
9. $\frac{6}{7}$
10. $1\frac{1}{10}$
11. $4\frac{7}{8}$
12. $\frac{5}{6}$
13. $3\frac{2}{3}$
14. $\frac{10}{49}$
15. $1\frac{1}{8}$
16. $1\frac{2}{7}$
17. $\frac{8}{15}$
18. 2
19. 40
20. $6\frac{5}{12}$
21. $\frac{6}{5}, \frac{2}{3}, \frac{5}{9}$
22. 20
23. $1\frac{5}{6}$ liters
24. 19 cups
25. $1,500
26. 123 sq ft
27. 44 sq in
28. $34\frac{58}{63}$ sq in

Chapter 8 Objectives Test

1. 0.004
2. $\frac{28}{10}$
3. 5.4
4. 5.21
5. $\frac{45}{1000}$
6. $\frac{4}{10}$
7. 102.93
8. $\frac{501}{100}$
9. 5.7814×10^1
10. 7.134×10^{-4}
11. 5.67×10^7
12. 1.406×10^{-2}
13. T
14. F
15. T
16. T
17. <
18. <
19. <

20. =
21. 4.5
22. 6
23. 15.51
24. 59.04

Chapter 8, Feedback Unit 1

1. 0.125
2. 0.01
3. 0.0037
4. 37.48
5. 9.8
6. 0.12
7. 0.398
8. 0.0483
9. a. thousandths
   b. ones
   c. tenths
   d. hundredths
   e. millionths
10. a. 2
    b. 3
    c. 7
    d. 0
    e. 1
11. $\frac{207}{100}$
12. $\frac{92}{1000}$
13. $\frac{10752}{100}$
14. $\frac{50}{10000}$

Chapter 8, Feedback Unit 2

1. 8,134,000
2. 0.7
3. 0.09132
4. 2.707
5. 8.35
6. 234,600
7. 0.08903
8. 7,500
9. 0.03402
10. 0.004312
11. 3.6×10^1
12. 7×10^4
13. 8×10^{-3}
14. 7.8×10^5
15. 9.2×10^{10}
16. 9.273×10^{-2}
17. 4.6829×10^5
18. 5.005×10^{-1}
19. 4.51×10^{-13}
20. 8×10^{11}

Chapter 8, Feedback Unit 3

1. T
2. T
3. F
4. T
5. T
6. F
7. F
8. T
9. T
10. F
11. >
12. <
13. >
14. =
15. >
16. >

Chapter 8, Feedback Unit 4

1. 1.89
2. 14.00
3. 0.46
4. 301.99
5. 26.300
6. 0.027
7. 12.479
8. 0.047
9. 14
10. 0.5
11. 243.5
12. 4.0494

Chapter 8 Mastery Test

1. 0.005
2. $\frac{18}{10}$
3. 4.2
4. 0.86
5. $\frac{56}{1000}$
6. $\frac{90}{100}$
7. 205.6
8. $\frac{408}{100}$
9. 4.1794×10^2
10. 6.107×10^{-3}
11. 8.3×10^8
12. 2.16×10^{-1}
13. T
14. F
15. T
16. T
17. =
18. <
19. >
20. <
21. 0.5
22. 1.27
23. 0.63
24. 125.633

Chapter 9 Objectives Test

1. 0.153
2. 3.588
3. 5,468.0
4. 5.2039
5. 1, 4, 3
6. 294 + 67 = 361
7. 0.009
8. 0.01796
9. 431.6
10. 9.159
11. 6.68
12. 0.608
13. 7.18
14. 7.4 gms
15. $149.90
16. 0.0058 interest per month

Chapter 9, Feedback Unit 1

1. 0.35
2. 0.17
3. 1.07
4. 10.175
5. 0.6252
6. 11.25
7. 0.509
8. 0.1622

9. 7.82
10. 2.0013
11. 65.108
12. 3,609.71
13. 8.0842%
14. 25.388 mg
15. 15.736 liters
16. $0.54745
17. 152.12 feet

Chapter 9, Feedback Unit 2

1. 0.21
2. 0.018
3. 6.321
4. 0.44
5. 8.0562
6. 0.715
7. 4.7
8. 0.0007
9. 0.0104
10. 152.22

Chapter 9, Feedback Unit 3

1. 1.11
2. 0.0037
3. 0.37
4. 53.37
5. 1.83
6. 0.38668
7. 0.0007
8. 11.5
9. 0.5538
10. 22.44
11. 8.7 gm
12. 2.183 liters
13. 18.1 feet

Chapter 9 Mastery Test

1. 1.3
2. 1.27
3. 0.101
4. 2.0712
5. A, B, C
6. 190 + 188 = 378
7. 0.002
8. 0.000087
9. 480.7
10. 99.031
11. 1.027
12. 4.9975
13. $1,101.75
14. 6.917 kgm
15. 3.98 liters
16. 6.72%
17. 83.786 meters
18. 8.46 meters

Chapter 10 Objectives Test

1. 150.096
2. 631.722
3. 209.8
4. 250,600
5. 20.72
6. 261.8 mg
7. $5,450
8. 22.2 grams
9. 21.5
10. 0.0059
11. 1.60
12. 30.81
13. 0.302
14. 43 stamps
15. $78.98

Chapter 10, Feedback Unit 1

1. 1
2. 1
3. 3
4. 1
5. 3
6. 5
7. 6
8. 0
9. 2
10. 3
11. 1
12. 4

Chapter 10, Feedback Unit 2

1. 0.306
2. 0.000078
3. 0.04875
4. 0.816296
5. 10.8
6. 0.05
7. 8.2
8. 49.7
9. 0.02781
10. 3.9

Chapter 10, Feedback Unit 3

1. 750.0
2. 36.34
3. 0.0776
4. 240.00
5. 8.137
6. 3,649.368
7. 0.05432
8. 0.51212
9. 2.8
10. 27.8
11. 3,278.4
12. 82.9

Chapter 10, Feedback Unit 4

1. $237.50
2. 62.4 kgm
3. $4.50
4. $2.88
5. 0.00225 kgm
6. 2.4
7. 800
8. 1 mgm
9. $2.26
10. 174 rats
11. $3,300
12. 2,596.66 sq yds
13. 66.585 sq in
14. 43.018 in
15. 2.5434 sq meters

Chapter 10, Feedback Unit 5

1. 0.00051
2. 0.025
3. 0.2
4. 0.0025
5. 5.005
6. 0.0814
7. 0.59125
8. 0.69
9. 0.375
10. 2.625

CHAPTER 10, FEEDBACK UNIT 6

1. 1.2
2. 512.5
3. 571.2
4. 500
5. 0.0063
6. 3.06

CHAPTER 10, FEEDBACK UNIT 7

1. 0.03
2. 6.67
3. 0.03
4. 1.24
5. 0.030
6. 0.038
7. 16.926
8. 937.111
9. 0.24
10. 0.57
11. 0.65
12. 0.16

CHAPTER 10, FEEDBACK UNIT 8

1. 904 tickets
2. 1.94 meters
3. $2.57
4. $8.40

CHAPTER 10 MASTERY TEST

1. 2.451
2. 7.52154
3. 9.9264
4. 7.89
5. 25,800 bushels
6. 143 mgm
7. $3,450
8. 0.48 gallons
9. 0.0259
10. 81.5
11. 1.15
12. 59.17
13. 0.167
14. 0.38
15. 1.71
16. 0.56
17. 0.55
18. 94 cards
19. $46.37
20. 205.942 sq ft
21. 976.26 sq in
22. 58.404 cm
23. 1.5386 sq m

CHAPTER 11 OBJECTIVES TEST

1. $\frac{1}{3}$
2. $\frac{2}{1}$
3. $\frac{11}{16}$
4. $\frac{43}{1}$
5. 12
6. 10
7. 40
8. $14\frac{2}{5}$
9. 48 pages
10. 40 days
11. 8.01 malfunctions
12. $231
13. 50 mg
14. $37\frac{1}{3}$ kgm

CHAPTER 11, FEEDBACK UNIT 1

1. yes
2. $\frac{6}{13}$
3. $\frac{13}{70}$
4. $\frac{8}{3}$
5. $\frac{4}{5}$
6. $\frac{120}{1}$
7. $\frac{9}{10}$
8. $\frac{315}{1}$
9. $\frac{34}{5}$
10. $\frac{40300}{551}$
11. $\frac{29}{1150}$
12. $\frac{47560}{23}$

CHAPTER 11, FEEDBACK UNIT 2

1. yes
2. yes
3. no
4. yes
5. no
6. yes
7. no
8. n x 16 = 8 x 2
9. 37 x 10 = 18 x n
10. 16 x n = 20 x 5
11. 7 x 2 = n x 1
12. 3 x 13 = n x 7

CHAPTER 11, FEEDBACK UNIT 3

1. b
2. a
3. 6
4. $1\frac{4}{17}$
5. 9
6. $3\frac{1}{3}$
7. $7\frac{2}{5}$
8. $\frac{4}{9}$
9. $2\frac{2}{7}$
10. $3\frac{1}{2}$

CHAPTER 11, FEEDBACK UNIT 4

1. $6\frac{1}{4}$
2. $4\frac{1}{2}$
3. $5\frac{5}{6}$
4. $\frac{6}{11}$
5. $4\frac{1}{5}$
6. $\frac{8}{15}$
7. $17\frac{1}{2}$
8. $3\frac{2}{3}$

Chapter 11, Feedback Unit 5

1. 15 cents
2. $1\frac{2}{5}$ tblspns
3. 150 problems
4. $4
5. $11\frac{2}{3}$ in
6. 100 kgm
7. $19.50
8. $11\frac{7}{11}$ cm

Chapter 11 Mastery Test

1. $\frac{1}{4}$
2. $\frac{2}{1}$
3. $\frac{4}{5}$
4. $\frac{8}{7}$
5. 4
6. 15
7. 30
8. 48
9. 40 min
10. $7\frac{1}{2}$ days
11. 67.5 days
12. $2,000
13. 325 particles of pollution
14. 56 kgm

Chapter 12 Objectives Test

1. $\frac{5}{100}$
2. $\frac{120}{100}$
3. $\frac{5.7}{100}$
4. 7%
5. 39%
6. .56
7. .085
8. 30%
9. 79%
10. 56%
11. 72%
12. b = 80, r = 55, p = 44
13. b = 500, r = 17, p = 85
14. 67%
15. 810
16. 52.22
17. 50%
18. 300
19. 33
20. $4,400
21. 10%
22. 1,500 kgm
23. 68.64
24. 5

Chapter 12, Feedback Unit 1

1. $\frac{17}{100}$
2. $\frac{49}{100}$
3. $\frac{57}{100}$
4. $\frac{143}{100}$
5. $\frac{7}{100}$
6. $\frac{8.4}{100}$
7. 100
8. 19%
9. 81%
10. 3%
11. 181%
12. 6.7%

Chapter 12, Feedback Unit 2

1. 0.27
2. 0.85
3. 0.56
4. 0.04
5. 0.37
6. 1.63
7. 0.92
8. 0.058
9. 0.16
10. 0.473
11. 26%
12. 89%
13. 8%
14. 63%
15. 240%
16. 52%
17. 145%
18. 22%
19. 4.3%
20. 141%
21. 1%
22. 2.5%

Chapter 12, Feedback Unit 3

1. 38%
2. 56%
3. 80%
4. 33%
5. 30%
6. 52%
7. 48%
8. 44%
9. 119%
10. 4%

Chapter 12, Feedback Unit 4

1. base, rate, part
2. of
3. %
4. 500, 53%, 265
5. 20, 85%, 17
6. 60, 125%, 75
7. 180, 80%, 144

Chapter 12, Feedback Unit 5

1. 32.52
2. 22%
3. 300
4. 750.96
5. 72%
6. 540
7. 26
8. 9%
9. 212
10. 78.47

Chapter 12, Feedback Unit 6

1. 544.04
2. 5%
3. 149
4. 26%
5. 2.82
6. 60
7. $2,400
8. 75 games
9. 43%
10. 32 students
11. 2.25 kgm
12. $30,000

Chapter 12 Mastery Test

1. $\frac{2}{100}$
2. $\frac{110}{100}$
3. $\frac{7.9}{100}$
4. 100
5. 31%
6. 5.2%
7. .55
8. .137
9. 20%
10. 58%
11. 63%
12. 61%
13. b = 70, r = 110, p = 77
14. b = 68, r = 75, p = 51
15. 49%
16. 212
17. 142.68
18. 50%
19. 300
20. 20
21. $6,000
22. 5%
23. 2300 kgm
24. 69.92
25. 8

Chapter 13 Objectives Test

1. -3
2. -8
3. -10
4. 5
5. -20
6. 24
7. -2
8. 5
9. 45
10. -5
11. 17
12. 4
13. 22
14. 15
15. 122° F
16. 100° C

Chapter 13, Feedback Unit 1

1. -10
2. 5
3. 9
4. 4
5. -2
6. -5
7. -6
8. 0

Chapter 13, Feedback Unit 2

1. -1
2. -11
3. -2
4. 3
5. -30
6. 36
7. 3
8. -16
9. -2
10. 2
11. 2
12. -5

Chapter 13, Feedback Unit 3

1. 47
2. 16
3. 4
4. 27
5. 38
6. 12
7. 61
8. 5

Chapter 13, Feedback Unit 4

1. 104° F
2. 149° F
3. 212° F
4. 32° F
5. 30° C
6. $15\frac{5}{9}$ ° C or 15.6°C
7. 0° C
8. -5°C

Chapter 13 Mastery Test

1. -9
2. 3
3. 2
4. -12
5. 10
6. -18
7. -2
8. 8
9. 30
10. -3
11. 15
12. 17
13. 79
14. 20
15. 95° F
16. 25° C

Chapter 14 Objectives Test

1. .2 meters
2. 34,500 meters
3. .256 liters
4. 1.25 kilograms
5. a. right
   b. acute
   c. obtuse

6. a. parallelogram
   b. rectangle
   c. square
   d. trapezoid
7. 14.27 feet
8. 39.36 sq cm
9. 95.1 inches
10. 21.1848 sq yards
11. 37.68 cm
12. 616 sq feet
13. 7.5 feet

Chapter 14, Feedback Unit 1

1. 100 centimeters
2. .01 meters
3. 1,000,000 meters
4. 1 meter
5. 0.1 meter
6. 1 liter
7. 1 kilometer
8. 1 kilogram
9. 16.5 grams
10. .55 kilometers

Chapter 14, Feedback Unit 2

1. a. right
   b. obtuse
   c. straight
   d. acute
2. 67°
3. 128°
4. 71°
5. a. obtuse
   b. equilateral
   c. acute
6. 14.7 feet
7. $6\frac{11}{24}$ inches
8. 16.96 sq feet
9. a. parallelogram
   b. square
   c. trapezoid
   d. rectangle
10. 34.9 feet
11. $20\frac{1}{2}$ meters
12. $42\frac{5}{6}$ inches
13. $22\frac{2}{3}$ yards
14. $19\frac{1}{2}$ sq feet
15. 14,500 sq cm or 1.45 sq meters
16. $7\frac{1}{3}$ inches
17. 282,600 sq meters
18. $14\frac{2}{5}$

Chapter 14 Mastery Test

1. 2.5 meters
2. 16,200 meters
3. .26 liters
4. .584 kilograms
5. a. acute
   b. obtuse
   c. right
6. a. rectangle
   b. parallelogram
   c. trapezoid
   d. square
7. 28.72 cm
8. 9.44 sq feet
9. 366 cm
10. 8,770 sq cm or 0.877 sq meters
11. 81.64 feet
12. 13.86 sq cm
13. 21.6 cm